T0350182

An Introduction to
ANALYSIS

An Introduction to
ANALYSIS

Jan Mikusiński

Piotr Mikusiński
University of Central Florida, USA

World Scientific

NEW JERSEY · LONDON · SINGAPORE · BEIJING · SHANGHAI · HONG KONG · TAIPEI · CHENNAI · TOKYO

Published by

World Scientific Publishing Co. Pte. Ltd.

5 Toh Tuck Link, Singapore 596224

USA office: 27 Warren Street, Suite 401-402, Hackensack, NJ 07601

UK office: 57 Shelton Street, Covent Garden, London WC2H 9HE

Library of Congress Cataloging-in-Publication Data
Names: Mikusiński, Piotr. | Mikusiński, Jan.
Title: An introduction to analysis / by Piotr Mikusinski
    (University of Central Florida, USA), Jan Mikusinski.
Description: New Jersey : World Scientific, 2017. | Includes index.
Identifiers: LCCN 2016052214 | ISBN 9789813202610 (hardcover : alk. paper)
Subjects: LCSH: Mathematical analysis. | Calculus. | Functions of real variables.
Classification: LCC QA300 .M54525 2017 | DDC 515--dc23
LC record available at https://lccn.loc.gov/2016052214

British Library Cataloguing-in-Publication Data
A catalogue record for this book is available from the British Library.

Printed in Singapore

To Grażyna

Preface

The precursor to this book is a book written by my father Jan Mikusiński. The title was *Wstęp do analizy matematycznej* (Introduction to Mathematical Analysis), it was published in Poland in 1957. In the word from the publisher we read "Students of mathematics and other disciplines that use mathematics (such as chemistry) will find in this book the most interesting and important ideas and problems of mathematical analysis, often introduced in an original way. It will shed a new light on certain concepts and theorems, their role and importance .... Graduate students, engineers, and researchers will find that a careful and active reading of it will deepen their understanding of mathematical analysis and improve the precision of their thinking." The book quickly became popular in Poland and is in use to this day.

In the early eighties my father decided to write an introductory book on mathematical analysis in English and invited me to work with him. The evolution included strengths of the original book while incorporating significant, new ideas. The result of the collaboration was *An Introduction to Analysis: From Number to Integral*, published by John Wiley & Sons in 1993. My father died in 1987, therefore he did not see the completion of our project.

The book has been used as a textbook for introductory analysis courses at several universities, including the University of Central Florida. Over the years I received much feedback from students and colleagues using the book in their classes. I am particularly indebted to Howard Sherwood, now a retired Professor of Mathematics of the University of Central Florida, for his comments and suggestions concerning simplifications and clarifications of some parts of the book.

The present, significantly revised edition includes improvements to the presentation of material and a major reorganization of the content. In the previous edition, the book started with an introduction of axioms of natural numbers. Then basic properties of natural numbers were carefully developed. Real numbers were introduced in the second chapter and their construction from natural numbers was presented. This followed the original plan of presenting a rigorous development of mathematical analysis starting from simple axioms of natural numbers and culminating with a presentation of the Lebesgue integral. This approach, while mathematically attractive, presented some pedagogical problems. It placed the nontrivial construction of the reals very early in the course. Many students do not have the mathematical maturity and sophistication necessary to fully understand the technicalities of the construction at the beginning of their first course in mathematical analysis. In the present edition we start from a system of axioms of real numbers and postpone the construction of the reals from natural numbers until the end of the book.

This book contains a rigorous exposition of calculus of a single real variable. In the first chapter we introduce axioms of real numbers and develop basic properties of the reals, with a section on countable and uncountable sets closing the chapter. Chapters 2, 3, and 4 discuss limits, continuity, and differentiability of functions. Sequences and series of numbers and functions are discussed in Chapters 5 and 6. Our presentation of partitions and rearrangements of absolutely convergent series is more elaborate than what we find in standard introductory textbooks. These properties are essential for our development of the Lebesgue integral.

Chapter 7 is devoted to elementary functions. We emphasize precise definitions of the functions and carefully derive their properties. The approach shows students how to combine differentiation and infinite series to construct important transcendental functions and derive many standard and impressive results of calculus. This gives students some feel for the value of the tools developed in previous chapters.

Chapters 8 and 9 are short and mostly contain results that ought to look familiar to students. Chapter 8 contains important theorems on limits of sequences and functions, including Abel's theorem. Chapter 9 is a review of the most important techniques of integration. The stated theorems are carefully proved and the discussed examples are often more challenging than those usually found in calculus textbooks.

Chapter 10 is probably the most distinctive feature of this book. It contains a direct treatment of the Lebesgue integral based solely on the concept of absolutely convergent series. This chapter in itself is equivalent to an entire book. In only a few pages the entire spectrum of the Lebesgue integral is revealed. A student with the standard introductory calculus experience is adequately prepared for the material of this chapter if only given time to adjust to precise arguments. The student who has grown accustomed to the precision of earlier chapters should be well prepared for the challenges presented by this one.

In Chapter 11 we present some topics that can be loosely called applications of integration. For instance, there is a short treatment of Fourier series. A section on the Riemann integral closes the chapter.

In the final chapter we present a construction of the real numbers from natural numbers. Instead of the standard Peano's axioms, a simpler approach to natural numbers is used; the axioms are formulated in terms of inequalities. Then the algebraic properties of natural numbers are carefully obtained from the axioms. The construction of the real numbers from the natural numbers, through integers and rational numbers, follows. A new approach to equivalence relations simplifies the construction of the real numbers. Only the most essential properties of numbers are developed, but enough details are given to show students the spirit of the process and some ideas of what they can legitimately use in a proof.

The exposition in this book is rigorous but light in tone. It is written so as to obtain maximum results with a minimum of work. Theorems are chosen that are central to the study of analysis yet with an eye on simplicity. The usual practice of the text is not to state the most general theorems but rather to find those that are simultaneously most simple to state and clearly exhibit fundamental themes and concepts. The text consistently renders a spare, central core of essential mathematics with just enough applications provided to show why the material is important. For example, some numerical calculations show how theoretical results can drastically simplify calculations.

In general, the level of presentation moves from simple to more complex. We find some very simple arguments in Chapter 1 and some quite sophisticated proofs in Chapter 10. Short dialogs appear throughout the book. They address commonly asked questions and misconceptions. They also give the reader a chance to pause for a moment from the formal language of mathematics.

The text is complemented by a large number of exercises and they are an important element of this text. A student should be able to do almost all of them while reading the book. Some exercises introduce new concepts. Those are not essential for understanding the following sections, but an instructor may decide to assign them as special projects. Routine exercises of the type usually given in a calculus course are given only sparingly. This book is written to follow a standard calculus course, not to replace it. Some problems are not very clearly stated and this is intentional. A precise formulation of the question is an essential part of the problem.

The book is designed as a text for a post-calculus course addressed to students majoring in mathematics or mathematics education. It can also be used for an honors calculus course or a guided study. It will provide students with a solid background for further studies in analysis, deepen their understanding of calculus, and provide sound training in rigorous mathematical proof. A shorter course may consist of Chapters 2, 3, 4, 5, 6, 8, 10, and selected sections from Chapter 11, with a possible inclusion of Sections 1.5 and 1.6. Some instructors may decide to skip some more technical proofs in Chapter 10. The standard prerequisite for this course is the calculus series. The authors assume that the reader is familiar with the basic set theoretical notation and basic principles of logic. Thus, some students may find it helpful to take a course like Logic and Proof before taking a course based on this book.

I would like to take this opportunity to express my gratitude to those who helped me at various stages of preparation of this version of the book. I am indebted to M. D. Taylor, Steve Goldman, and Meredith Clements for their comments on different parts of the manuscript. I am grateful to the World Scientific Publishing team, including Rochelle Kronzek, Suqi Pan, Rok Ting Tan and Janice Sim for their support and for making the final stage of this process quite painless. Finally, I would like to thank the TeX-LaTeX Stack Exchange community for helping me on several occasions with my LaTeX problems.

Piotr Mikusiński
Orlando, Florida
December, 2016

Contents

Chapter 1

REAL NUMBERS

The foundation of Mathematical Analysis is the real numbers. While the set of real numbers is relatively simple, it provides nontrivial examples for most important ideas in analysis. A good understanding of real numbers is a necessary prerequisite for understanding the concepts of differential and integral calculus as well as more advanced topics in analysis.

In this chapter we introduce real numbers using the axiomatic approach. As primitive notions we take the set of real numbers, the operations of addition and multiplication, and the inequality. We then formulate ten axioms that completely describe the set of real numbers. Everything else in this book can be derived from these ten axioms.

Most facts in this chapter will look familiar. We prove these properties of real numbers not to verify their correctness, but to show that they follow from the axioms.

The set of all real numbers will be denoted by $\mathbb{R}$. In what follows, real numbers will be referred to simply as numbers.

1.1 Addition

These are the fundamental properties of addition of numbers:

1$^+$ $a + b = b + a$;
2$^+$ $(a + b) + c = a + (b + c)$;
3$^+$ The equation $a + x = b$ is solvable.

Equalities **1$^+$** and **2$^+$** hold for all numbers a, b, and c. Property **3$^+$** says that for every pair of numbers a and b there exists a number x satisfying $a + x = b$. We do not prove properties **1$^+$**, **2$^+$**, and **3$^+$**, so they can be considered as axioms.

Later we will assume, for the sake of logical completeness, that there are at least two distinct real numbers. Why that assumption is desirable will be dealt with when the need arises. Proposition 1.1.1 asserts the existence of a number with a very special property. But without an understanding tacitly assumed but not explicitly mentioned, the statement is not quite true and the proposed proof is not quite valid. See if you can find the chink in the logic and the remedy.

1.1.1. *There exists one and only one number x such that $b + x = b$ for all numbers b.*

Proof. The existence of a number x satisfying $b + x = b$ follows from $\mathbf{3}^+$. However we may only assert that for every b there exists some number x satisfying the required equation, but we do not know whether that number is the same for all b. This must be proved.

Let a denote an arbitrary number. There exists a number x satisfying

$$a + x = a. \tag{1.1}$$

We will prove that this same number x satisfies $b + x = b$ for every number b. In fact, from $\mathbf{3}^+$ it follows that there is a number y such that $a + y = b$, or, equivalently,

$$b = a + y. \tag{1.2}$$

By (1.2), $\mathbf{2}^+$, $\mathbf{1}^+$, $\mathbf{2}^+$, (1.1), and (1.2), we conclude that

$$b + x = (a + y) + x = a + (y + x) = a + (x + y) = (a + x) + y = a + y = b.$$

We thus have proved that the number x satisfying $a + x = a$ for some arbitrary a, satisfies the same equation for any a. We still have to prove that this is the only number with this property. Assume that there are two such numbers and denote them by x_1 and x_2. Then we have simultaneously

$$x_1 + x_2 = x_1 \quad \text{and} \quad x_2 + x_1 = x_2.$$

Consequently, by $\mathbf{1}^+$, we have

$$x_1 = x_1 + x_2 = x_2 + x_1 = x_2.$$

This proves that there is only one number with the required property. $\qquad \square$

Proposition 1.1.1 is interesting for the reason that it allows one to distinguish from the set of all real numbers exactly one real number with the following property: If we add that number to any number a, then as the result we obtain the same number a. The number with this particular property is called *zero* and is denoted by 0. In view of $\mathbf{1}^+$ we have

$$0 + a = a + 0 = a. \tag{1.3}$$

1.1.2. *If $a + x = b$ and $a + y = b$, then $x = y$.*

In other words: The solution of $a + x = b$ is unique for each pair of numbers a and b.

Proof. If $a + x = b$ and $a + y = b$, then

$$a + x = a + y. \tag{1.4}$$

By $\mathbf{3}^+$ there is a number z such that

$$a + z = 0. \tag{1.5}$$

From (1.3), (1.5), $\mathbf{2}^+$, $\mathbf{1}^+$, (1.4), $\mathbf{1}^+$, $\mathbf{2}^+$, (1.5), and (1.3), we obtain

$$x = x + 0 = x + (a + z) = (x + a) + z = (a + x) + z$$
$$= (a + y) + z = (y + a) + z = y + (a + z) = y + 0 = y.$$

$\square$

Note that 1.1.2 implies the following useful property, which is often called the *cancellation law for addition*:

1.1.3. *If $a + x = a + y$, then $x = y$.*

The unique solution x of the equation $a + x = b$ is called the *difference* of a and b and is denoted by $b - a$. From this definition it follows that

1.1.4. $a + (b - a) = b$.

By introducing the difference we have defined a new operation which associates a number $b - a$ with a pair of numbers a and b. This operation is called *subtraction*. Proposition 1.1.2 asserts the uniqueness of subtraction and property $\mathbf{3}^+$ its feasibility.

1.1.5. $a - a = 0$.

Proof. By the definition of 0 we have $a + 0 = a$ and, by 1.1.4, $a + (a - a) = a$. Hence $a - a = 0$, by 1.1.2. $\square$

1.1.6. $(b - a) + a = b$ *and* $(b + a) - a = b$.

Proof. The first of these identities follows at once from 1.1.4, by $\mathbf{1}^+$. In view of 1.1.4 we also may write $a + ((b+a)-a) = b+a$. Hence, by $\mathbf{1}^+$, we get $((b+a)-a)+a = b+a$, and finally $(b + a) - a = b$, by 1.1.3. $\square$

The number $0 - a$ is denoted simply by $-a$ and is called the *opposite* of a. We thus have

$$0 - a = -a. \tag{1.6}$$

1.1.7. $a + (-a) = 0$.

Proof. Substitute $b = 0$ in 1.1.4 and use (1.6). $\square$

1.1.8. $b + (-a) = b - a$.

Proof. From 1.1.7 we have $(a+(-a))+b = 0+b = b$ and consequently $a+(b+(-a)) = b$. Comparing this with 1.1.4 we see that $b + (-a) = b - a$, by 1.1.2.   □

1.1.9. $-(-b) = b$.

Proof. Substituting b for a in 1.1.7 we get $b + (-b) = 0$ and therefore, by $\mathbf{1^+}$, $(-b)+b = 0$. On the other hand, substituting $-b$ for a into the same equation 1.1.7 we obtain $(-b) + (-(-b)) = 0$. Consequently $-(-b) = b$, by 1.1.2.   □

EXERCISES 1.1

(1) Does the set $\{0\}$ with the usual definition of addition of numbers satisfy $\mathbf{1^+}$, $\mathbf{2^+}$, $\mathbf{3^+}$?

(2) Does the empty set with the usual notion of addition of numbers satisfy $\mathbf{1^+}$, $\mathbf{2^+}$, $\mathbf{3^+}$?

(3) In the proof of Proposition 1.1.1, find the chink in the logic and the remedy.

(4) Denote $\mathbb{Z}_3 = \{0, 1, 2\}$. Define addition $\oplus$ in $\mathbb{Z}_3$ by the table below. Does $(\mathbb{Z}_3, \oplus)$ satisfy $\mathbf{1^+}$, $\mathbf{2^+}$, $\mathbf{3^+}$?

$\oplus$	0	1	2
0	0	1	2
1	1	2	0
2	2	0	1

(5) Denote $\mathbb{Z}_4 = \{0, 1, 2, 3\}$. Define addition $\oplus$ in $\mathbb{Z}_4$ by the table below. Does $(\mathbb{Z}_4, \oplus)$ satisfy $\mathbf{1^+}$, $\mathbf{2^+}$, $\mathbf{3^+}$?

$\oplus$	0	1	2	3
0	0	1	2	3
1	1	2	3	0
2	2	3	0	1
3	3	0	1	2

(6) Prove the following identities:

    (a) $a - 0 = a$.

    (b) $a + (b - c) = (a + b) - c$.

    (c) $a - (b - c) = (a + c) - b$.

    (d) $(a - b) + (c - d) = (a + c) - (b + d)$.

    (e) $-(a + b) = -a - b$.

    (f) $-(b - a) = a - b$.

1.2   Multiplication

Multiplication assigns to each pair of numbers a and b another number ab, called the *product* of a and b. Multiplication has properties similar to those of addition:

1* $ab = ba$;

2* $(ab)c = a(bc)$;

3* The equation $ax = b$ is solvable whenever $a \neq 0$.

Property **3*** differs from the analogous property of addition by the additional assumption that a is different from 0. The necessity of this restriction is discussed in Section 1.3.

1.2.1. *There exists one and only one number x such that, $bx = b$ for all numbers b.*

Proof. Let a denote an arbitrary number different from 0. In view of **3*** there exists a number x such that

$$ax = a. \tag{1.7}$$

We are going to show that the same number x also satisfies $bx = b$ for all b.

By **3*** there exists a number y satisfying $ay = b$, or

$$b = ay. \tag{1.8}$$

Hence

$$bx = (ay)x = a(yx) = a(xy) = (ax)y = ay = b.$$

We have proved existence of a number with the desired property. It remains to show its uniqueness. If there were two such numbers, x_1 and x_2, then we would have

$$x_1 x_2 = x_1 \quad \text{and} \quad x_2 x_1 = x_2.$$

By **1*** the second equation can be written as $x_1 x_2 = x_2$ and thus it follows that $x_1 = x_2$. This proves that the number is unique. $\square$

Proposition 1.2.1 distinguishes a number with the property that each number b multiplied by that number gives b as the result. This particular number is called the *unit* and denoted by 1. In view of **1*** and 1.2.1 we have

$$1 \cdot a = a \cdot 1 = a.$$

In expressions with specific numbers like the one above, it is customary to write $a \cdot b$ instead of ab.

1.2.2. *If $ax_1 = b$ and $ax_2 = b$, and $a \neq 0$, then $x_1 = x_2$.*

In other words: *The solution of the equation $ax = b$ with $a \neq 0$ is unique.*

Proof. If $ax_1 = b$ and $ax_2 = b$, then

$$ax_1 = ax_2. \tag{1.9}$$

By **3*** there is a number y such that

$$ay = 1. \tag{1.10}$$

From (1.9), (1.10), **1*** and **2***, we find successively

$$(ax_1)y = (ax_2)y,$$
$$(x_1a)y = (x_2a)y,$$
$$x_1(ay) = x_2(ay),$$
$$x_1 \cdot 1 = x_2 \cdot 1,$$
$$x_1 = x_2.$$

$\square$

As in the case of addition, the above property implies the cancellation law for multiplication:

1.2.3. *If $ax = ay$ and $a \neq 0$, then $x = y$.*

In view of **3*** and 1.2.2, given a pair of numbers a and b such that $a \neq 0$, there exists exactly one number x such that $ax = b$. This number is called the *quotient* of the numbers b and a with *denominator a* and *numerator b*. This number is denoted by $\frac{b}{a}$. Since the number $\frac{b}{a}$ satisfies the equation $ax = b$, we have

$$a\frac{b}{a} = b \quad \text{for } a \neq 0. \tag{1.11}$$

Note that the above defines a new operation which assigns to every pair of numbers a and b such that $a \neq 0$, the number $\frac{b}{a}$. This operation is called *division*.

1.2.4. $\dfrac{a}{a} = 1$ *for $a \neq 0$.*

Proof. For $b = a$, (1.11) implies $a\frac{a}{a} = a$. On the other hand, $a \cdot 1 = a$. Hence, by 1.2.2, we obtain $\frac{a}{a} = 1$. $\square$

1.2.5. $c\dfrac{b}{a} = \dfrac{b}{a}c = \dfrac{bc}{a}.$

Proof. The first equality is an immediate consequence of **1***. To establish the second equality, we multiply both sides of (1.11) by c to get $(a\frac{b}{a})c = bc$, and hence

$$a\left(\frac{b}{a}c\right) = bc \tag{1.12}$$

by **2***. Since formula (1.11) holds for each number b it thus remains true when b is replaced by bc:

$$a\frac{bc}{a} = bc. \tag{1.13}$$

Now (1.12) and (1.13) implies $\frac{b}{a}c = \frac{bc}{a}$, by 1.2.2. $\square$

1.2.6. $\dfrac{b}{a}a = b$ *and* $\dfrac{ba}{a} = b$ *for $a \neq 0$.*

Proof. The first identity follows from (1.11) and **1***. If in (1.11) we write ba instead of b, then we get $a\frac{ba}{a} = ba = ab$. Hence the identity $\frac{ba}{a} = b$ follows by 1.2.2. $\quad\square$

The number $\frac{1}{a}$, for $a \neq 0$, is called the *reciprocal* of a.

1.2.7. $c\dfrac{1}{a} = \dfrac{c}{a}$ *for* $a \neq 0$.

Proof. From 1.2.5 with $b = 1$ we get $c\frac{1}{a} = \frac{c \cdot 1}{a}$ and hence $c\frac{1}{a} = \frac{c}{a}$. $\quad\square$

EXERCISES 1.2

(1) Does the set $\{0\}$ satisfy **1***, **2***, **3***?

(2) Is $0 = 1$ or is $0 \neq 1$? Explain.

(3) Denote $\mathbb{Z}_3 = \{0, 1, 2\}$. Define multiplication $\otimes$ in $\mathbb{Z}_3$ by the table below. Does $(\mathbb{Z}_3, \otimes)$ satisfy **1***, **2***, **3***?

$\otimes$	0	1	2
0	0	0	0
1	0	1	2
2	0	2	1

(4) Show that $\frac{1}{2} = 2$ in $(\mathbb{Z}_3, \otimes)$.

(5) Denote $\mathbb{Z}_4 = \{0, 1, 2, 3\}$. Define multiplication $\otimes$ in $\mathbb{Z}_4$ by the table below. Does $(\mathbb{Z}_4, \otimes)$ satisfy **1***, **2***, **3***?

$\otimes$	0	1	2	3
0	0	0	0	0
1	0	1	2	3
2	0	2	0	2
3	0	3	2	1

(6) Denote $\mathbb{Y} = \{0, 1, 2\}$. Define addition and multiplication in $\mathbb{Y}$ by the following tables:

$\oplus$	0	1	2
0	0	1	2
1	1	2	0
2	2	0	1

$\otimes$	0	1	2
0	2	0	1
1	0	1	2
2	1	2	0

Does $\mathbb{Y}$ with addition $\oplus$ and multiplication $\otimes$ satisfy **1$^+$**, **2$^+$**, **3$^+$**, **1***, **2***, and **3***? Which of the three elements in $\mathbb{Y}$ equals $\frac{1}{2}$?

(7) Does it follow from axioms **1$^+$**, **2$^+$**, **3$^+$**, **1***, **2***, and **3*** that $ab \neq 0$ whenever $a \neq 0$ and $b \neq 0$?

(8) Prove that $\frac{1}{a}\frac{1}{c} = \frac{1}{ac}$, provided $a \neq 0$, $c \neq 0$, and $ac \neq 0$.

(9) Prove that $\frac{1}{\left(\frac{1}{a}\right)} = a$, provided $a \neq 0$ and $\frac{1}{a} \neq 0$.

(10) In the proof of 1.2.5, both sides of (1.11) are multiplied by c to get $\left(a\frac{b}{c}\right)c = bc$. Why can we multiply both sides of an equality by the same number and then assert that the resulting quantities are equal?

1.3 Distributivity

We now complete the list of properties of addition and multiplication with a property that links the two operations.

$+*$ $a(b + c) = ab + ac$

This property is called the *distributivity of multiplication over addition*. In view of 1^* we may also write $(b + c)a = ba + ca$.

1.3.1. $a(b - c) = ab - ac$ *and* $(b - c)a = ba - ca$.

Proof. In view of 1.1.4 we have

$$ac + (ab - ac) = ab \tag{1.14}$$

and

$$c + (b - c) = b. \tag{1.15}$$

Multiplying both sides of (1.15) by a we get $a(c + (b - c)) = ab$ and then applying $+*$ we get

$$ac + a(b - c) = ab. \tag{1.16}$$

From (1.14) and (1.16) we get $a(b - c) = ab - ac$, by 1.1.2. This proves the first identity. The second identity follows by 1^*.   □

1.3.2. $0 \cdot a = a \cdot 0 = 0$.

Proof. From 1^* we have $0 \cdot a = a \cdot 0$. For $c = b$ we obtain, from the first identity in 1.3.1, $a(b - b) = ab - ab$ and hence $a \cdot 0 = 0$, by 1.1.5.   □

From 1.3.2 it follows that the equation $ax = b$ has no solution if $a = 0$ and $b \neq 0$. This explains the necessity of the restriction $a \neq 0$ in 3^*. On the other hand, if $a = 0$ and $b = 0$, then every number x satisfies $ax = b$. This explains the necessity of the restriction $a \neq 0$ in 1.2.2.

1.3.3. $(-a)b = -(ab)$ *and* $(-a)(-b) = ab$.

Proof. Since

$$(-a)b = (0 - a)b = 0 \cdot b - ab = 0 - ab = -(ab),$$

we have $(-a)b = -(ab)$. Using this equality we may write

$$(-a)(-b) = -(a(-b)) = -(a(0 - b)) = -(a \cdot 0 - ab)$$
$$= -(0 - ab) = -(-(ab)) = ab,$$

by 1.3.1, 1.3.2, and 1.1.9.   □

1.3.4. *If $a \neq 0$ and $b \neq 0$, then $ab \neq 0$.*

Proof. It suffices to prove that if $a \neq 0$ and $ab = 0$, then $b = 0$. Indeed, since $a \cdot 0 = 0$ (by 1.3.2), $b = 0$ follows from 1.2.2. $\qquad\square$

1.3.5. *If $a \neq 0$ and $b \neq 0$, then $\dfrac{b}{a} \neq 0$.*

Proof. It suffices to prove that, if $a \neq 0$ and

$$\frac{b}{a} = 0, \tag{1.17}$$

then $b = 0$. Multiplying (1.17) by a we get $\frac{b}{a}a = 0 \cdot a$. Hence $b = 0$, by 1.2.6 and 1.3.2. $\qquad\square$

1.3.6. $\dfrac{b}{a}\dfrac{d}{c} = \dfrac{bd}{ac}$ *for $a \neq 0$ and $c \neq 0$.*

Proof. Since

$$(ac)\left(\frac{b}{a}\frac{d}{c}\right) = \left((ac)\frac{b}{a}\right)\frac{d}{c} = \left(a\left(c\frac{b}{a}\right)\right)\frac{d}{c}$$
$$= \left(a\left(\frac{b}{a}c\right)\right)\frac{d}{c} = \left(\left(a\frac{b}{a}\right)c\right)\frac{d}{c} = \left(a\frac{b}{a}\right)\left(c\frac{d}{c}\right),$$

we have

$$(ac)\left(\frac{b}{a}\frac{d}{c}\right) = bd.$$

On the other hand, we may write

$$(ac)\frac{bd}{ac} = bd,$$

because ac is different from zero, by 1.3.4. Hence $\frac{b}{a}\frac{d}{c} = \frac{bd}{ac}$, by 1.2.2. $\qquad\square$

1.3.7. $\dfrac{b}{a} = \dfrac{bc}{ac}$ *for $a \neq 0$ and $c \neq 0$.*

Proof. From 1.2.4 and 1.3.6 we obtain

$$\frac{b}{a} = \frac{b}{a} \cdot 1 = \frac{b}{a}\frac{c}{c} = \frac{bc}{ac}.$$

$\qquad\square$

1.3.8. $\dfrac{b}{a} + \dfrac{c}{a} = \dfrac{b+c}{a}$ *and* $\dfrac{b}{a} - \dfrac{c}{a} = \dfrac{b-c}{a}$ *for $a \neq 0$.*

Proof. We have

$$\frac{b}{a} + \frac{c}{a} = b\frac{1}{a} + c\frac{1}{a} = (b+c)\frac{1}{a} = \frac{b+c}{a},$$

which proves the first identity. The proof of the second identity is similar. $\qquad\square$

Identities in 1.3.7 and 1.3.8 allow us to add and subtract fractions with arbitrary denominators different from zero. Indeed,

$$\frac{b}{a} + \frac{d}{c} = \frac{bc}{ac} + \frac{ad}{ac} = \frac{bc + ad}{ac}$$

and

$$\frac{b}{a} - \frac{d}{c} = \frac{bc}{ac} - \frac{ad}{ac} = \frac{bc - ad}{ac}.$$

1.3.9. $\dfrac{c}{\left(\frac{b}{a}\right)} = c\dfrac{a}{b}$ *for* $a \neq 0$ *and* $b \neq 0$.

Proof. By 1.3.4, we have $ab \neq 0$ and thus

$$1 = \frac{ab}{ab} = \frac{ba}{ab} = \frac{b}{a} \cdot \frac{a}{b}$$

and hence

$$\left(\frac{b}{a} \cdot \frac{a}{b}\right) c = 1 \cdot c = c,$$

and

$$\frac{b}{a}\left(c\frac{a}{b}\right) = c. \tag{1.18}$$

Since $\frac{b}{a} \neq 0$, we may write

$$\frac{b}{a} \cdot \frac{c}{\left(\frac{b}{a}\right)} = c, \tag{1.19}$$

by (1.11). From (1.18) and (1.19), by using 1.2.2, we obtain the desired identity. $\square$

EXERCISES 1.3

(1) Prove that $\dfrac{\left(\frac{d}{c}\right)}{\left(\frac{b}{a}\right)} = \dfrac{ad}{bc}$ for $a \neq 0$, $b \neq 0$, $c \neq 0$.

(2) Does $(\mathbb{Y}, \oplus, \otimes)$ defined in Exercise 1.2.6 satisfy $+*$?

(3) Does $(\mathbb{Z}_3, \oplus, \otimes)$ satisfy $+*$? (see Exercises 1.1.4 and 1.2.4.)

(4) Does $(\mathbb{Z}_4, \oplus, \otimes)$ satisfy $+*$? (see Exercises 1.1.5 and 1.2.5.)

1.4   Inequalities

All properties of real numbers considered so far can be called algebraic. They concern the algebraic operations of addition and multiplication. We now include a non-algebraic property of being *positive*. To indicate that a is positive we write $0 < a$. We assume the following properties:

$\mathbf{1}^<$ For every number a, one and only one of the following relations holds:

$$a = 0 \quad \text{or} \quad 0 < a \quad \text{or} \quad 0 < -a.$$

$\mathbf{2}^<$ If $0 < a$ and $0 < b$, then $0 < a + b$ and $0 < ab$.

Any number a such that $0 < -a$ is called *negative*. Property $\mathbf{1}^<$ says that every number is either zero, positive, or negative. Property $\mathbf{2}^<$ tells us that the sum and the product of two positive numbers are positive.

1.4.1. *If* $0 < a$ *and* $0 < b$, *then* $0 < \dfrac{b}{a}$.

Proof. If we had $\frac{b}{a} = 0$, then we would get $b = \frac{b}{a}a = 0 \cdot a = 0$, which contradicts $\mathbf{1}^<$. If we had $0 < -\frac{b}{a}$, it would mean by 1.3.3 that $0 < \left(-\frac{b}{a}\right)a = -b$, which again contradicts $\mathbf{1}^<$. Consequently, we must have $0 < \frac{b}{a}$. $\qquad\square$

1.4.2. *If* $a \neq 0$, *then* $0 < aa$.

Proof. If $a \neq 0$, then $0 < a$ or $0 < -a$, by $\mathbf{1}^<$. If $0 < a$, then $0 < aa$, by $\mathbf{2}^<$. If $0 < -a$, then $0 < (-a)(-a) = aa$, by $\mathbf{2}^<$ and 1.3.3. $\qquad\square$

1.4.3. $0 < 1$.

Proof. By 1.4.2 we have $0 < 1 \cdot 1 = 1$. (See Exercise 1.4.1.) $\qquad\square$

If $0 < b - a$, we say that a *is less than* b and write $a < b$. We can also say that b *is greater than* a and write $b > a$. By this convention the symbol $0 < b$ can be read in two ways: "the number b is positive" or "the number b is greater than zero", because $0 < b - 0$.

From the definition of inequality and from $\mathbf{1}^<$ we obtain the following important property:

1.4.4 (Trichotomy). *For each pair of numbers* a *and* b, *one and only one of the following relations holds:*

$$a = b \quad or \quad a < b \quad or \quad b < a.$$

1.4.5 (Transitivity). *If* $a < b$ *and* $b < c$, *then* $a < c$.

Proof. The inequalities $a < b$ and $b < c$ mean that $0 < b - a$ and $0 < c - b$. Therefore, in view of $\mathbf{2}^<$, we obtain $0 < (b - a) + (c - b) = c - a$. $\qquad\square$

Instead of "$a < b$ and $b < c$" we often write "$a < b < c$."

1.4.6. *If* $a < b$, *then* $a + c < b + c$.

Proof. Since $b - a = (b + c) - (a + c)$, we have $0 < (b + c) - (a + c)$, which means that $a + c < b + c$. $\qquad\square$

1.4.7. *If* $a < b$ *and* $c < d$, *then* $a + c < b + d$.

Proof. This is a direct consequence of $\mathbf{2}^<$ and the definition of inequality. $\qquad\square$

1.4.8. *If* $a < b$ *and* $0 < c$, *then* $ac < bc$.

Proof. If $a < b$, then $0 < b - a$ and $0 < (b - a)c$ by $\mathbf{2}^<$. Hence $0 < bc - ac$, which means that $ac < bc$. $\qquad\square$

1.4.9. *If $a < b$, $c < d$ and $0 < b$, $0 < c$, then $ac < bd$.*

Proof. From 1.4.8 it follows that $ac < bc$ and $bc < bd$. Hence, by transitivity, $ac < bd$. $\qquad\square$

Note that in general, $a < b$ and $c < d$ does not imply $ac < bd$.

We write $a \leq b$ to indicate that either $a = b$ or $a < b$. As an immediate consequence of 1.4.4 and this notational convention we obtain the following property: *For each pair of real numbers a and b, one and only one of the following relations holds:*

$$a \leq b \quad \text{or} \quad b < a.$$

1.4.10. *If $a \leq b$ and $b \leq a$, then $a = b$.*

Proof. Since the inequalities $a < b$ and $b < a$ exclude each other, by 1.4.4, only one possibility remains, namely $a = b$. $\qquad\square$

EXERCISES 1.4

(1) In the proof of 1.4.3, use was made of 1.4.2. But in order to legitimately use 1.4.2, an understanding, tacitly made but not explicitly mentioned, must have been employed. What was that understanding?

(2) Provide a complete proof for 1.4.4.

(3) Provide a complete proof for 1.4.7.

(4) Show that, if $0 < \varepsilon$, then $m - \varepsilon < m$ for any real number m.

(5) Show that, in general, $a < b$ and $c < d$ does not imply $ac < bd$.

(6) Let $a, b \in \mathbb{Z}_3$. We will write $a < b$ if the same inequality holds for the natural numbers a and b. Does $(\mathbb{Z}_3, \oplus, \otimes, <)$ satisfy $\mathbf{1}^<$, $\mathbf{2}^<$? (See Exercises 1.1.4 and 1.2.4.)

(7) Let $a, b \in \mathbb{Z}_4$. We will write $a < b$ if the same inequality holds for the natural numbers a and b. Does $(\mathbb{Z}_4, \oplus, \otimes, <)$ satisfy $\mathbf{1}^<$, $\mathbf{2}^<$? (See Exercises 1.1.5 and 1.2.5.)

(8) Is it possible to define in $\mathbb{Z}_3$ a relation "$<$" so that $\mathbf{1}^<$, $\mathbf{2}^<$ would be satisfied? How about $\mathbb{Z}_4$?

(9) Let X be a set with addition $+$, multiplication $\cdot$, and order $<$, such that all conditions $\mathbf{1}^+$, $\mathbf{2}^+$, $\mathbf{3}^+$, $\mathbf{1}^*$, $\mathbf{2}^*$, $\mathbf{3}^*$, $+_*$, $\mathbf{1}^<$, $\mathbf{2}^<$ are satisfied. Prove that, if X has at least two elements, then it has infinitely many elements.

(10) Does the set $\{0\}$ satisfy all conditions $\mathbf{1}^+$, $\mathbf{2}^+$, $\mathbf{3}^+$, $\mathbf{1}^*$, $\mathbf{2}^*$, $\mathbf{3}^*$, $+_*$, $\mathbf{1}^<$, $\mathbf{2}^<$?

(11) Does the set $\mathbb{N}$ of all natural numbers satisfy all conditions $\mathbf{1}^+$, $\mathbf{2}^+$, $\mathbf{3}^+$, $\mathbf{1}^*$, $\mathbf{2}^*$, $\mathbf{3}^*$, $+_*$, $\mathbf{1}^<$, $\mathbf{2}^<$?

(12) Does the set $\mathbb{Q}$ of all rational numbers satisfy all conditions $\mathbf{1^+}$, $\mathbf{2^+}$, $\mathbf{3^+}$, $\mathbf{1^*}$, $\mathbf{2^*}$, $\mathbf{3^*}$, $\mathbf{+_*}$, $\mathbf{1^<}$, $\mathbf{2^<}$?

1.5 Bounded sets

A set of numbers is said to be *bounded from below* by a number m, if all its elements are greater than or equal to m. The number m is then called a *lower bound* of that set. If m is a lower bound of a set, then each number less than m is another lower bound of the same set. For instance the set of all positive numbers is bounded from below. All negative numbers as well as zero are its lower bounds. The number 0 is the greatest lower bound. It is also the greatest lower bound of the set of all non-negative numbers. In this case the greatest lower bound belongs to the set, whereas in the case of the set of all positive numbers it does not.

Definition 1.5.1. A number m is called the *greatest lower bound* of a given set P, if
(a) m is a lower bound for P, that is, $m \leq x$ for each $x \in P$
and
(b) m is the greatest of the lower bounds for P, that is, if $n \leq x$ for each $x \in P$, then $m \geq n$.

Notice that a set can have only one greatest lower bound. Indeed, assuming that m_1 and m_2 are the greatest lower bounds of P, we have $m_1 \leq m_2$ and $m_2 \leq m_1$, so $m_1 = m_2$, which proves the uniqueness.

A question arises, whether every set bounded from below possesses a greatest lower bound. This does not follow from the properties of real numbers considered so far (see Exercise 1.5.1). If we want to use this property, it must be adopted as an additional axiom.

D Every nonempty set bounded from below has a greatest lower bound.

This axiom is often called the Dedekind Axiom. Dedekind (Richard Dedekind, 1831–1916) was the first to construct a model of real numbers. All together we get the following Axioms of The Real Numbers:

$\mathbf{1^+}$ $a + b = b + a$;
$\mathbf{2^+}$ $(a + b) + c = a + (b + c)$;
$\mathbf{3^+}$ The equation $a + x = b$ is solvable;
$\mathbf{1^*}$ $ab = ba$;
$\mathbf{2^*}$ $(ab)c = a(bc)$;
$\mathbf{3^*}$ The equation $ax = b$ is solvable whenever $a \neq 0$;
$\mathbf{+_*}$ $a(b + c) = ab + ac$;

1$^<$ For every number a, one and only one of the following relations holds:

$$a = 0 \quad \text{or} \quad 0 < a \quad \text{or} \quad 0 < -a;$$

2$^<$ If $0 < a$ and $0 < b$, then $0 < a + b$ and $0 < ab$;

D Every nonempty set bounded from below has a greatest lower bound.

We will refer to this collection of ten axioms as $\mathfrak{R}$.

S: Do these axioms completely characterize the set of real numbers?

T: People trained in formal logic would say no, because both the empty set and the set consisting of the single element 0 satisfy all ten of the axioms. And yet neither is the set of all real numbers.

S: What if we assume that the set contains at least two elements?

T: Then everything is fine. The existence of two numbers implies, by means of the adopted axioms, the existence of all real numbers. But we have to be very careful with the words "two elements" which have to be understood in the colloquial sense: "one element and another one." A careful reader could notice that we have used the existence of two different real numbers in the proofs of 1.1.1 and 1.4.3. In the first case we need an element b in order to have an equation which must then have 0 as its solution. In the second case we use the fact that $1 \neq 0$, which follows from the existence of two different elements.

S: So the existence of at least two elements is to be adopted as an additional axiom of real numbers.

T: This is one possibility. Or else we may say that the set of real numbers is the set containing at least two elements and satisfying the axioms in $\mathfrak{R}$.

S: I like this better, because then the number of axioms remains 10, and it is nice to have exactly as many axioms as fingers.

A set is said to be *bounded from above* if all its elements are less than or equal to a number n.

Definition 1.5.2. By the *least upper bound* of a given set P we mean a number m such that:

(a) m is an upper bound for P, that is, $x \leq m$ for each $x \in P$

and

(b) m is the least of the upper bounds for P, that is, if $x \leq n$ for each $x \in P$, then $m \leq n$.

Using axiom **D** one can easily prove property

D' Every nonempty set bounded from above has a least upper bound.

To see that **D** implies **D'** and, conversely, **D'** implies **D**, it suffices to consider the set P' of all elements opposite to those from P, that is, a number x belongs to P' if and only if $-x$ belongs to P. Consequently, axiom **D** can be replaced by **D'** and we will obtain an equivalent set of axioms. In view of this equivalence, we will use the same name, Dedekind Axiom, to mean either of the two versions.

In order to avoid possible misunderstanding it should be strongly emphasized that if we say "the set P has a greatest lower bound m" (or "a least upper bound m"), then m need not belong to P. This only means that a number with the required properties exists.

Here is a nice application of the Dedekind Axiom.

Theorem 1.5.3 (Archimedean Property). *For any positive real number ε there is a natural number n such that $n\varepsilon > 1$.*

Proof. We argue by contradiction. If there is no n such that $n\varepsilon > 1$, then the set $S = \{n\varepsilon : n \in \mathbb{N}\}$ is bounded by 1. By the Dedekind Axiom, S has a least upper bound m. Then $n\varepsilon \leq m$ for all $n \in \mathbb{N}$. Hence, $(n - 1)\varepsilon \leq m - \varepsilon$ for all $n \in \mathbb{N}$. But in this case we would have $n\varepsilon \leq m - \varepsilon$ for all $n \in \mathbb{N}$, contradicting the definition of m. $\qquad\square$

From the Archimidean Property we easily obtain the following useful property of the real numbers.

Corollary 1.5.4. *There is a rational number between any two distinct numbers.*

Proof. Let a and b be arbitrary numbers such that $a < b$. Since $b - a > 0$, there exists a natural number n such that $n(b - a) > 1$. Then there must be an integer m such that $na < m < nb$. Otherwise, for some integer m we would have

$$m \leq na < nb \leq m + 1$$

and consequently

$$n(b - a) = nb - na \leq m + 1 - m = 1.$$

For any integer m such that $na < m < nb$ we have

$$a < \frac{m}{n} < b.$$

$\qquad\square$

EXERCISES 1.5

(1) The set $\mathbb{Q}$ of all rational numbers with addition and multiplication defined as usual satisfy all conditions 1^+, 2^+, 3^+, 1^*, 2^*, 3^*, $+_*$, $1^<$, $2^<$ (see Exercise 2.4.12). Show that $\mathbb{Q}$ does not satisfy condition **D**. This will show that condition **D** cannot be deduced from the other nine in $\mathfrak{R}$.

(2) Provide details of the proof of equivalence of **D** and **D'**.

(3) Find greatest lower bounds and least upper bounds of the following sets:

(a) $\left\{1, \dfrac{1}{2}, \dfrac{1}{3}, \dfrac{1}{4}, \cdots\right\}$,

(b) $\left\{0, -\frac{1}{2}, \frac{2}{3}, -\frac{3}{4}, \frac{4}{5}, -\frac{5}{6}, \ldots \right\}$.

(4) Find the least upper bound of the set of all rational numbers whose square is less than 2.

(5) Consider the set $X = \{a_1, a_2, a_3, \ldots\}$ where a_n's are defined inductively:

$$a_1 = 1, \quad a_{n+1} = \frac{3a_n + 4}{2a_n + 3} \quad \text{for } n = 1, 2, 3, \ldots$$

Show that X is bounded from above and from below. Can you find the least upper bound of X?

(6) Using Crolollary 1.5.4, show that there are infinitely many rational numbers between any two distinct numbers.

1.6   Countable and uncountable sets

Suppose we have two boxes of matches: one with white matches and the other one with red. Is there a way for a person who cannot count, to determine which box contains more matches?

Let's take one match from each box and put them aside. From the remaining matches, again take one match from each box and put them aside too. Continuing this process, either we will exhaust both boxes simultaneously or one of the boxes will become empty while the other box will still contain at least one match. The first case occurs if and only if the number of matches is the same in both boxes. In the second case, the box that became empty had fewer matches and the other more. Note that we do not have to count the matches.

Using the described idea, we can check whether two sets have the same quantity of elements without counting them. Two sets are called *equipotent* if, to every element from the first set we can assign exactly one element from the second set in such a way that each element of the second set is assigned to exactly one element of the first set. We then say that there is a bijection between those sets.

Obviously, every set is equipotent with itself. Moreover, if a set A is equipotent with a set B, then B is equipotent with A. Equipotency is also transitive, that is, if a set A is equipotent with a set B and B is equipotent with a set C, then A is equipotent with C. Thus equipotency is an equivalence relation.

In order to establish a correspondence between elements of equipotent sets, we do not necessarily need to join elements in pairs; it suffices to give a law which says how the elements should be joined. Here, however, facts may come to light which seem paradoxical at first glance. For instance the set of all natural numbers is equipotent with the set of all even natural numbers. To establish a bijection between these sets, we associate with every natural number the number that is exactly twice as large, thus even. This is an example of a set which is equipotent with one of its proper subsets. This property can be used as a definition of an infinite set: a nonempty set is called *infinite* if it is equipotent with a proper subset of itself. On the other hand, a nonempty set is called *finite* if it is not equipotent with any proper subset of itself. Finite sets can be also defined as follows: a set A is finite if it is empty or there exists a natural number n such that A is equipotent with the set $\{1, 2, \ldots, n\}$. In this case, we can write $A = \{a_1, a_2, \ldots, a_n\}$.

A set which is equipotent with the set of all natural numbers is called *countable*. Elements of such a set can be numbered so that to each element of the set there corresponds a natural number and each natural number corresponds to one and only one element from the set. All elements of a countable set can thus be arranged into an infinite sequence. In other words, if A is a countable set, we can write $A = \{a_1, a_2, a_3, \ldots\}$. Conversely, if all elements of a set can be arranged into an infinite sequence such that every element appears in the sequence exactly once, then the set is countable. Hence the union of a countable set and a finite set or another countable set is countable. It is also clear that a subset of a countable set is either countable or finite.

Theorem 1.6.1. *The union of a countable family of countable sets is countable.*

Proof. Let $Z_1, Z_2, Z_3, \ldots$ be a sequence of countable sets. The elements of each set Z_n can be arranged into an infinite sequence, so we can write

$$Z_n = \{a_{n,1}, a_{n,2}, a_{n,3}, \ldots\}.$$

Consequently, the elements of the union can be arranged into an infinite matrix

$$
\begin{matrix}
a_{1,1} & a_{1,2} & a_{1,3} & \cdots \\
a_{2,1} & a_{2,2} & a_{2,3} & \cdots \\
a_{3,1} & a_{3,2} & a_{3,3} & \cdots \\
\vdots & \vdots & \vdots & \\
a_{n,1} & a_{n,2} & a_{n,3} & \cdots \\
\vdots & \vdots & \vdots &
\end{matrix}
$$

Now we want to arrange all entries of this matrix into a single sequence. We can accomplish this by first splitting entries of the matrix into finite groups such that the n-th group consists of all entries $a_{i,j}$ such that $i + j = n + 1$. Thus, in the n-th group we will have

$$a_{n,1}, a_{n-1,2}, a_{n-2,3}, \ldots, a_{2,n-1}, a_{1,n}.$$

These entries form the n-th diagonal of the matrix.

By writing down the first diagonal, then the second diagonal, and so on, we will arrange all entries of the matrix into a single sequence:

$$a_{1,1}, a_{2,1}, a_{1,2}, a_{3,1}, a_{2,2}, a_{1,3}, \ldots$$

In this sequence, some elements of the union may appear several times if some Z_n's have elements in common. In such a case, we have to remove each element which is equal to one of the preceding elements. In this way we obtain an arrangement of the union in a sequence, which proves that the union is countable. $\square$

The following diagram represents the idea of the above proof in a more intuitive and easier to remember way. To arrange all entries of the infinite matrix into a single sequence, we follow the indicated path.

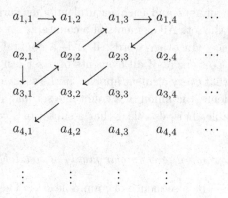

Theorem 1.6.2. *The set $\mathbb{Q}$ of all rational numbers is countable.*

Proof. Let, for $n = 1, 2, 3, \ldots$, Z_n denote the set of all rational numbers that can be written with n in the denominator, that is,

$$Z_n = \left\{ \frac{0}{n}, \frac{1}{n}, \frac{-1}{n}, \frac{2}{n}, \frac{-2}{n}, \ldots \right\}.$$

Clearly, each set Z_n is countable and we have

$$\mathbb{Q} = \bigcup_{n=1}^{\infty} Z_n.$$

Thus the set $\mathbb{Q}$ is countable by Theorem 1.6.1. □

Theorem 1.6.3. *The set $\mathbb{R}$ of all real numbers is not countable.*

In the proof of the above theorem we will use the fact that every real number between 0 and 1 can be uniquely represented in the form of an infinite sequence of digits

$$0.d_1 d_2 d_3 \ldots \tag{1.20}$$

such that $d_n \neq 0$ for some index n (to exclude 0) and there are arbitrarily large indices n for which $d_n \neq 9$ (to exclude sequences that have nothing but 9's from some point on.) The representation of a real number in the form (1.20) is called a *decimal expansion*. Any such sequence represents a real number between 0 and 1. (Decimal expansions of real numbers will be discussed in more detail in Section 5.7.)

Proof. We argue by contradiction. Suppose that the set $\mathbb{R}$ is countable. Then the set of all real numbers between 0 and 1 must be countable and thus its elements can be arranged into a sequence $a_1, a_2, a_3, \ldots$. Let

$$a_n = 0.d_{n,1}d_{n,2}d_{n,3}\cdots$$

be the unique decimal expansion of a_n. Now we consider a real number whose decimal expansion is

$$0.e_1e_2e_3\ldots, \tag{1.21}$$

where $e_n \neq d_{n,n}$ and $e_n \neq 9$. Note that (1.21) represents a real number between 0 and 1, though it is not among the numbers in the sequence $a_1, a_2, a_3, \ldots$. Thus the assumption that the set $\mathbb{R}$ is countable leads to a contradiction. $\square$

A set that is neither finite nor countable is called *uncountable*. Real numbers which are not rational numbers are called *irrational numbers*. Note that Theorems 1.6.2 and 1.6.3 imply that the set of irrational numbers is uncountale. Moreover, we obtain the following useful theorem.

Theorem 1.6.4. *There are infinitely many irrational numbers between any two distinct numbers.*

Proof. If $a < b$, then

$$x \longleftrightarrow \frac{x - a}{b - a}$$

establishes a bijection between the set of all real numbers between a and b and the set of all real numbers between 0 and 1. Since there are uncountably many numbers between 0 and 1, there are uncountably many numbers between a and b. $\square$

EXERCISES 1.6

(1) Prove that the union of a finite set and a countable set is countable.

(2) Prove that the union of a finite family of countable sets is countable.

(3) Let A and B be countable sets. Prove that the set $A \times B$ is countable. ($A \times B$ denotes the set of all ordered pairs (a, b) with $a \in A$ and $b \in B$.)

(4) Prove that every infinite subset A of a countable set B is countable.

(5) Is the set of all finite sequences of rational numbers countable?

(6) Is the set of all infinite sequences of rational numbers countable?

(7) Is the set of all infinite sequences of natural numbers countable?

(8) Is the set of all infinite binary sequences countable?

(9) A number $\alpha \in \mathbb{R}$ is called *algebraic* if it is a root of a polynomial with integer coefficients, that is, there are integers $\lambda_0, \lambda_1, \ldots, \lambda_n$, not all zero, such that $\lambda_0\alpha^n + \lambda_1\alpha^{n-1} + \cdots + \lambda_{n-1}\alpha + \lambda_n = 0$. Is the set of all algebraic numbers countable?

Chapter 2

LIMITS OF FUNCTIONS

2.1 Functions

The concept of a function, arguably the most important concept in mathematics, is also a very important tool in describing phenomena of the surrounding world in which there is a dependence between entities. For instance, how does the distance traveled depend on the elapsed time? How does the pressure of a gas depend on its temperature? Such questions suggest that the concept of a function stems naturally from that of variable entities, called variables for short. However the concept of a variable is difficult to define accurately and all attempts to do so seem to obscure the concept of a function rather than clarify it. In some books a function is defined as a correspondence or a rule. But in fact such definitions only replace one word by another. The reader is assumed not to know what a function is but is supposed to know what a correspondence is. It seems that it would be better to say that everybody knows what a function is and therefore we do not have to define it. On the other hand we do not want to base a mathematical theory on such an assumption, because it would sooner or later cause some problems. There are important examples in the history of mathematics of such unfortunate cases.

So what do we do if we want to have a precise definition of a function? The modern concept of a function is based on the concept of a relation, that is, a set of ordered pairs.

A *function f* from a set X to a set Y is a collection of ordered pairs (x, y) such that

- (a) $x \in X$ and $y \in Y$,
- (b) For every $x \in X$ there exists a $y \in Y$ such that $(x, y) \in f$,
- (c) If $(x, y_1) \in f$ and $(x, y_2) \in f$, then $y_1 = y_2$.

S: Maybe the above formal definition of a function is a very rigorous one but it seems rather artificial. We often say that $y = x^2$ or $f(x) = x^2$ is a function. According to the above definition it is not, because $f(x) = x^2$ is not a collection of ordered pairs.

T: Strictly speaking this is true. On the other hand it is easy to see that the formula $f(x) = x^2$ describes a collection of pairs: a pair $(x, y) \in f$ if and only if $y = f(x)$, that is, if and only if $y = x^2$. Using symbols, we can write $f = \{(x, x^2) : x \in \mathbb{R} \}$.

S: I like the usual method better. The formula $f(x) = x^2$ tells me all I need to know and it is simpler. One has to agree that, for example, $\sin \pi = 0$ looks much better than $(\pi, 0) \in \sin$. And what about $(2, 4) \in^2$?

T: We may like the usual method better for specifying which particular function we are talking about, but that method does not tell us what a function is. Our definition overcomes that difficulty.

We are not going to change the way we specify particular functions. It is important, however, to know that the concept of a function can be rigorously defined in terms of primitive notions. In this case the primitive notion is actually the concept of a set because an ordered pair can be defined as $(a, b) = \{\{a\}, \{a, b\}\}$.

The formal definition of a function is, in some sense, natural and necessary. It is very convenient when a function can be given by a short formula like $f(x) = x^2$ or $f(x) = \sin x$, but what if there is no such formula? If you conduct an experiment in which you want to find a relation between two quantities, then the result of the experiment is a set of pairs. For instance, you drop a small ball from a height h and measure the time t it takes to hit the ground. The data from your experiment is going to be a table; for example:

h feet	1	1.5	2	2.5	3	3.5	4	4.5	5
t seconds	0.25	0.31	0.35	0.40	0.43	0.47	0.50	0.53	0.56

This is actually a set of pairs, and thus it represents a function. Of course, at this point we could try to find a formula which would describe the correspondence. Such a formula would amount to an extension of the correspondence and would represent a larger collection of ordered pairs. A table representation of a function is often used. Its main disadvantage is that it can contain only a finite number of entries.

In the formal definition of a function f there are three elements: a set X, a set Y, and a collection of pairs. The set X is called the *domain* of f. Whenever $(x, y) \in f$, we call y *the value of f at x*, or just *a value of f*. The set of all elements of Y which are values of f is called the *range* of f. In other words, $y \in Y$ is an element of the range of f if there is an $x \in X$ such that $f(x) = y$.

In order to completely specify a function, it is not enough to give a formula which describes the correspondence; we need to specify its domain as well. The range is determined by the formula and the domain. For instance, if $f(x) = x^2$ and the domain is the set of all real numbers, then the range is the set of all non-negative numbers. If $f(x) = x^2$ and the domain is the set of all numbers which are less than -2, then the range is the set of all numbers which are greater than 4.

Instead of saying "f is a function with domain X" we can simply say "f is a function on X". The notation "$f : X \to Y$" is very useful. It means "f is a function with domain X whose range is a subset of Y".

In calculus texts we are often asked to find the domain of a function. To be more precise we should ask what is the largest set of real numbers for which the rule for the function is well-defined. The problem is more or less an arithmetical one. For example, the function $f(x) = x^2$ is defined for all real numbers; $f(x) = \frac{1}{x}$ is defined for all numbers except $x = 0$; $f(x) = \sqrt{x}$ is defined for all non-negative numbers. Whenever a rule for a function is given and its domain is not specified, we understand that its domain is intended to be the largest collection of real numbers for which the rule makes sense. If we are asked to find the range of a function, we mean the largest possible set of values of the function.

The most important property of a function is the one described in (c): for every x in the domain of f, there exists exactly one y such that $f(x) = y$. For example $f(x) = x^2$ describes a function because every real number has exactly one square. It is possible, however, that the same y is assigned to different elements of the domain: $1^2 = 1$ and $(-1)^2 = 1$.

Instead of saying "the function defined by the formula $f(x) = x^2$" we often say "the function $f(x) = x^2$" or even "the function x^2". The last abbreviation is not quite correct, for x^2 is not a function, but a number associated with x, namely, the value of the function $f(x) = x^2$ at x. In situations when doubt could arise, it is better to use another symbol, for example, $\{x^2\}$. Similarly, the function defined by the formula $f(x) = 1/x$ would be denoted by $\{1/x\}$. It may happen that we do not know a general formula for a function or, for some reason, we do not want to use it. The number associated with x can be denoted by $f(x)$ and the function itself by $\{f(x)\}$ or just by f. If we simultaneously have to use other functions we may use other letters like g or h.

A systematic use of braces $\{\ \}$ to distinguish a function from its values is cumbersome in practice. We therefore will drop the braces if there is no fear of misunderstanding.

Some functions are traditionally denoted by special symbols. For instance, the function f which assigns to x the greatest integer less than or equal to x is often denoted by $f(x) = \lfloor x \rfloor$. In the following chapters we will also discuss functions denoted by: ln, sin, cos, tan, cot, arctan, arcsin, arccos.

EXERCISES 2.1

(1) Guess an algebraic formula for the function described approximately by the table given in this section.

(2) Find the domains of the following functions:

$\quad$ (a) $f(x) = \sqrt{3 - 5x - 2x^2}$,
$\qquad\qquad$ (c) $f(x) = 1/\sqrt{x^2}$,
$\quad$ (b) $f(x) = 1/(x^2 - 4)$,
$\qquad\qquad$ (d) $f(x) = \sqrt[4]{[1/(x^2 + 1)]} - 1$.

(3) Find the ranges of the following functions:

$\quad$ (a) $f(x) = 1 + x^2$,
$\qquad\qquad$ (e) $f(x) = (1 + x)^3$,
$\quad$ (b) $f(x) = (1 + x)^2$,
$\qquad\qquad$ (f) $f(x) = \lfloor x \rfloor$,
$\quad$ (c) $f(x) = 1/(1 + x^2)$,
$\qquad\qquad$ (g) $f(x) = \lfloor x^2 + 7 \rfloor$.
$\quad$ (d) $f(x) = 1 + x^3$,

(4) Do the following expressions define functions?

$\quad$ (a) $f(x) = $ the bigger of the numbers $x + 7$ and $\frac{1}{2}x$,
$\quad$ (b) $f(x) = $ the smaller of the numbers x^2 and $2x$,
$\quad$ (c) $f(x) = $ a number whose square equals $x^2 + 1$,
$\quad$ (d) $f(x) = $ a number whose cube equals $x + 1$,
$\quad$ (e) $f(x) = $ a number which multiplied by x gives 11.

(5) Is the set of all functions $f : \{1, 2\} \to \mathbb{N}$ countable? How do you know?
(6) Is the set of all functions $f : \mathbb{N} \to \{1, 2\}$ countable? How do you know?

2.2 Geometric interpretation of numbers and functions

In the past mathematical analysis was often based on geometric proofs. In the modern approach, geometry is completely eliminated as a proving instrument in analysis. Nevertheless, connecting analysis with geometry is very important, because geometry provides intuitive interpretation of many ideas and offers precious hints in various investigations. The reader is encouraged to find geometric interpretations whenever possible.

$\quad$ In this book the geometry will be used only to illustrate the text. It will be treated rather intuitively, without observing logical accuracy.

$\quad$ Numbers can be interpreted geometrically as points on a straight line in such a way that every real number corresponds to one and only one point on the line, and conversely, every point on the line corresponds to exactly one real number. This can be done in infinitely many ways. We first chose an arbitrary point on the line and assign to it the number 0. Then we chose another point (traditionally it is a point to the right of the one assigned to zero) and call it 1. The distance from 0 to 1 can be used to assign numbers to the remaining points. If a point is to the right of 0 then we assign to it the positive real number equal to the distance from zero to this point, where the distance from 0 to 1 is the unit. The same rule applies to points to the left of 0 with the only difference being that negative numbers are used. A line for which this has been done will be called a *real axis* or a *real line* (see Fig. 2.1).

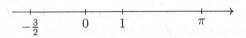

Fig. 2.1   The real axis.

We often use a geometrical way of speaking. For instance, we say "the value of the function f at the point x" when we mean "the value of the function f at the number x".

By an *open interval* (a, b) we mean the set of all numbers x such that $a < x < b$. If we include the numbers a and b, then we obtain a *closed interval* $[a, b]$. Thus the closed interval $[a, b]$ is the set of all numbers x such that $a \le x \le b$. The numbers a and b are called the *end-points* of the interval; the numbers x satisfying the inequalities $a < x < b$ are called the *interior points* of the interval. Every point in the open interval (a, b) is an interior point.

Open and closed intervals have simple geometric interpretations. They represent line segments between points a and b: an open interval is represented by a segment without ends and a closed interval with ends (see Fig. 2.2).

Fig. 2.2   Open and closed intervals.

We will also use *infinite intervals*: (a, ∞), $(-\infty, b)$, and $(-\infty, \infty)$. The first one denotes the set of all real numbers greater than a, the second one the set of all numbers less than b, and the third one the set of all real numbers. Other intervals often used are the so-called *half-open* or *half-closed* intervals. Below we list all possible types of intervals.

$$x \in (a, b) \quad \text{means } a < x < b,$$
$$x \in (a, b] \quad \text{means } a < x \le b,$$
$$x \in [a, b) \quad \text{means } a \le x < b,$$
$$x \in [a, b] \quad \text{means } a \le x \le b,$$
$$x \in (a, \infty) \quad \text{means } x > a,$$
$$x \in [a, \infty) \quad \text{means } x \ge a,$$
$$x \in (-\infty, b) \quad \text{means } x < b,$$
$$x \in (-\infty, b] \quad \text{means } x \le b,$$

and

$$(-\infty, \infty) = \mathbb{R}.$$

Given two nonparallel real axes, with each point in the plane we may associate two points on the axes, which are its projections. In this way to each point of the plane there correspond two numbers, and conversely, to each pair of numbers there corresponds a point of the plane. Unless specifically mentioned otherwise, we always assume that the axes are perpendicular and the intersection point is 0 on both lines. The intersection is called the *origin* and is denoted by O. A plane with two perpendicular axes is called a *coordinate plane* (see Fig. 2.3).

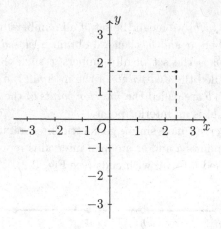

Fig. 2.3   The coordinate plane.

Since a function is a set of pairs of numbers it can be represented by a set of points on the plane. This set of points is called the *graph* of that function. A graph can be sketched by first constructing a table of the function and then plotting the points on the plane. If the points are dense enough, then we complete the sketch by guessing its shape. We cannot expect great accuracy but it usually gives us an idea what the graph looks like. In the past, technicians often used a French curve to complete the sketch. It did not necessarily improve the accuracy; it only gave a better appearance. Nowadays, computers can generate graphs of functions with practically arbitrary accuracy. Figure 2.4 contains a few examples of graphs.

EXERCISES 2.2

(1) Match each function with its graph in Figure 2.4: $a(x) = \lfloor x \rfloor$; $b(x) = x^2$; $c(x) = \frac{1}{x}$; $d(x) = x - \lfloor x \rfloor$.

(2) Sketch graph of the following functions:

    (a) $a(x) = \lfloor x^2 \rfloor$ for x in $[0, 3]$;

    (b) $b(x) = \lfloor -x^2 \rfloor$ for x in $[0, 3]$;

    (c) $c(x) = -\lfloor x^2 \rfloor$ for x in $[0, 3]$;

    (d) $d(x) =$ the distance from x to 3 for x in $[-5, 5]$;

(e) $e(x) =$ the average distance from x to -1 and 1 for x in $[-5, 5]$;

(f) $f(x) = \left\lfloor \frac{\lfloor x \rfloor + 1}{3} \right\rfloor$ for x in $[0, 10]$;

(g) $g(x) = x/\left((x+2)^{\lfloor 2 - x^2 \rfloor} - 2\right)^2$ for x in $[-1, 1]$.

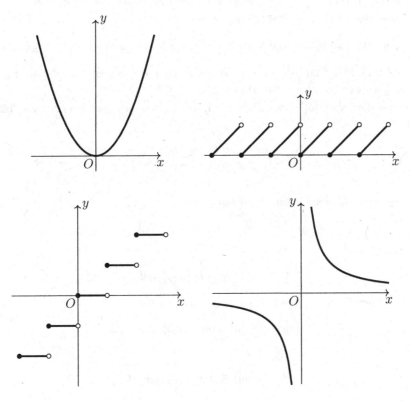

Fig. 2.4   Examples of graphs.

2.3   The absolute value

The *absolute value* of a non-negative number is the number itself, and the absolute value of a negative number is the opposite of that number, that is,

$$|a| = \begin{cases} a, & \text{if } a \geq 0, \\ -a, & \text{if } a < 0. \end{cases} \tag{2.1}$$

Note that the absolute value of a number is always a non-negative number. The absolute value of a number different from zero is always a positive number. For instance, we have $|5| = 5$, $|0| = 0$, $|-7| = 7$.

2.3.1. $|a| = |-a|$ *for any number a.*

Proof. The assertion is obvious if $a = 0$. If $a > 0$, then $|a| = a$ and $|-a| = -(-a) = a$. If $a < 0$, then $|a| = -a$ and $|-a| = -a$, because $-a > 0$.   □

2.3.2. $-|a| \leq a \leq |a|$ *for any number a.*

Proof. If $a \geq 0$, then $-a \leq a$ and, by (2.1), we have $-|a| \leq a = |a|$. If $a < 0$, then $a \leq -a$ and, by (2.1), $-|a| = a \leq |a|$.   □

2.3.3. *For any numbers a and b we have* $|a| \leq b$ *if and only if* $-b \leq a \leq b$.

Proof. Assume first that $-b \leq a \leq b$. If $0 \leq a$, then $a \leq b$ gives us $|a| \leq b$; if $a < 0$, then $-b \leq a$ gives us $-b \leq -|a|$, thus again $|a| \leq b$.

Assume now that $|a| \leq b$. Then we have $-b \leq -|a|$, and hence, by 2.3.2, we obtain

$$-b \leq -|a| \leq a \leq |a| \leq b.$$

□

2.3.4. *For any numbers a and b we have*

$$|a| - |b| \leq |a + b| \leq |a| + |b|$$

and

$$|a| - |b| \leq |a - b| \leq |a| + |b|.$$

Proof. Adding the corresponding parts of inequalities

$$-|a| \leq a \leq |a| \quad \text{and} \quad -|b| \leq b \leq |b|$$

we get

$$-(|a| + |b|) \leq a + b \leq |a| + |b|.$$

Hence

$$|a + b| \leq |a| + |b| \tag{2.2}$$

by 2.3.3. Replacing b by $-b$ in (2.2) we get

$$|a - b| \leq |a| + |b| \tag{2.3}$$

which implies

$$|a| = |(a + b) - b| \leq |a + b| + |b|,$$

and therefore

$$|a| - |b| \leq |a + b|. \tag{2.4}$$

Finally, replacing b by $-b$ in (2.4) we obtain

$$|a| - |b| \leq |a - b| \tag{2.5}$$

Inequalities in 2.3.4 are abbreviated forms of (2.2), (2.3), (2.4), and (2.5).   □

2.3.5. $|ab| = |a|\,|b|$ *for any numbers a and b.*

Proof. The equality

$$|ab| = |a||b|, \quad \text{for } a \geq 0 \text{ and } b \geq 0,$$

is obvious. In the remaining cases, we make use of the above equality and 2.3.1 to obtain

$$|ab| = |-ab| = |a(-b)| = |a|\,|-b| = |a|\,|b|, \quad \text{for } a \geq 0 \text{ and } b < 0;$$
$$|ab| = |-ab| = |(-a)b| = |-a|\,|b| = |a|\,|b|, \quad \text{for } a < 0 \text{ and } b \geq 0;$$
$$|ab| = |(-a)(-b)| = |-a|\,|-b| = |a|\,|b|, \quad \text{for } a < 0 \text{ and } b \geq 0.$$

$\square$

2.3.6. $\left|\frac{a}{b}\right| = \frac{|a|}{|b|}$ *for any number a and any number $b \neq 0$.*

Proof. If $b \neq 0$, then from 2.3.5 we have $\left|b\frac{a}{b}\right| = |b|\left|\frac{a}{b}\right|$, that is, $|a| = |b|\left|\frac{a}{b}\right|$. This gives the desired identity. $\square$

EXERCISES 2.3

(1) Graph the following functions:

    (a) $a(x) = |x|$ for x in $[-5,5]$,
    (b) $b(x) = ||x| - 1|$ for x in $[-5,5]$,
    (c) $c(x) = |||x| - 1| - 1|$ for x in $[-5,5]$,
    (d) $d(x) = |x+1|/(|x|+1)$ for x in $[-5,5]$.

(2) Prove that if $b - 1 < a < b + 1$ then $|a| < |b| + 1$.

(3) On the real axis, graph the sets of numbers satisfying the following conditions:

    (a) $|x - 3| < 2$,   (d) $|x - a| < 1$,
    (b) $|x + 2| \leq 3$,   (e) $|x + b| \geq 4$,
    (c) $0 < \left|x - \frac{1}{2}\right| \leq 2$,   (f) $|x^2 - 4| < \frac{1}{3}$.

(4) Prove that, for any $x, y \in \mathbb{R}$,

    (a) $\max\{x, y\} = \frac{1}{2}(x + y + |x - y|)$,
    (b) $\min\{x, y\} = \frac{1}{2}(x + y - |x - y|)$.

(5) Let a and b be real numbers. Give a geometric interpretation for $|a - b| < 4$.

(6) Let x denote a real number. Give a geometric interpretation for $|x + 2| \geq 3$.

(7) Let x, a and δ be real numbers with $\delta > 0$. Give a geometric interpretation for $0 < |x - a| < \delta$.

(8) Let P_1, P_2, P_3 be the vertices of a triangle and let $d(P_i, P_j)$ denote the distance from P_i to P_j. Then,

$$d(P_1, P_3) \leq d(P_1, P_2) + d(P_2, P_3)$$

is called *the triangle inequality*. Draw a picture of the triangle and explain the geometric significance of the inequality. Also, in light of this, draw another illustrative picture and explain why the inequality $|a + b| \leq |a| + |b|$ is also called the triangle inequality.

2.4   Limit of a function

The function $f(x) = (x - 2)^2$ (see Fig. 2.5) assumes the value 0 at $x = 2$. If x is only a little greater than 2, then the value of the function differs only a little from 0.

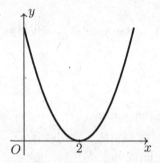

Fig. 2.5   Graph of $y = (x - 2)^2$.

Now consider the function defined by

$$f(x) = \frac{-(x - 2)^2}{|x - 2| + (x - 2)},$$

(see Fig. 2.6). If $x \leq 2$, then $|x - 2| = -(x - 2)$ and the denominator equals zero. Thus the function is not defined at $x = 2$. It is defined only for $x > 2$ because then $|x - 2| = x - 2$ and the value of the function is $-\frac{1}{2}(x - 2)$. Therefore, if x is only a little greater than 2, then the value of the function differs only a little from 0. This shows that the discussed property is not connected with the value of the function at the point 2, but rather with points on the right of that point. If x is close to 2 on the right side of 2, then the value of the function at x is close to 0. In fact, the value of the function f can be made as close to 0 as one pleases by taking x sufficiently close to 2 on the right side of 2. One can also say that values of the function tend to zero when x approaches 2 from the right. The following definition expresses this situation in the precise language of analysis.

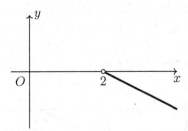

Fig. 2.6 Graph of $y = \frac{-(x-2)^2}{|x-2|+(x-2)}$.

Definition 2.4.1. We say that the number l is the *right-hand limit* of a function f at a point a, if for every positive number ε there exists a positive number δ such that for every $x \in (a, a + \delta)$ we have $|f(x) - l| < \varepsilon$.

If the right-hand limit of f at a is l, we write

$$\lim_{x \to a+} f(x) = l \quad \text{or} \quad f(a+) = l.$$

The arbitrary positive ε makes precise "as close as one pleases" while the δ makes precise "sufficiently close."

S: I don't like this definition. It seems to be unnecessarily complicated. I understand the examples before the definition, but it is not easy to see the connection.

T: Only beginners have this problem. Probably the difficulty arises from the use of multiple quantifiers.

S: What are quantifiers?

T: The expressions *for every* and *there exists* are called quantifiers. In our definition we have three quantifiers: *for every*, then *there exists*, and finally again *for every*. The order of these quantifiers is essential. For ε we can take any positive number, whereas the choice of δ depends on ε. Quantifiers play a very important role in the language of mathematics. Other synonyms are also used. For instance, instead of *for every* one can say *for each*, *for all*, or *for any*. Similarly, *there exists* can be replaced by *there is*, *for some*, or *for certain*. The quantifier *for every* is often omitted. For example, if we say *for $a < x < a + \delta$ we have* $|f(x) - l| < \varepsilon$, we actually mean *for every x such that $a < x < a + \delta$ we have* $|f(x) - l| < \varepsilon$.

A function that has a limit at a point may behave in a more complicated manner than in the examples given above. For instance, the right-hand limit at 0 of the function

$$f(x) = x \left\lfloor \frac{2}{x} \right\rfloor + \frac{x}{2} - 2$$

is 0, and it assumes both positive and negative values in any right neighborhood of zero (see Fig. 2.7).

Fig. 2.7   Graph of $y = x \lfloor \frac{2}{x} \rfloor + \frac{x}{2} - 2$.

The left-hand limit is defined similarly to the right-hand limit.

Definition 2.4.2. We say that the number l is the *left-hand limit* of a function f at a point a, if for every positive number ε there exists a positive number δ such that for every $x \in (a - \delta, a)$ we have $|f(x) - l| < \varepsilon$.

If the left-hand limit of f at a is l, we write

$$\lim_{x \to a-} f(x) = l \quad \text{or} \quad f(a-) = l.$$

If both the left-hand limit and right-hand limit exist and are equal to l, then we say that the function has the limit l at a.

Definition 2.4.3. We say that the number l is the *limit* of a function f at a point a, if for every positive number ε there exists a positive number δ such that for every x satisfying $0 < |x - a| < \delta$ we have $|f(x) - l| < \varepsilon$.

If the limit of f at a is l, we write

$$\lim_{x \to a} f(x) = l.$$

For example,

$$\lim_{x \to a} x = a.$$

If $f(x) = c$ for $a \le x \le b$, then $\lim_{x \to a+} f(x) = c$, $\lim_{x \to b-} f(x) = c$, and $\lim_{x \to d} f(x) = c$ for $a < d < b$. In other words: *a constant function $f(x) = c$ has the limit c at every point.*

It is very important to remember that a function need not be defined at a point to have a limit at that point. In the two examples considered above that is exactly the case. The function

$$f(x) = \frac{-(x - 2)^2}{|x - 2| + (x - 2)}$$

is not defined at $x = 2$, but

$$\lim_{x \to 2+} \frac{-(x - 2)^2}{|x - 2| + (x - 2)} = 0.$$

Similarly, the function

$$f(x) = x \left\lfloor \frac{2}{x} \right\rfloor + \frac{x}{2} - 2$$

is not defined at $x = 0$, but

$$\lim_{x \to 0+} \left(x \left\lfloor \frac{2}{x} \right\rfloor + \frac{x}{2} - 2 \right) = 0.$$

For a function f to have a right-hand limit at a it is necessary for f to be defined in an interval of the form $(a, a + \delta)$, where δ is some positive number. In other words, an interval of the form $(a, a + \delta)$ has to be contained in the domain of f. Similarly, to have a left-hand limit at a, f has to be defined in an interval of the form $(a - \delta, a)$, for some $\delta > 0$. Finally, to have a two-sided limit at a, f has to be defined in both intervals $(a - \delta, a)$ and $(a, a + \delta)$ for some positive δ. Of course, being defined in those intervals is not sufficient for existence of a limit. Consider, for example, the function

$$f(x) = \frac{1}{x} - \left\lfloor \frac{1}{x} \right\rfloor.$$

This function is defined for all $x \neq 0$, but does not have a limit at 0, not even one-sided limits. The partial graph of the function, shown in Fig. 2.8, indicates the behavior of the function near 0. In Example 2.4.7 we give a formal proof of the fact that the right-hand limit of this function does not exist.

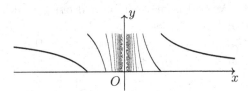

Fig. 2.8 Graph of $f(x) = \frac{1}{x} - \left\lfloor \frac{1}{x} \right\rfloor$.

The intervals $(a, a + \delta)$ and $(a - \delta, a)$ are called a *right neighborhood* of a and a *left neighborhood* of a, respectively. The set $(a - \delta, a) \cup (a, a + \delta)$ is often called a *deleted neighborhood* of a. Note that the inequalities $0 < |x - a| < \delta$ in Definition 2.4.3 describe such a deleted neighborhood. The role of the inequality $0 < |x - a|$ is to exclude $x = a$.

Example 2.4.4. We are going to use Definition 2.4.3 to prove that

$$\lim_{x \to 1} \frac{1}{x^2 + 1} = \frac{1}{2}.$$

Let ε be an arbitrary positive number. (We cannot specify ε. We have to present an argument which works for every positive ε.) We need to find a positive number δ such that

$$\left| \frac{1}{x^2 + 1} - \frac{1}{2} \right| < \varepsilon \quad \text{whenever } 0 < |x - 1| < \delta.$$

(Finding δ actually means expressing it in terms of ε. It is difficult to guess what δ is going to have the desired property. But since $\left|\frac{1}{x^2+1} - \frac{1}{2}\right| = 0$ when $x = 1$, it seems likely that we can rewrite $\left|\frac{1}{x^2+1} - \frac{1}{2}\right|$ in such a way that $|x - 1|$ will be a factor. That rewrite may give us a hint concerning a reasonable next step.) For any number x, we have

$$\left|\frac{1}{x^2+1} - \frac{1}{2}\right| = \left|\frac{1-x^2}{2(x^2+1)}\right| = \left|\frac{(1-x)(1+x)}{2(x^2+1)}\right| = \left|\frac{1+x}{2(x^2+1)}\right| |x-1|. \tag{2.6}$$

Now lets find a number α such that $\left|\frac{1+x}{2(x^2+1)}\right| \leq \alpha$. It is easy to see that, for any x, $2 \leq \left|2(x^2+1)\right|$, so

$$\left|\frac{1+x}{2(x^2+1)}\right| = \frac{|1+x|}{|2(x^2+1)|} \leq \frac{|1+x|}{2}. \tag{2.7}$$

Notice that

$$|1+x| < 3 \tag{2.8}$$

whenever $|x - 1| < 1$. This is because $|x - 1| < 1$ means $-1 < x - 1 < 1$ which implies that $1 < x + 1 < 3$. Now, from (2.6) and (2.7), and (2.8) we have

$$\left|\frac{1}{x^2+1} - \frac{1}{2}\right| \leq \frac{3}{2}|x - 1|$$

whenever $|x - 1| < 1$. We will have accomplished our goal if $|x - 1| < \frac{2}{3}\varepsilon$. Since we want to be sure that $|x - 1|$ is less than both 1 and $\frac{3}{2}\varepsilon$, we choose δ to be the smaller of 1 and $\frac{2}{3}\varepsilon$. Now, for every x such that $0 < |x - 1| < \delta$, we have $|x - 1| < 1$ and $|x - 1| < \frac{2}{3}\varepsilon$ so that

$$\left|\frac{1}{x^2+1} - \frac{1}{2}\right| \leq \frac{3}{2}|x - 1| < \frac{3}{2} \cdot \frac{2}{3}\varepsilon = \varepsilon.$$

The proof is now complete.

To find the limit of a function at a point we often use the following theorem.

Theorem 2.4.5. *If two functions are identical in a deleted neighborhood of a point a and one of them has the limit l at a, then the other one also has the limit l at a. More precisely, if f and g are both defined and equal in $(a - \gamma, a)$ and in $(a, a + \gamma)$ for some $\gamma > 0$, and $\lim_{x \to a} f(x) = l$, then $\lim_{x \to a} g(x) = l$.*

Proof. Assume that f and g are both defined and equal in $(a - \gamma, a)$ and in $(a, a + \gamma)$, for some $\gamma > 0$. In other words,

$$f(x) = g(x) \quad \text{whenever } 0 < |x - a| < \gamma.$$

Now let ε be an arbitrary positive number. Since $\lim_{x \to a} f(x) = l$, there exists a positive number δ such that for every x satisfying $0 < |x - a| < \delta$ we have $|f(x) - l| < \varepsilon$. Note that, if $0 < \delta' < \delta$, then for every x satisfying $0 < |x - a| < \delta'$ we still have $|f(x) - l| < \varepsilon$. For this reason we can assume, without loss of generality, that $\delta < \gamma$. Now, if $0 < |x - a| < \delta$, then $0 < |x - a| < \gamma$, and consequently $f(x) = g(x)$. Therefore, if $0 < |x - a| < \delta$, then

$$|g(x) - l| = |f(x) - l| < \varepsilon.$$

Since ε is an arbitrary positive number, this proves that $\lim_{x \to a} g(x) = l$. $\quad\square$

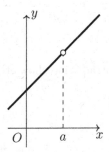

Fig. 2.9   Graph of $y = \frac{x^2 - a^2}{x - a}$.

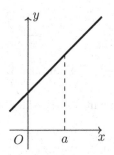

Fig. 2.10   Graph of $y = x + a$.

For instance, the functions

$$f(x) = \frac{x^2 - a^2}{x - a} \quad \text{and} \quad g(x) = x + a,$$

(see Fig. 2.9 and Fig. 2.10), are identical everywhere except for $x = a$. Thus, instead of calculating the limit of the first function, we can find the limit of the second one. It is easily seen that

$$\lim_{x \to a} (x + a) = 2a.$$

Hence also

$$\lim_{x \to a} \frac{x^2 - a^2}{x - a} = 2a.$$

Theorem 2.4.6. *If a function has a limit at a point, then that limit is unique. In other words, if $\lim_{x \to a} f(x) = l$ and $\lim_{x \to a} f(x) = m$, then $l = m$. The same is true for one-sided limits.*

Proof. Our proof is a proof by contradiction. Assume $\lim_{x \to a} f(x) = l$ and $\lim_{x \to a} f(x) = m$ and suppose that $l \neq m$. Let $\varepsilon = \frac{1}{2}|l - m|$. Note that this ε is a positive number. Since $\lim_{x \to a} f(x) = l$, there exists a $\delta_l > 0$ such that

$$|f(x) - l| < \varepsilon \quad \text{whenever } 0 < |x - a| < \delta_l.$$

Similarly, since $\lim_{x \to a} f(x) = m$, there exists a $\delta_m > 0$ such that

$$|f(x) - m| < \varepsilon \quad \text{whenever } 0 < |x - a| < \delta_m.$$

Define δ to be the smaller of δ_m and δ_l. If $0 < |x - a| < \delta$, then

$$\varepsilon = \frac{1}{2}|l - m|$$

$$= \frac{1}{2}|l - f(x) + f(x) - m|$$

$$\leq \frac{1}{2}(|l - f(x)| + |f(x) - m|)$$

$$< \frac{1}{2}(\varepsilon + \varepsilon) = \varepsilon,$$

since $0 < |x - a| < \delta_l$ and $0 < |x - a| < \delta_m$. This contradiction shows that we must have $l = m$.

The argument for one-sided limits is very similar. $\qquad\square$

Example 2.4.7. We intend to show that $\lim_{x \to 0+} \left(\frac{1}{x} - \lfloor \frac{1}{x} \rfloor \right)$ does not exist by using

$$\lim_{x \to a} \frac{1}{x} = \frac{1}{a}, \quad \text{for any real number } a > 0, \tag{2.9}$$

and

$$\lim_{x \to \frac{1}{n}+} \left(\frac{1}{x} - \left\lfloor \frac{1}{x} \right\rfloor \right) = 1, \quad \text{for any natural number } n, \tag{2.10}$$

whose proofs are left as exercises (see Exercises 2.4.3 and 2.4.4). Our proof that $\lim_{x \to 0+} \left(\frac{1}{x} - \lfloor \frac{1}{x} \rfloor \right)$ does not exist is motivated by the fact that in every right neighborhood of 0 there are points where our function $f(x) = \frac{1}{x} - \lfloor \frac{1}{x} \rfloor$ has the value 0 and other points where it has values arbitrarily close to 1 (see Fig. 2.8). Our proof is also a proof by contradiction. To this end, suppose $\lim_{x \to 0+} f(x) = l$ for some number l. Then there is some $\delta > 0$ such that for every $x \in (0, \delta)$ we have

$$|f(x) - l| < \frac{1}{3}. \tag{2.11}$$

By Corollary 1.5.4, there exists a rational number β such that $0 < \beta < \delta$. By the Archimedean property (Theorem 1.5.3), there exists a natural number n such that $n > \frac{1}{\beta}$. Thus $0 < \frac{1}{n} < \delta$. It follows from (2.10) that there is some $x_0 \in (\frac{1}{n}, \delta)$ such that

$$|f(x_0) - 1| < \frac{1}{3}. \tag{2.12}$$

Now, using the fact that $\frac{1}{n}, x_0 \in (0, \delta)$, n is a natural number, (2.11), (2.12), and the triangle inequality, we conclude that

$$1 = \left| 1 - f\left(\frac{1}{n}\right) \right|$$

$$= \left| 1 - f(x_0) + f(x_0) - f\left(\frac{1}{n}\right) \right|$$

$$\leq |1 - f(x_0)| + \left| f(x_0) - f\left(\frac{1}{n}\right) \right|$$

$$< \frac{1}{3} + \left| f(x_0) - {}'k + k - f\left(\frac{1}{n}\right) \right|$$

$$\leq \frac{1}{3} + |f(x_0) - k| + \left| k - f\left(\frac{1}{n}\right) \right|$$

$$\leq \frac{1}{3} + \frac{1}{3} + \frac{1}{3} = 1,$$

a contradiction. Thus, $\lim_{x \to 0+} \left(\frac{1}{x} - \lfloor \frac{1}{x} \rfloor\right)$ does not exist.

EXERCISES 2.4

(1) Use Definition 2.4.3 to prove that $\lim_{x \to a} x = a$ for an arbitrary number a.

(2) Use Definition 2.4.3 to prove that $\lim_{x \to a}(x + a) = 2a$ for an arbitrary number a.

(3) Use Definition 2.4.3 to prove that $\lim_{x \to a} \frac{1}{x} = \frac{1}{a}$ for any real number $a > 0$.

(4) Use Definition 2.4.1 to prove that $\lim_{x \to \frac{1}{n}+} \left(\frac{1}{x} - \lfloor \frac{1}{x} \rfloor\right) = 1$, for any natural number n.

(5) Use Definition 2.4.3 to prove that $\lim_{x \to 0} |x| = 0$.

(6) Use Definition 2.4.1 to prove that $\lim_{x \to 0+}(x^2 + x + 1) = 1$.

(7) Use Definition 2.4.2 to prove that $\lim_{x \to 1-} \frac{1}{x+1} = \frac{1}{2}$.

(8) Use Definition 2.4.3 to prove that $\lim_{x \to -2} \frac{1}{x^2-1} = \frac{1}{3}$.

(9) Use Definitions 2.4.1 and 2.4.2 to prove that $\lim_{x \to 0+} \frac{x}{|x|} = 1$ and $\lim_{x \to 0-} \frac{x}{|x|} = -1$.

(10) Use Definitions 2.4.1 and 2.4.2 to prove that $\lim_{x \to n-} \lfloor x \rfloor = n - 1$ and $\lim_{x \to n+} \lfloor x \rfloor = n$, for any integer n.

(11) Use Definition 2.4.3 to prove that $\lim_{x \to 0} |x| \neq 1$.

(12) Use Definition 2.4.3 to prove that $\lim_{x \to 0}(x^2 + x + 1) \neq 2$.

(13) Use Definition 2.4.3 to prove that $\lim_{x \to 0} \frac{\sqrt{x^2-x+2}}{\sqrt{x-2}+x-2} \neq 0$.

(14) Show that $\lim_{x \to 0} \frac{|x|}{x}$ does not exist.

(15) Show that $\lim_{x \to 1+} \frac{1}{x-1}$ does not exist.

(16) Show that $\lim_{x \to -2} \lfloor x \rfloor$ does not exist.

(17) Let

$$f(x) = \begin{cases} 1, & \text{if } x \text{ is a rational number,} \\ 0, & \text{if } x \text{ is an irrational number.} \end{cases}$$

Show that $\lim_{x \to a} f(x)$ does not exist at any $a \in \mathbb{R}$.

(18) Let

$$f(x) = \begin{cases} x, & \text{if } x \text{ is a rational number,} \\ 0, & \text{if } x \text{ is an irrational number.} \end{cases}$$

Find all $a \in \mathbb{R}$ for which $\lim_{x \to a} f(x)$ exists.

(19) Prove Theorem 2.4.6 for one-sided limits.

(20) Prove that $\lim_{x \to a} f(x) = l$ implies $\lim_{x \to a} |f(x)| = |l|$. Is the converse true?

(21) Prove that $\lim_{x \to 0} f(x) = l$ if and only if $\lim_{x \to 0} f(-x) = -l$. Is the same true for one-sided limits?

2.5   Algebraic operations on limits

In the following theorem the sign "lim" means any of the three symbols $\lim_{x \to a+}$, $\lim_{x \to a-}$, or $\lim_{x \to a}$ (but the same in the whole theorem).

Theorem 2.5.1. *If* $\lim f(x) = l$ *and* $\lim g(x) = m$, *then*

(a) $\lim (f(x) + g(x)) = l + m$,

(b) $\lim (f(x) - g(x)) = l - m$,

(c) $\lim f(x) g(x) = lm$.

Moreover, if $m \neq 0$, *then*

(d) $\lim \frac{f(x)}{g(x)} = \frac{l}{m}$.

In order to prove the theorem we first establish the following lemma.

Lemma 2.5.2. *Let h be a function defined in an interval (a, b). The following conditions are equivalent:*

(a) *For every number $\varepsilon > 0$ there exists a number $\delta > 0$ such that for every $x \in (a, a + \delta)$ we have $|h(x)| < \varepsilon$,*

(b) *There exists a number $M > 0$ such that for every number $\varepsilon > 0$ there exists a number $\delta > 0$ such that for every $x \in (a, a + \delta)$ we have $|h(x)| < M\varepsilon$.*

It is important that the constant M does not depend on ε. Before approaching the proof note that the only difference between the last condition and Definition 2.4.1 is the introduction of the constant M. The meaning of the lemma will become clear in the proof of Theorem 2.5.1. Obviously, the lemma remains true if the interval $(a, a + \delta)$ is replaced by $(b - \delta, b)$.

Proof of Lemma 2.5.2. Clearly (a) implies (b), since we can take $M = 1$.

Now assume that M is such that for every $\varepsilon > 0$ there is $\delta > 0$ such that for every $x \in (a, a + \delta)$ we have $|h(x)| < M\varepsilon$. Then, in particular, it is true for $\frac{\varepsilon}{M}$ instead of ε, and hence $|h(x)| < M \frac{\varepsilon}{M} = \varepsilon$. $\qquad \square$

Proof of Theorem 2.5.1. By the triangle inequality, we have

$$|f(x) + g(x) - (l + m)| = |f(x) - l + g(x) - m|$$
$$\leq |f(x) - l| + |g(x) - m|. \tag{2.13}$$

For every $\varepsilon > 0$ there exist numbers $\delta_1 > 0$ and $\delta_2 > 0$ such that

$$|f(x) - l| < \varepsilon \quad \text{for all } x \in (a, a + \delta_1),$$
$$|g(x) - m| < \varepsilon \quad \text{for all } x \in (a, a + \delta_2). \tag{2.14}$$

Let δ denote the smaller of the numbers δ_1 and δ_2. Then, by (2.13), we have

$$|f(x) + g(x) - (l + m)| < 2\varepsilon \quad \text{for all } x \in (a, a + \delta).$$

This shows that the function $f + g$ has the right-hand limit at a equal to $l + m$, by Lemma 2.5.2 (with $h(x) = f(x) + g(x) - (l + m)$ and $M = 2$).

Similarly

$$|f(x) - g(x) - (l - m)| \leq |f(x) - l| + |g(x) - m|$$

and the rest of the proof follows as before.

The proofs of (c) and (d) are slightly more difficult. First notice that $f(x)g(x) - lm = (f(x) - l)g(x) + l(g(x) - m)$.

There is a number $\delta_0 > 0$ such that $|g(x) - m| < 1$ for all $x \in (a, a + \delta_0)$, and thus

$$|g(x)| = |g(x) - m + m| \leq |g(x) - m| + |m| < 1 + |m|.$$

Hence, by 2.3.5 and the triangle inequality,

$$|f(x)g(x) - lm| = |(f(x) - l)g(x) + l(g(x) - m)|$$
$$\leq |f(x) - l|\,|g(x)| + |l|\,|g(x) - m|$$
$$< \varepsilon\,(1 + |m|) + |l|\,\varepsilon = (1 + |m| + |l|)\,\varepsilon$$

for all $x \in (a, a + \delta)$, where δ is the least of the numbers δ_0, δ_1, and δ_2 (where δ_1 and δ_2 are such that (2.14) holds).

To prove (d) we write

$$\frac{f(x)}{g(x)} - \frac{l}{m} = \frac{m(f(x) - l) - l(g(x) - m)}{mg(x)}.$$

There is a number δ_0 such that $|g(x) - m| < \frac{|m|}{2}$, and thus, by 2.3.4,

$$|m| - |g(x)| \leq |m - g(x)| = |g(x) - m| < \frac{|m|}{2}$$

so that $|g(x)| > \frac{|m|}{2}$ for all $x \in (a, a + \delta_0)$. Hence

$$\left| \frac{f(x)}{g(x)} - \frac{l}{m} \right| = \left| \frac{m(f(x) - l) - l(g(x) - m)}{mg(x)} \right|$$
$$\leq \frac{|m|\,|f(x) - l| + |l|\,|g(x) - m|}{|m|\,|g(x)|}$$
$$< \frac{|m|\varepsilon + |l|\varepsilon}{\frac{1}{2}m^2} = \frac{2(|m| + |l|)}{m^2}\,\varepsilon$$

for all $x \in (a, a + \delta)$, where δ is the least of numbers δ_0, δ_1, and δ_2.

For the left-hand limit, the proof is the same; we only have to replace the intervals $(a, a + \delta)$, $(a, a + \delta_0)$, $(a, a + \delta_1)$, and $(a, a + \delta_2)$ by the intervals $(a - \delta, a)$, $(a - \delta_0, a)$, $(a - \delta_1, a)$, and $(a - \delta_2, a)$, respectively. Since Theorem 2.5.1 is true for left-hand and right-hand limits, it is evidently true for two-sided limits, when both one sided limits exist and are equal. $\qquad\square$

Note that Theorem 2.5.1 can be thought of as a theorem about commutativity of certain operations. For example, it says that the order of operations of addition of functions and the limit of functions can be interchanged. We can find the limits first and then add the results or add the functions first and then find the limit and we will end up with the same result. Theorems of these type are important in mathematics. Not all operations commute and changing the order of non-commuting operations leads to serious errors.

It is important to remember that the statements in Theorem 2.5.1 hold only if both limits $\lim f(x)$ and $\lim g(x)$ exist. For example, the theorem says that, if both limits $\lim_{x \to a} f(x)$ and $\lim_{x \to a} g(x)$ exist, then the limit $\lim_{x \to a}(f(x) + g(x))$ exits and we have $\lim_{x \to a} f(x) + \lim_{x \to a} g(x) = \lim_{x \to a}(f(x) + g(x))$. This does not mean that existence of the limit $\lim_{x \to a}(f(x) + g(x))$ implies existence of the limits $\lim_{x \to a} f(x)$ and $\lim_{x \to a} g(x)$.

Theorem 2.5.1 allows us to find limits of some more complicated functions.

Example 2.5.3.
$$\lim_{x \to 1} \frac{2x}{x^2 + x - 1} = 2.$$

Since $\lim_{x \to 1} x = 1$, we have $\lim_{x \to 1} x^2 = 1$, by (c), $\lim_{x \to 1}(x^2 + x) = 2$, by (a), and thus $\lim_{x \to 1}(x^2 + x - 1) = 1$, by (b) and the fact that $\lim_{x \to 1} 1 = 1$. Since $\lim_{x \to 1} 2 = 2$, we have $\lim_{x \to 1} 2x = 2$, by (c). Finally, by (d), we obtain the desired result. We can summarize this argument in the following form
$$\lim_{x \to 1} \frac{2x}{x^2 + x - 1} = \frac{(\lim_{x \to 1} 2)(\lim_{x \to 1} x)}{(\lim_{x \to 1} x)(\lim_{x \to 1} x) + (\lim_{x \to 1} x) - (\lim_{x \to 1} 1)}$$
$$= \frac{2 \cdot 1}{1 \cdot 1 + 1 - 1} = 2.$$
With some practice, it should not be necessary to write down the middle term.

To better appreciate Theorem 2.5.1 you can try showing that $\lim_{x \to 1} \frac{2x}{x^2 + x - 1} = 2$ using Definition 2.4.3 instead.

Example 2.5.4. For $x \neq 0$ we have
$$\lim_{h \to 0} \frac{\frac{1}{x+h} - \frac{1}{x}}{h} = -\frac{1}{x^2}.$$
In this example the letter h plays the same role as x in the first example and x is an arbitrary constant number different from zero. For $h \neq 0$, $x \neq 0$, and $x + h \neq 0$, we have
$$\frac{\frac{1}{x+h} - \frac{1}{x}}{h} = \frac{-1}{x(x + h)}.$$

Applying Theorem 2.5.1 we easily find that the function on the right-hand side has the limit $-\frac{1}{x^2}$ at $h = 0$. Thus the same limit is assigned to the function on the left-hand side.

EXERCISES 2.5

(1) Find the following limits:

(a) $\lim\limits_{x \to 1} \dfrac{x^3 + x}{x^4 + x^2 + 1}$,

(b) $\lim\limits_{x \to -1} \dfrac{x^2 + x}{x + 1}$,

(c) $\lim\limits_{x \to 2} \dfrac{x^2 - 4}{x + 2}$,

(d) $\lim\limits_{x \to 0} \dfrac{x^2}{|x|}$,

(e) $\lim\limits_{h \to 0} \dfrac{(x + h)^2 - x^2}{h}$,

(f) $\lim\limits_{h \to 0} \dfrac{\frac{1}{(x+h)^2} - \frac{1}{x^2}}{h}$,

(g) $\lim\limits_{h \to 0-} \dfrac{|x + h| - |x|}{h}$,

(h) $\lim\limits_{h \to 0+} \dfrac{|x + h| - |x|}{h}$.

(2) Use the Induction Principle to prove that, if the limits $\lim_{x \to a} f_1(x)$, $\lim_{x \to a} f_2(x) \ldots$, $\lim_{x \to a} f_n(x)$ all exist, then the limit

$$\lim_{x \to a} (f_1(x) + f_2(x) + \cdots + f_n(x))$$

exists and we have

$$\lim_{x \to a} (f_1(x) + f_2(x) + \cdots + f_n(x)) = \lim_{x \to a} f_1(x) + \lim_{x \to a} f_2(x) + \cdots + \lim_{x \to a} f_n(x).$$

(3) Prove that, if the limit $\lim_{x \to a} f(x)$ exists, then the limit $\lim_{x \to a} (f(x))^3$ exists, and we have

$$\lim_{x \to a} (f(x))^3 = (\lim_{x \to a} f(x))^3.$$

2.6   Limits and inequalities

In the preceding section we were investigating how limits "interact" with the four algebraic operations. Now we consider questions concerning limits and inequalities.

Theorem 2.6.1. *If $f(x) \leq g(x)$ in some neighborhood of a and both limits $\lim_{x \to a} f(x)$ and $\lim_{x \to a} g(x)$ exist, then $\lim_{x \to a} f(x) \leq \lim_{x \to a} g(x)$. The same property holds for one-sided limits.*

Proof. We present a proof for the right-hand limit. We assume that $f(x) \leq g(x)$ for all $x \in (a, a + \gamma)$ for some $\gamma > 0$. We argue by contradiction to show that $\lim_{x \to a+} f(x) \leq \lim_{x \to a+} g(x)$. Suppose that

$$\lim_{x \to a+} f(x) - \lim_{x \to a+} g(x) > 0.$$

Then, by Theorem 2.5.1 (b),

$$\lim_{x \to a^+} (f(x) - g(x)) = l > 0$$

and there exists a number $\delta > 0$ such that

$$|f(x) - g(x) - l| < \frac{l}{2} \quad \text{for all } x \in (a, a + \delta). \tag{2.15}$$

Without loss of generality, we can assume that $\delta < \gamma$, so that $f(x) \leq g(x)$ for all $x \in (a, a + \delta)$. Note that 2.15 can be written as

$$\frac{l}{2} < f(x) - g(x) < \frac{3l}{2} \quad \text{for all } x \in (a, a + \delta).$$

But this means that $f(x) - g(x) > 0$ for all $x \in (a, a+\delta)$, contrary to the hypothesis.

The proof for the left-hand limit is obtained from the above by replacing $(a, a+\delta)$ by $(a-\delta, a)$. The results for one-sided limits imply the result for two-sided limits. $\square$

It is important to remember that in Theorem 2.6.1 the weak inequality cannot be replaced by the strong inequality. For instance, if $f(x) = -x$ and $g(x) = x$, then $f(x) < g(x)$ for $x > 0$, but in spite of this we have $\lim_{x \to 0+} f(x) = \lim_{x \to 0+} g(x) = 0$.

The following theorem is often called the "squeeze theorem" or the "sandwich theorem." It says that, if a function is squeezed between two functions that have the same limit at a point, then it must have the same limit at that point.

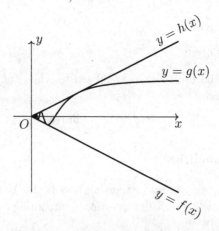

Fig. 2.11   Illustration of the squeeze theorem.

Theorem 2.6.2. *If $f(x) \leq g(x) \leq h(x)$ in some neighborhood of a and*

$$\lim_{x \to a} f(x) = \lim_{x \to a} h(x) = l,$$

then

$$\lim_{x \to a} g(x) = l.$$

The same property holds for one-sided limits.

Proof. It may seem that Theorem 2.6.2 follows at once from Theorem 2.6.1. However, we must first show that $\lim_{x \to a} g(x)$ exists. As before, we first prove the result for the right-hand limit.

Assume that $f(x) \leq g(x) \leq h(x)$ for all $x \in (a, a + \gamma)$, for some $\gamma > 0$. Then, for every $\varepsilon > 0$ there exist numbers δ_1 and δ_2 such that

$$f(x) - l > -\varepsilon \quad \text{for all } x \in (a, a + \delta_1)$$

and

$$h(x) - l < \varepsilon \quad \text{for all } x \in (a, a + \delta_2).$$

Since

$$f(x) - l \leq g(x) - l \leq h(x) - l \quad \text{for all } x \in (a, a + \gamma),$$

we have

$$-\varepsilon < g(x) - l < \varepsilon \quad \text{for all } x \in (a, a + \delta),$$

where δ is the least of the numbers δ_1, δ_2, and γ. This proves Theorem 2.6.2 for the right-hand limit. We extend the result to other types of limits in the usual way. □

EXERCISES 2.6

(1) Show that $\lim_{x \to 0+} \left(1 - x \left\lfloor \frac{1}{x} \right\rfloor \right) = 0$.

(2) Show that the limit $\lim_{x \to 0} \left(\lfloor x \rfloor - \lfloor -x \rfloor \right)$ does not exist, but
$\lim_{x \to 0} x \left(\lfloor x \rfloor - \lfloor -x \rfloor \right) = 0$.

(3) Show that $\lim_{x \to 0+} \left(x \left\lfloor \frac{2}{x} \right\rfloor + \frac{x}{2} - 2 \right) = 0$.

(4) Find the limit $\lim_{x \to 0} f(x)$
if $f(x) = \begin{cases} 1 + x^2, & \text{if } x \text{ is a rational number,} \\ 1 - x^3, & \text{if } x \text{ is an irrational number.} \end{cases}$

(5) Prove: If $a \leq f(x) \leq b$ in some heigborhood of x_0 and $\lim_{x \to x_0} f(x)$ exists, then $a \leq \lim_{x \to x_0} f(x) \leq b$.

(6) Let $f \colon \mathbb{R} \to \mathbb{R}$ satisfy the *Cauchy functional equation*, that is,

$$f(x + y) = f(x) + f(y) \quad \text{for all } x, y \in \mathbb{R}.$$

Suppose f has a limit at 0. Find that limit and prove that f has a limit at every other point.

(7) True or false: If $f(x) \leq g(x)$ and the limit $\lim_{x \to a} g(x)$ exists, then the limit $\lim_{x \to a} f(x)$ exists and we have $\lim_{x \to a} f(x) \leq \lim_{x \to a} g(x)$?

2.7 Limit at infinity

If in the function $f(x) = \frac{1}{x}$ we substitute greater and greater positive numbers for x, then the values of the function become closer and closer to 0. In fact we can make $f(x)$ as close as we please to 0 merely by requiring that x is sufficiently large. For this reason, we say that the function f has the limit 0 at infinity.

Definition 2.7.1. Let f be a function defined on the interval (a, ∞) for some number a. We say that f *tends to a limit l at infinity*, denoted by $\lim_{x \to \infty} f(x) = l$, if for every $\varepsilon > 0$ there exists a number x_0 such that $|f(x) - l| < \varepsilon$ for all $x > x_0$.

For example:

$$\lim_{x \to \infty} \frac{1}{x} = 0.$$

The next theorem shows that the limit at infinity can be converted to a one-sided limit at 0.

Theorem 2.7.2. $\lim_{x \to \infty} f(x) = l$ *if and only if* $\lim_{x \to 0+} f\left(\frac{1}{x}\right) = l$.

Proof. Let f be a function defined on the interval (a, ∞) for some number a. Without loss of generality, we can assume that $a > 0$. We define a new function

$$g(x) = f\left(\frac{1}{x}\right) \quad \text{for } x \in \left(0, \frac{1}{a}\right).$$

Note that g is well-defined on the interval $(0, \frac{1}{a})$.

If $\lim_{x \to \infty} f(x) = l$, then for every $\varepsilon > 0$ there exists a positive number x_0 such that

$$|f(x) - l| < \varepsilon \quad \text{for all } x > x_0. \tag{2.16}$$

If $\delta = \frac{1}{x_0}$, then for all $x \in (0, \delta)$ we have $\frac{1}{x} > x_0$ and, by (2.16),

$$|g(x) - l| = \left| f\left(\frac{1}{x}\right) - l \right| < \varepsilon. \tag{2.17}$$

Thus, for every $\varepsilon > 0$ there exists $\delta > 0$ such that (2.17) holds for all $x \in (0, \delta)$. This proves that $\lim_{x \to \infty} f(x) = l$ implies

$$\lim_{x \to 0+} f\left(\frac{1}{x}\right) = \lim_{x \to 0+} g(x) = l.$$

Now we assume that $\lim_{x \to 0+} f\left(\frac{1}{x}\right) = l$. Then, for every $\varepsilon > 0$, there exists a number $\delta > 0$ such that (2.17) holds for all $x \in (0, \delta)$. Using the same notation as before, $x \in (0, \delta)$ is equivalent to $x > x_0$ and (2.17) is equivalent to (2.16). Hence, for every $\varepsilon > 0$ there exists a number x_0 such that (2.16) holds for all $x > x_0$. Thus we have proved that $\lim_{x \to 0+} f\left(\frac{1}{x}\right) = l$ implies $\lim_{x \to \infty} f(x) = l$. $\qquad \square$

For example,

$$\lim_{x \to \infty} \frac{x^2 - 2x}{2x^2 - 3x - 1} = \lim_{x \to 0+} \frac{\frac{1}{x^2} - \frac{2}{x}}{\frac{2}{x^2} - \frac{3}{x} - 1} = \lim_{x \to 0+} \frac{1 - 2x}{2 - 3x - x^2} = \frac{1}{2}.$$

Corollary 2.7.3. *If* $\lim_{x \to \infty} f(x) = l$ *and* $\lim_{x \to \infty} g(x) = m$, *then*

(a) $\lim_{x \to \infty} (f(x) + g(x)) = l + m$,

(b) $\lim_{x \to \infty} (f(x) - g(x)) = l - m$,

(c) $\lim\limits_{x\to\infty} f(x)g(x) = lm.$

Moreover, if $m \neq 0$, then

(d) $\lim\limits_{x\to\infty} \dfrac{f(x)}{g(x)} = \dfrac{l}{m}.$

Proof. By Theorem 2.7.2, this is an immediate consequence of Theorem 2.5.1. $\square$

We will also need the *limit at negative infinity* which can be defined as follows:

Definition 2.7.4. We say that a function f *tends to a limit l at* $-\infty$, denoted by $\lim_{x\to-\infty} f(x) = l$, if for every $\varepsilon > 0$ there exists a number x_0 such that $|f(x)-l| < \varepsilon$ for all $x < x_0$.

The limit at $-\infty$ can be converted to a limit at ∞ or to a one-sided limit at 0:

Theorem 2.7.5.

(a) $\lim\limits_{x\to-\infty} f(x) = \lim\limits_{x\to\infty} f(-x).$

(b) $\lim\limits_{x\to-\infty} f(x) = l$ *if and only if* $\lim\limits_{x\to0-} f\left(\dfrac{1}{x}\right) = l.$

The proofs of these two properties are left as exercises.

EXERCISES 2.7

(1) Prove Theorem 2.7.5.

(2) Discuss the limit $\lim\limits_{x\to\infty} \dfrac{a_n x^n + \cdots + a_1 x + a_0}{b_m x^m + \cdots + b_1 x + b_0}.$

(3) Discuss the limit $\lim\limits_{x\to-\infty} \dfrac{a_n x^n + \cdots + a_1 x + a_0}{b_m x^m + \cdots + b_1 x + b_0}.$

2.8   Monotone functions

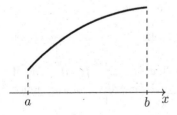

Fig. 2.12   An example of a function increasing on (a,b).

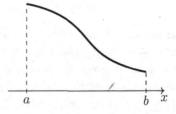

Fig. 2.13   An example of a function decreasing on (a,b).

A function is called increasing (see Fig. 2.12) if its value increases whenever x increases. A function is called decreasing (see Fig. 2.13) if its value decreases whenever x increases. More precisely:

Definition 2.8.1. A function f is *increasing* on an interval if for any two points $x_1 < x_2$ in that interval we have $f(x_1) < f(x_2)$. A function f is *decreasing* on an interval if for any two points $x_1 < x_2$ in that interval we have $f(x_1) > f(x_2)$.

From the above definition it follows that a function that is increasing or decreasing on an interval assumes each of its values exactly once on that interval.

Functions can be increasing on some intervals and decreasing on some other intervals. For instance, the function $f(x) = x^2$ is increasing on $(0, \infty)$ and decreasing on $(-\infty, 0)$.

Definition 2.8.2. A function f is called *non-increasing* in an interval if for every pair of points $x_1 < x_2$ in that interval we have $f(x_1) \geq f(x_2)$. A function f is called *non-decreasing* in an interval if for every pair of points $x_1 < x_2$ in that interval we have $f(x_1) \leq f(x_2)$. If a function is non-increasing or non-decreasing, we say it is *monotone*.

Every increasing function is non-decreasing, but not conversely. Similarly, every decreasing function is non-increasing. If f is increasing, $-f$ is decreasing. If f is non-decreasing, $-f$ is non-increasing. Also, if f is decreasing or non-increasing, $-f$ is increasing or non-decreasing, respectively.

Definition 2.8.3. We say that a function f is *bounded* if there exists a constant M such that $|f(x)| < M$ for all points x in the domain of f.

The following important theorem is a consequence of the Dedekind Axiom of the real numbers.

Theorem 2.8.4. *If a function f is monotone and bounded on an interval (a, ∞), then the limit $\lim_{x \to \infty} f(x)$ exists.*

Proof. Let Z be the set of values which f admits in (a, ∞), that is,

$$Z = \{y : y = f(x) \text{ for some } x > a\}.$$

The boundedness of f means that the set Z is bounded. Let M be the least upper bound of Z and m its greatest lower bound.

For every $\varepsilon > 0$ there exists a number $x_0 > a$ such that $M - \varepsilon < f(x_0) \leq M$. If f is non-decreasing, then we also have $M - \varepsilon < f(x) \leq M$ for every $x > x_0$. Hence $|f(x) - M| < \varepsilon$ for all $x > x_0$, proving $\lim_{x \to \infty} f(x) = M$.

If f is non-increasing, then $\lim_{x \to \infty} f(x) = m$. Indeed, for every $\varepsilon > 0$ there exists a number x_0 such that $m \leq f(x_0) < m + \varepsilon$. Then $m \leq f(x) < m + \varepsilon$ for all $x > x_0$, proving $\lim_{x \to \infty} f(x) = m$. $\qquad \square$

The following theorem can be easily obtained as a corollary of Theorems 2.7.5 and 2.8.4.

Theorem 2.8.5. *For a monotone and bounded function f defined on an interval $(-\infty, a)$ the limit $\lim_{x \to -\infty} f(x)$ exists.*

EXERCISES 2.8

(1) Prove Theorem 2.8.5.
(2) Justify the following statements.

   (a) Every increasing function is non-decreasing, but not conversely.
   (b) Every decreasing function is non-increasing, but not conversely.
   (c) If f is increasing, then $-f$ is decreasing.
   (d) If f is non-decreasing, then $-f$ is non-increasing.
   (e) If f is decreasing, then $-f$ is increasing.
   (f) If f is non-increasing, then $-f$ is non-decreasing.

(3) Show that the product of two increasing functions is increasing. Show that the same is true for decreasing, non-increasing, and non-decreasing functions.
(4) If f is increasing and g is non-decreasing, what can we say about fg?
(5) If both f and g are increasing functions and $g > 0$, what can we say about $\frac{f}{g}$?
(6) Show that the sum of two increasing functions is increasing. Show that the same is true for decreasing, non-increasing, and non-decreasing functions.
(7) If f is increasing and g is non-decreasing, what can we say about $f + g$?
(8) If both f and g are increasing functions, what can we say about $f - g$?
(9) In the proof of Theorem 2.8.4 we claim that "For every $\varepsilon > 0$ there exists a number $x_0 > a$ such that $M - \varepsilon < f(x_0) \leq M$." Justify that claim.
(10) In the proof of Theorem 2.8.4 we claim that "for every $\varepsilon > 0$ there exists a number x_0 such that $m \leq f(x_0) < m + \varepsilon$." Justify that claim.

Chapter 3

CONTINUOUS FUNCTIONS

3.1 Continuity of a function at a point

Consider the function

$$f(x) = \frac{2x}{x^2 + x - 1}.$$

In Example 2.5.3 we have shown, using Theorem 2.5.1, that $\lim_{x \to 1} f(x) = 2$. The same number can be obtained in a much simpler way by evaluating the function at $x = 1$:

$$f(1) = \frac{2 \cdot 1}{1^2 + 1 - 1} = 2.$$

Is there, then, any difference between finding the value of a function at a point x_0 and finding the limit of that function at x_0? Or are these just an easy and a hard way of solving the same problem?

The first important difference between the limit at a point and the value at a point is that a function need not be defined at a point to have a limit there, as we have seen in Section 2.4. Conversely, a function defined at a point does not have to have a limit at that point. Consider, for example, the function f defined by

$$f(x) = \frac{x}{\left((x+2)^{\lfloor 2 - x^2 \rfloor} - 2 \right)^2}.$$

We have $f(0) = 0$, but f has no limit at 0.

The value of a function at a point x_0 is a property of the function at that point, whereas the limit of the function at x_0 is a property of the function at *different*, nearby points. The equality of the value of a function at a point and the limit of that function at that point is a desirable property called continuity. The following definitions treat this notion in more careful detail.

Definition 3.1.1.

(a) If $\lim_{x \to x_0+} f(x) = f(x_0)$, then f is called *right-hand continuous* at x_0.

(b) If $\lim_{x \to x_0-} f(x) = f(x_0)$, then f is called *left-hand continuous* at x_0.

Note that the condition $\lim_{x \to x_0+} f(x) = f(x_0)$ requires f to be defined in an interval $[x_0, x_0 + \delta)$ for some $\delta > 0$. Similarly, the condition $\lim_{x \to x_0+} f(x) = f(x_0)$ requires f to be defined in an interval $(x_0 - \delta, x_0]$ for some $\delta > 0$.

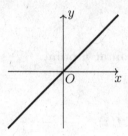

Fig. 3.1   Graph of $y = x$.

The function $f(x) = x$ (see Fig. 3.1) is right-hand and left-hand continuous at each point x_0, because $\lim_{x \to x_0+} x = x_0$ and $\lim_{x \to x_0-} x = x_0$ for every real number x_0. The function $f(x) = \lfloor x \rfloor$ (see Fig. 3.2) is right-hand continuous at each point, but it is not left-hand continuous at integral points. The function $f(x) = \frac{|x|}{x}$ (see Fig. 3.3) is neither left-hand continuous nor right-hand continuous at zero, because f is not defined there. If we enlarge the domain of f and define $f(0) = 1$, then f becomes right-hand continuous at 0. If we define $f(0) = -1$, then f becomes left-hand continuous at 0.

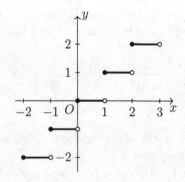

Fig. 3.2   Graph of $y = \lfloor x \rfloor$.

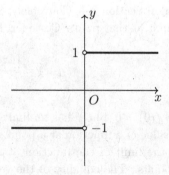

Fig. 3.3   Graph of $y = |x|/x$.

Definition 3.1.2. If $\lim_{x \to x_0} f(x) = f(x_0)$, then we say that f is *continuous* at x_0. If f is continuous at every point, we simply say that f is continuous.

Note that a function f is continuous at a point x_0 if and only if f is right-hand and left-hand continuous at x_0.

Functions $f(x) = c$ (a constant function) or $f(x) = x$ are continuous at every point. The function $f(x) = \frac{x^2-4}{x-2}$ is not continuous at $x = 2$ because it is not defined at that point. However, we can make the function continuous by enlarging its domain and defining $f(2) = 4$. On the other hand the function $f(x) = \frac{|x|}{x}$ (see Fig. 3.3) cannot be made continuous at zero by enlarging its domain, because the limit at that point does not exist; the left-hand limit is different from the right-hand limit.

Theorem 3.1.3. *If functions f and g are continuous at x_0, then their sum, difference, and product are continuous at x_0. If, in addition, $g(x_0) \neq 0$, then the quotient $\frac{f}{g}$ is continuous at x_0. The same is true for the right-hand and the left-hand continuity.*

Proof. Assume that f and g are right-hand continuous, that is,

$$\lim_{x \to x_0+} f(x) = f(x_0) \quad \text{and} \quad \lim_{x \to x_0+} g(x) = g(x_0).$$

Then, in view of Theorem 2.5.1, we have

$$\lim_{x \to x_0+} (f(x) + g(x)) = f(x_0) + g(x_0),$$

$$\lim_{x \to x_0+} (f(x) - g(x)) = f(x_0) - g(x_0),$$

$$\lim_{x \to x_0+} f(x)g(x) = f(x_0)g(x_0),$$

and, if $g(x_0) \neq 0$,

$$\lim_{x \to x_0+} \frac{f(x)}{g(x)} = \frac{f(x_0)}{g(x_0)}.$$

This proves the theorem for right-hand continuity. To obtain the proof for left-hand continuity it suffices to replace x_0+ by x_0-. Combining both results we obtain the assertion for continuity. $\qquad \square$

For natural numbers n the powers a^n are defined by induction:

$$a^1 = a, \quad a^{n+1} = a \cdot a^n \quad \text{for} \quad n \in \mathbb{N}.$$

Thus we have $a^2 = a \cdot a$, $a^3 = a \cdot a^2 = a \cdot a \cdot a$, $a^4 = a \cdot a^3 = a \cdot a \cdot a \cdot a$, and so on.

Theorem 3.1.4. *For every $n \in \mathbb{N}$ the function $f(x) = x^n$ is continuous at every point.*

Proof. The assertion is true for $n = 1$, because the function $f(x) = x$ is continuous at every point. If the function $f(x) = x^n$ is continuous at $x = x_0$ for some $n \in \mathbb{N}$, then the function $f(x) = x \cdot x^n = x^{n+1}$ is continuous at $x = x_0$ by Theorem 3.1.3. Thus the theorem follows by the induction principle. $\qquad \square$

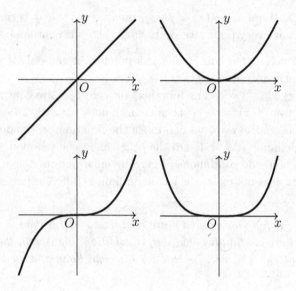

Fig. 3.4   Graphs of $y = x$, $y = x^2$, $y = x^3$, and $y = x^4$.

The power x^n can easily be extended to all integers n by letting

$$x^0 = 1 \text{ for every } x$$

and

$$x^{-n} = \frac{1}{x^n} \text{ for } n \in \mathbb{N} \text{ and } x \neq 0.$$

EXERCISES 3.1

(1) Let $f(x) = \frac{x^2+x-6}{x-2}$ for $x \neq 2$. How should f be defined at $x = 2$ so that f becomes a continuous function?

(2) Prove that the function $f(x) = \lfloor x \rfloor$ is right-hand continuous at each point, but it is not left-hand continuous at integral points.

(3) Prove that the function

$$f(x) = \frac{x}{\left((x+2)^{\lfloor 2-x^2 \rfloor} - 2\right)^2}$$

is not continuous at $x = 0$.

(4) Prove: A function f is continuous at x_0 if and only if for every $\varepsilon > 0$ there exists $\delta > 0$ such that $|f(x) - f(x_0)| < \varepsilon$ whenever $|x - x_0| < \delta$.

(5) Find an example of a function $f \colon \mathbb{R} \to \mathbb{R}$ that is not continuous at any point even though f^2 is continuous at every point.

(6) Show that, if f is continuous at a point, then $|f|$ is continuous at that point.

(7) Let f and g be continuous at x_0. Show that the functions

$$M(x) = \max\{f(x), g(x)\} \quad \text{and} \quad m(x) = \min\{f(x), g(x)\},$$

are continuous at $x = x_0$.

(8) Prove that the function

$$f(x) = \begin{cases} 1, & \text{if } x \text{ is a rational number,} \\ 0, & \text{if } x \text{ is an irrational number} \end{cases}$$

is not continuous at any point.

(9) Prove that the function

$$f(x) = \begin{cases} x, & \text{if } x \text{ is a rational number,} \\ 0, & \text{if } x \text{ is an irrational number} \end{cases}$$

is discontinuous (that is, not continuous) at every point except 0 where it is continuous.

(10) Prove that Dirichlet's function

$$f(x) = \begin{cases} \frac{1}{q}, & \text{if } x = \frac{p}{q} \text{ is a rational number in lowest terms,} \\ 0, & \text{if } x \text{ is an irrational number} \end{cases}$$

is continuous at every irrational point and discontinuous at every rational point.

(11) Find a function that is continuous at two distinct points a and b and discontinuous everywhere else.

(12) Decide whether the following are possible:

(a) f is continuous at x_0, g is not continuous at x_0, and $f+g$ is continuous at x_0.

(b) Neither f nor g is continuous at x_0, but $f + g$ is continuous at x_0.

(c) f is continuous at x_0, g is not continuous at x_0, and fg is continuous at x_0.

(d) Neither f nor g is continuous at x_0, but fg is continuous at x_0.

(13) Let f be a continuous function on $[a, b]$. Prove that if $f(x) = 0$ for every rational x in $[a, b]$ then $f(x) = 0$ for every x in $[a, b]$.

(14) Let f and g be two continuous functions on $[a, b]$. Prove that if f and g are equal at rational points of $[a, b]$, then f and g are identical on $[a, b]$.

(15) True or false: If f and g are two continuous functions on $[a, b]$ and f and g are equal at irrational points of $[a, b]$, then f and g are identical on $[a, b]$?

(16) Suppose that function $f \colon \mathbb{R} \to \mathbb{R}$ satisfies Cauchy's functional equation

$$f(x + y) = f(x) + f(y) \quad \text{for all } x, y \in \mathbb{R}$$

and has a limit at 0. Prove that f is a continuous function, that is, f is continuous at every real number x_0.

(17) Find all continuous functions $f \colon \mathbb{R} \to \mathbb{R}$ satisfying Cauchy's functional equation $f(x + y) = f(x) + f(y)$ for all $x, y \in \mathbb{R}$.

(18) Prove that, if $a > 0$, then $a^n > 0$ for all integers n. Is the same true for $a < 0$?

(19) A function f is called a *monomial function* if $f(x) = ax^n$ for some $a \in \mathbb{R}$ and some $n \in \mathbb{N}$. Show that every monomial function is continuous at every point.

(20) A function f is called a *polynomial function* if it is a linear combination of monomials. In other words, f is a polynomial function if

$$f(x) = a_n x^n + a_{n-1} a^{n-1} + \cdots + a_1 x + a_0,$$

where $a_n, a_{n-1}, \ldots, a_1, a_0$ are arbitrary real numbers and $n \in \mathbb{N}$. Show that every polynomial function is continuous at every point.

(21) A function f is called a *rational function* if it can be written in the form $f(x) = \frac{P(x)}{Q(x)}$, where P and Q are polynomial functions. Show that every rational function is continuous at every point of its domain.

(22) Prove that, if f is continuous at $x = a$, then the function $g(x) = (f(x))^n$ is continuous at $x = a$ for any $n \in \mathbb{N}$.

3.2 Functions continuous on an interval

We say that a function is continuous on an open interval (a, b) if it is continuous at each point of that interval. We say that a function is continuous on a closed interval $[a, b]$ if it is continuous on the open interval (a, b) and, moreover, it is right-hand continuous at a and left-hand continuous at b.

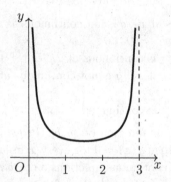

Fig. 3.5   Graph of $y = \frac{1}{x(3-x)}$ for $x \in (0, 3)$.

For instance the function

$$f(x) = \frac{1}{x(3 - x)}$$

(see Fig. 3.5) is continuous on $(0, 3)$ but it is not continuous on $[0, 3]$, because it fails to be defined for both $x = 0$ and $x = 3$. Moreover, it is not possible to define this function at 0 and 3 to make it continuous on $[0, 3]$. The function is continuous on $[1, 2]$ and, in fact, on every interval $[\varepsilon, 3 - \delta]$ where ε and δ are arbitrary positive numbers such that $\varepsilon < 3 - \delta$.

In general, if f is continuous on an interval, then it is continuous on any subinterval.

Theorem 3.2.1.

(a) *If a function is right-hand continuous at a point a and $f(a) < k$, then $f(x) < k$ for every $x \in (a, a + \delta)$ for some $\delta > 0$.*

(b) *If a function is right-hand continuous at a point a and $f(a) > k$, then $f(x) > k$ for every $x \in (a, a + \delta)$ for some $\delta > 0$.*

(c) *If a function is left-hand continuous at a point a and $f(a) < k$, then $f(x) < k$ for every $x \in (a - \delta, a)$ for some $\delta > 0$.*

(d) *If a function is left-hand continuous at a point a and $f(a) > k$, then $f(x) > k$ for every $x \in (a - \delta, a)$ for some $\delta > 0$.*

The above properties have simple geometric interpretations; see Fig. 3.6 and Fig. 3.7.

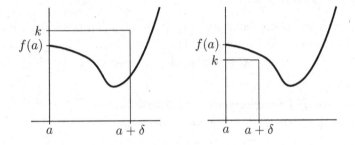

Fig. 3.6　Illustration of Theorem 3.2.1 for right-hand continuity.

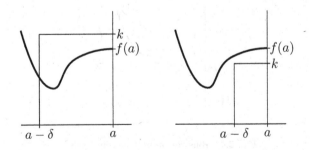

Fig. 3.7　Illustration of Theorem 3.2.1 for left-hand continuity.

Proof of Theorem 3.2.1. Assume that f is right-hand continuous at a and that $f(a) < k$. Then for every $\varepsilon > 0$ there exists a number $\delta > 0$ such that

$$|f(x) - f(a)| < \varepsilon \quad \text{for } a < x < a + \delta.$$

In particular, if $\varepsilon < k - f(a)$ we have

$$|f(x) - f(a)| < k - f(a) \quad \text{for } a < x < a + \delta.$$

Hence $f(x) - f(a) < k - f(a)$ and consequently $f(x) < k$ for $a < x < a + \delta$. This completes the proof of part (a). The proofs of the remaining parts are similar with obvious modifications. □

Corollary 3.2.2.

(a) *If a function is continuous at a point a and $f(a) < k$, then $f(x) < k$ for every $x \in (a - \delta, a + \delta)$ for some $\delta > 0$.*

(b) *If a function is continuous at a point a and $f(a) > k$, then $f(x) > k$ for every $x \in (a - \delta, a + \delta)$ for some $\hat{\delta} > 0$.*

Proof. If a function is continuous at a point a, then it is right-hand continuous and left-hand continuous at that point and thus the above properties are immediate consequences of Theorem 3.2.1. □

The graph of a function continuous in an interval can be drawn without lifting the pencil. Such an intuitive understanding of continuity allows us to foresee the following theorem. This is one of the most important properties of continuous functions.

Theorem 3.2.3. *If f is continuous in an interval $[a, b]$ and $f(a) < 0$ and $f(b) > 0$, then there exists a point x_0 in (a, b) such that $f(x_0) = 0$.*

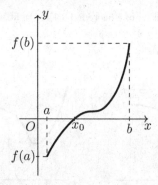

Fig. 3.8   Illustration of Theorem 3.2.3.

Proof. Let Z be the set of all numbers $x \in [a, b]$ such that $f(x) < 0$. The set is not empty because $a \in Z$. Moreover, it is bounded from above by b. Let x_0 be the least upper bound of Z. In view of Theorem 3.2.1, we have $a < x_0 < b$. We are going to prove that $f(x_0) = 0$. Suppose $f(x_0) < 0$. Then, by Corollary 3.2.2, $f(x) < 0$ for every x in some interval $(x_0 - \delta, x_0 + \delta)$, and thus x_0 cannot be the least upper bound of Z. Similarly, if $f(x_0) > 0$, then $f(x) > 0$ for every x in some interval $(x_0 - \delta, x_0 + \delta)$, contradicting the definition of x_0. Therefore $f(x_0) = 0$. $\qquad\square$

The function $f(x) = x^5 + x - 1$ is continuous in the interval $[0, 1]$ and admits values 1 and -1 at its ends. Therefore there exists a number $x_0 \in (0, 1)$ such that $x_0^5 + x_0 - 1 = 0$. While Theorem 3.2.3 ensures existence of solutions of some equations, it does not help in finding the solutions. Finding a solution is often much more difficult.

In practice, instead of Theorem 3.2.3, we often need the more general theorem called the Intermediate Value Theorem.

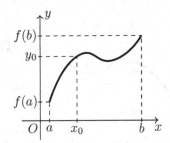

Fig. 3.9   Illustration of the intermediate value property.

Theorem 3.2.4. *(Intermediate Value Theorem). A function which is continuous on a closed interval assumes in its interior all values between the initial and the end value. More precisely, if f is continuous in $[a, b]$ and $f(a) \neq f(b)$ then for every number y_0 between $f(a)$ and $f(b)$ there exists a number $x_0 \in (a, b)$ such that $f(x_0) = y_0$.*

Proof. If $f(a) < y_0 < f(b)$, then the continuous function $g(x) = f(x) - y_0$ is negative at a and positive at b. There thus exists $x_0 \in (a, b)$ such that $g(x_0) = 0$. Consequently, $f(x_0) = y_0$. If $f(b) < f(a)$, the proof is similar. $\qquad\square$

EXERCISES 3.2

(1) Provide the details to complete the proof of Theorem 3.2.1.
(2) True or false: If a function is continuous at a point a and $f(a) \leq k$, then $f(x) \leq k$ for every $x \in (a - \delta, a + \delta)$ for some $\delta > 0$?
(3) Provide the details to complete the proof of Theorem 3.2.4 for the case when $f(b) < f(a)$.
(4) Show that the equation $x^5 - 3x^3 - x + 1 = 0$ has at least two solutions in the interval $[0, 2]$.
(5) Let $f : [a, b] \to [a, b]$ be a continuous function. Prove that there exists $c \in [a, b]$ such that $f(c) = c$.

(6) Let f and g be continuous functions on $[a, b]$. Show that, if $f(a) \leq g(a)$ and $f(b) \geq g(b)$, then there exists $c \in [a, b]$ such that $f(c) = g(c)$.

(7) Let $f : [a, b] \to \mathbb{Q}$ be a continuous function where $\mathbb{Q}$ is the set of all rational numbers. Prove that f is a constant function.

(8) Show that the range of a continuous function on $[a, b]$ is either a one element set or an uncountable set.

3.3   Bounded functions

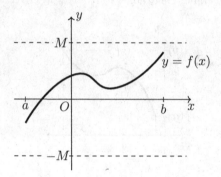

Fig. 3.10   A function bounded on $[a, b]$.

In this section we show that a function continuous on a closed bounded interval is bounded on that interval.

Definition 3.3.1. We say that a function f is *bounded* on some interval if there exists a constant M such that $|f(x)| < M$ for every point x in that interval.

The function $f(x) = \frac{2}{x} - 1$ (see Fig. 3.11) is bounded on the interval $(1, 3)$, because $|f(x)| < 1$ for all points in $(1, 3)$. However, it is not bounded on $(0, 3)$, because for any number $M > 0$ we have $f\left(\frac{2}{M+2}\right) = M + 1$, and $\frac{2}{M+2} \in (0, 3)$ for all positive $M \in \mathbb{R}$.

Theorem 3.3.2. *A function continuous on a closed bounded interval is bounded on that interval.*

Proof. Assume the function f is continuous on the closed interval $[a, b]$. Since $f(a) - 1 < f(a) < f(a) + 1$, it follows from Theorem 3.2.1 that we have

$$f(a) - 1 < f(x) < f(a) + 1 \quad \text{for all } x \in [a, c_0],$$

for some $c_0 > a$. Thus f is bounded on $[a, c_0]$.

Let Z be the set of all numbers $c \leq b$ such that f is bounded on the interval $[a, c]$. Since $c_0 \in Z$, Z is not empty. Let x_0 be the least upper bound of Z. Clearly

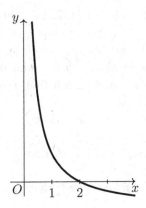

Fig. 3.11 Graph of $y = \frac{2}{x} - 1$ for $x \in (0, 4)$.

$a < x_0 \leq b$. We want to show that $x_0 = b$. Suppose, to the contrary, that $x_0 < b$. Since

$$f(x_0) - 1 < f(x_0) < f(x_0) + 1,$$

it follows from Corollary 3.2.2 that there exists a $\delta > 0$ such that

$$f(x_0) - 1 < f(x) < f(x_0) + 1 \quad \text{for all } x \in (x_0 - \delta, x_0 + \delta).$$

Hence f is bounded on the closed interval $[x_0 - \frac{\delta}{2}, x_0 + \frac{\delta}{2}]$. Since $x_0 - \frac{\delta}{2} \in Z$, f is bounded on the interval $[a, x_0 - \frac{\delta}{2}]$. But, if f is bounded on $[a, x_0 - \frac{\delta}{2}]$ and on $[x_0 - \frac{\delta}{2}, x_0 + \frac{\delta}{2}]$, then f is bounded on $[a, x_0 + \frac{\delta}{2}]$ contradicting the definition of x_0. This shows that we must have $x_0 = b$ which means that f is bounded on $[a, b]$. $\square$

EXERCISES 3.3

(1) Prove the following property of bounded functions used in the proof of Theorem 3.3.2: If f is bounded on $[a, b]$ and on $[b, c]$, then f is bounded on $[a, c]$.

(2) Prove that, if $f(x) \leq g(x) \leq h(x)$ for every $x \in [a, b]$ and the functions f and h are bounded on $[a, b]$, then g is bounded on $[a, b]$.

(3) Give an example of an unbounded function on a closed bounded interval $[a, b]$.

(4) True or false: If f is bounded on every closed subinterval of (a, b), then f is bounded on (a, b)?

(5) Let $\lim_{x \to x_0} f(x) = 0$ and let g be a bounded function in some interval $[x_0 - \delta, x_0 + \delta]$. Show that $\lim_{x \to x_0} f(x)g(x) = 0$.

(6) Prove: If f and g are bounded on $[a, b]$, then $f + g$ and fg are bounded on $[a, b]$.

(7) True or false: If f and g are bounded on $[a, b]$ and $g(x) \neq 0$ for all $x \in [a, b]$, then f/g is bounded on $[a, b]$?

(8) True or false: If f is bounded on $[a, b]$, g is continuous on $[a, b]$ and $g(x) \neq 0$ for all $x \in [a, b]$, then f/g is bounded on $[a, b]$?

(9) True or false: If f and g are bounded on (a, b), g is continuous on (a, b) and $g(x) \neq 0$ for all $x \in (a, b)$, then f/g is bounded on (a, b)?

(10) Can you find functions f and g such that f is bounded, g is unbounded, and $f + g$ is bounded?

(11) Can you find functions f and g such that f and g are unbounded, but $f + g$ is bounded?

(12) Can you find functions f and g such that f is bounded, g is unbounded, and fg is bounded?

(13) Can you find functions f and g such that f and g are unbounded, but fg is bounded?

(14) Find all functions f that are bounded on $\mathbb{R}$ and satisfy Cauchy's functional equation

$$f(x + y) = f(x) + f(y) \quad \text{for all } x, y \in \mathbb{R}.$$

3.4 Minimum and maximum of a function

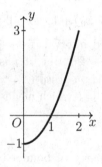

Fig. 3.12 Graph of $y = x^2 - 1$ for $x \in [0, 2]$.

The least value of the function $f(x) = x^2 - 1$ in the interval $[0, 2]$ is -1 and the greatest value is 3. In the open interval $(0, 2)$ the function admits all the values in the interval $(-1, 3)$ and only those values. Thus in the open interval $(0, 2)$ the defined function has neither a greatest nor least value.

The greatest value of a function is called its *maximum* and the least value its *minimum*. Each of them can be called an *extremum*. Existence of a minimum and a maximum in a closed bounded interval is one of the fundamental properties of continuous functions.

Theorem 3.4.1. *A function continuous on a closed bounded interval has a minimum and a maximum on that interval.*

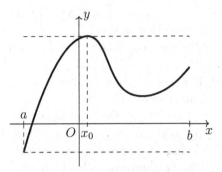

Fig. 3.13 A function with a minimum at a and a maximum at x_0.

Proof. Let f be a function continuous on an interval $[a, b]$ and let W be the set of all values of f on $[a, b]$, that is, $W = \{f(x) : x \in [a, b]\}$. By Theorem 3.3.2, W is a bounded set. Let K be the least upper bound of W. We want to show that there is an $x_0 \in [a, b]$ such that $f(x_0) = K$. Suppose, on the contrary, that $f(x) < K$ for all x in $[a, b]$. Let

$$g(x) = \frac{1}{K - f(x)}.$$

Then $g(x) > 0$ for all $x \in [a, b]$ and g is continuous on $[a, b]$. By Theorem 3.3.2, there is a positive number M such that $g(x) < M$ for every $x \in [a, b]$. Hence,

$$\frac{1}{K - f(x)} < M,$$

or, equivalently,

$$f(x) < K - \frac{1}{M}$$

for every x in $[a, b]$. But this contradicts the definition of K. Therefore, we must have $f(x_0) = K$ for some x_0 in $[a, b]$, which means that K is the maximum of f in $[a, b]$.

To prove that f also admits a minimum it is enough to apply the first part of the proof to the function $-f$. $\qquad\square$

EXERCISES 3.4

(1) Complete the proof of Theorem 3.4.1 by proving that a function continuous on a closed bounded interval has a minimum in that interval.

(2) Give an example of a function on a closed bounded interval that does not have a maximum in that interval.

(3) Give an example of a function that is bounded on a closed bounded interval that has neither a maximum nor a minimum on that interval.

(4) Give an example of a bounded continuous function on an unbounded interval that does not have a maximum on that interval.

(5) Give an example of a continuous function on a bounded interval that does not have a maximum on that interval.

(6) True or false: If M is the maximum value of f, then $|M|$ is the maximum value of $|f|$?

(7) True or false: If M is the maximum value of f and λ is a constant, then λM is the maximum value of λf?

(8) True or false: If M_1 and M_2 are the maximum values of f_1 and f_2, respectively, then $M_1 + M_2$ is the maximum value of $f_1 + f_2$?

(9) True or false: If M_1 and M_2 are the maximum values of f_1 and f_2, respectively, then $M_1 M_2$ is the maximum value of $f_1 f_2$?

(10) True or false: If f_1 and f_2 both have maximum and minimum values on a closed interval, then $f_1 f_2$ also has both a maximum value and a minimum value on that interval?

3.5   Composition of functions

Consider two functions f and g such that the range of g is contained in the domain of f. Then we can define a new function $h(x) = f(g(x))$ which is called the *composition* of f and g. The composition of f and g is denoted by $f \circ g$.

For example, if $f(x) = |x|$ and $g(x) = x^2 - x$, then $f(g(x)) = |x^2 - x|$ and $g(f(x)) = x^2 - |x|$, since $|x|^2 = x^2$. If $f(x) = \sqrt{x}$ and $g(x) = x^2 + 1$, then $f(g(x)) = \sqrt{x^2 + 1}$ and $g(f(x)) = x + 1$. Notice that in this example $f(g(x))$ is defined for all values of x, while $g(f(x))$ is defined only for nonnegative values of x, because the domain of f is $[0, \infty)$. When defining a composition of functions, we always have to pay attention to the domain and the range of the composed functions. For instance, if $f(x) = \sqrt{x}$ and $g(x) = -x^2 - 1$, then the composition $f(g(x))$ is not defined for any $x \in \mathbb{R}$.

Theorem 3.5.1. *If f is a function continuous at z_0 and $\lim_{x \to a} g(x) = z_0$, then*

$$\lim_{x \to a} f(g(x)) = f(z_0). \tag{3.1}$$

The assertion also holds, if we replace $x \to a$ by $x \to a+$ or $x \to a-$.

Proof. Let ε be an arbitrary positive number. From continuity of f at z_0 it follows that there exists a number $\delta > 0$ such that

$$|f(z) - f(z_0)| < \varepsilon \quad \text{whenever } |z - z_0| < \delta.$$

If $\lim_{x \to a+} g(x) = z_0$, then there exists a number $\gamma > 0$ such that

$$|g(x) - z_0| < \delta \quad \text{whenever } a < x < a + \gamma.$$

Consequently, for every $\varepsilon > 0$ there exists $\gamma > 0$ such that

$$|f(g(x)) - f(z_0)| < \varepsilon \quad \text{whenever } a < x < a + \gamma,$$

which means that $\lim_{x \to a+} f(g(x)) = f(z_0)$.

The proof for the left-hand limit is similar. Combining both results we obtain the assertion for the two-sided limit.   □

Equality (3.1) can be also written as

$$\lim_{x \to a} f(g(x)) = f(\lim_{x \to a} g(x)),$$

and similarly for the right-hand and the left-hand limits. The above equality says that the operations of taking a limit and evaluation of a function can be interchanged if the function is continuous. For example,

$$\lim_{x \to a} |x^2 - b| = |\lim_{x \to a} (x^2 - b)| = |a^2 - b|,$$

because the absolute value is a continuous function. On the other hand for the function $f(x) = \lfloor x \rfloor$ a similar calculation could be false as illustrated by the fact that

$$\left\lfloor \lim_{x \to 0} (-x^2) \right\rfloor = \lfloor 0 \rfloor = 0$$

whereas

$$\lim_{x \to 0} \lfloor -x^2 \rfloor = -1,$$

because $\lfloor -x^2 \rfloor = -1$ for $0 < x < 1$ and for $-1 < x < 0$. Changing the order in which the operations are performed leads to different results because the function $f(x) = \lfloor x \rfloor$ is not continuous at 0.

Theorem 3.5.1 implies that the composition of two continuous functions is continuous. The next theorem makes this statement more precise.

Theorem 3.5.2. *If g is continuous at a, and f is continuous at $g(a)$, then the function $f \circ g$ is continuous at a.*

Proof. In view of Theorem 3.5.1 we have $\lim_{x \to a} f(g(x)) = f(g(a))$ which means that the function $f \circ g$ is continuous at a. □

The following simple corollary is often useful.

Corollary 3.5.3. *A function f is continuous at a point a if and only if*

$$\lim_{x \to 0} f(a + x) = f(a).$$

Proof. Since the function $g(x) = a + x$ is continuous everywhere and f is continuous at $a = g(0)$, the function $h(x) = f(a + x)$ is continuous at 0, by Theorem 3.5.2. Consequently

$$\lim_{x \to 0} f(x + a) = \lim_{x \to 0} h(x) = h(0) = f(a).$$

To prove the converse, first note that f can be considered as the composition of the functions $h(x) = f(a+x)$ and $g(x) = x-a$, that is, $f = h \circ g$. If $\lim_{x \to 0} f(a+x) = f(a)$, then h is continuous at 0. Since g is continuous everywhere and $0 = g(a)$, the composition $f = h \circ g$ is continuous at a, by Theorem 3.5.2. □

EXERCISES 3.5

(1) An example in this section shows that $f \circ g$ need not equal $g \circ f$. How about the associativity of composition: $f \circ (g \circ h) = (f \circ g) \circ h$?

(2) Does composition distribute over addition, either from the left or the right? In other words is either of the following true:

$$f \circ (g + h) = f \circ g + f \circ h \quad \text{or} \quad (g + h) \circ f = g \circ f + h \circ f?$$

(3) Does there exist a function f such that $f \circ g = g \circ f = g$ for any function g?

(4) Is it possible to find f and g so that f is continuous, g is not continuous, but $f \circ g$ is continuous?

(5) Is it possible to find f and g so that f is continuous, g is not continuous, but $g \circ f$ is continuous?

(6) Is it possible to find f and g so that neither f nor g is continuous, but $f \circ g$ is continuous?

(7) Is it possible to find f that is not continuous, but $f \circ f$ is continuous?

(8) Show that the composition of two increasing functions is an increasing function.

(9) What can we say about the composition of two decreasing functions?

(10) What can we say about the composition of an increasing function and a decreasing function?

(11) What can we say about the composition of an increasing function and a non-decreasing function?

3.6   Inverse functions

Let f be a function defined on a set A. Assume that for any two distinct points x_1 and x_2 of A the values $f(x_1)$ and $f(x_2)$ are different. Let B be the set of all values of f. In other words, B is the range of f. Then for every number y from B there is exactly one number x in A such that

$$f(x) = y. \tag{3.2}$$

Thus the relation defines a function

$$x = g(y) \tag{3.3}$$

whose domain is B and the range is A. The function g is called the *inverse* of f. The inverse of f is denoted by f^{-1}. This should not be confused with $\frac{1}{f}$.

If f is regarded as a set of ordered pairs, then the inverse function is obtained by interchanging the order of elements in each pair. It is easily seen then that the concept of inverse function is symmetric: if g is the inverse of f, then f is the inverse of g. The graph of f is the set of points $(x, f(x))$ and the graph of the inverse function to f is the set of points $(f(x), x)$. Consequently, the graphs of inverse functions are symmetric with respect to the line $y = x$.

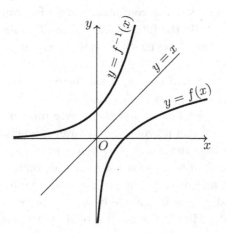

Fig. 3.14   Graphs of a function and its inverse.

Substituting (3.2) in (3.3) we get

$$g(f(x)) = x \quad \text{for all} \quad x \in A.$$

Similarly, substituting (3.3) in (3.2) we get

$$f(g(x)) = x \quad \text{for all} \quad x \in B.$$

In other words, the composition of a function f and its inverse is the identity function on the domain of the inner function. For instance, if

$$f(x) = \frac{1}{1+x} \quad \text{and} \quad g(x) = \frac{1-x}{x},$$

then

$$f(g(x)) = \frac{1}{1 + \frac{1-x}{x}} = x \quad \text{for all} \quad x \neq 0,$$

and

$$g(f(x)) = \frac{1 - \frac{1}{1+x}}{\frac{1}{1+x}} = x \quad \text{for all} \quad x \neq -1.$$

Not every function has an inverse. If at two distinct points x_1 and x_2 in the domain of a function f we have $f(x_1) = f(x_2)$, then f cannot have an inverse. A function which takes different values at different points is called a *one-to-one* function. Functions that are one-to-one are *invertible*, which means that they have inverses. Every increasing or decreasing function is invertible.

Theorem 3.6.1. *If a function f is increasing and continuous on an interval A (its domain), then the set B of all values of f (the range of f) is also an interval and the inverse of f is an increasing and continuous function on B. The theorem remains true if the word "increasing" is everywhere replaced by "decreasing".*

Proof. Assume that a function f is continuous on an interval A and let B be the set of all values of f on A. By the Intermediate Value Theorem (Theorem 3.2.4), if y_1 and y_2 belong to B, then f assumes each value between y_1 and y_2, which shows that B is an interval.

If y_1 and y_2 belong to B and $y_1 < y_2$, then there exist two points x_1 and x_2 in A such that $f(x_1) = y_1$ and $f(x_2) = y_2$. Thus $f(x_1) < f(x_2)$. If f is increasing, we must have $x_1 < x_2$, which proves that the inverse function is increasing. If f is decreasing, then $x_1 > x_2$ which proves that the inverse function is decreasing.

It remains to prove that the inverse function is continuous. (To help see what is happening in this proof, draw a picture of what the graph of g might look like on its domain B and then add more to the picture as the proof progresses.) Assume that f is increasing and denote its inverse by g. Let y_0 be an interior point of B. Then the point $x_0 = g(y_0)$ lies in the interior of A. Given any $\varepsilon > 0$ we choose in A two numbers x_1 and x_2 such that

$$x_0 - \varepsilon < x_1 < x_0 < x_2 < x_0 + \varepsilon. \tag{3.4}$$

Let $y_1 = f(x_1)$ and $y_2 = f(x_2)$. Then we have $y_1 < y_0 < y_2$. We can find a number $\delta > 0$ such that

$$y_1 < y_0 - \delta < y_0 + \delta < y_2.$$

For $|y - y_0| < \delta$, we have $y_1 < y < y_2$ and $g(y_1) < g(y) < g(y_2)$, that is, $x_1 < g(y) < x_2$. Consequently, in view of (3.4),

$$x_0 - \varepsilon < g(y) < x_0 + \varepsilon$$

whenever $|y - y_0| < \delta$, which proves the continuity of g at y_0.

The proof is not complete. We must consider what happens if y_0 is an end point of the interval B. Also we must consider what happens if f is a decreasing function. We leave these parts of the proof as exercises. $\square$

EXERCISES 3.6

(1) Complete the proof of Theorem 3.6.1.

    (a) Prove continuity of g at y_0 when y_0 is an end point of the interval B.

    (b) Prove continuity of g when f is decreasing.

(2) Find an example of a function f for which the inverse f^{-1} equals $\frac{1}{f}$?

(3) Prove: The composition of two invertible functions is invertible and we have

$$(f \circ g)^{-1} = g^{-1} \circ f^{-1}.$$

(4) True or false: If f and g are invertible, then $f + g$ is invertible?

(5) True or false: If f and g are invertible, then fg is invertible?

(6) Prove: If f is an invertible function and $f(x) \neq 0$ for all x in the domain of f, then the function $g(x) = \frac{1}{f(x)}$ is invertible.

Chapter 4

DERIVATIVES

4.1 Ascent and velocity

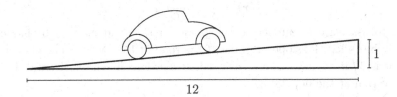

Fig. 4.1 A road with a constant slope.

The road drawn in Fig. 4.1 ascends by 1 meter for each 12 meters passed horizontally. We say that the slope of the road is $\frac{1}{12}$. In order to find the slope we do not have to take a 12 meter long segment of the road. For instance, if we note that for every 6 meters in the horizontal direction the road raises by $\frac{1}{2}$ of a meter, then the slope can be found by dividing $\frac{1}{2}$ by 6 which again gives $\frac{1}{12}$. The slope does not depend on the choice of the chosen segment of the road. Note that also the choice of meters to measure the length is unessential. If feet were used we would get exactly the same result.

The situation changes if the inclination of the road is different in different places, as in Fig. 4.2. The average slope for the 12 meter segment marked on the picture can be found by dividing 4 by 12, which gives us $\frac{1}{3}$. On the other hand, if we consider only the second half of the segment, the average slope is $\frac{1}{6}$.

It is clear from the picture that the road is steepest at the beginning of the road and it gradually gets less and less steep. The steepness of such road can easily be discussed qualitatively, that is, comparisons can be made about whether it is steeper at one point or another; just as one can say this pillow is softer than that one. But can the steepness of such a road be expressed quantitatively, that is, can we assign a number at each point of the road that measures its steepness at that point? If so, it is that number we call the *slope* of the road at that point.

Assume that the shape of the road on Fig. 4.2 is approximately represented by

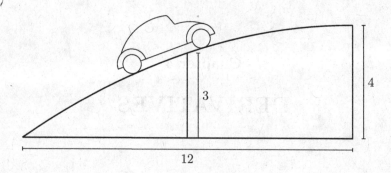

Fig. 4.2   A road with a varying slope.

the function

$$f(x) = \frac{2}{3}x - \frac{1}{36}x^2.$$

We wish to associate a number to the point on the road at $x = 6$ that could reasonably be called the slope or grade of the road at that point. The elevation at that point is 3. If we move along the road to the right by the horizontal distance h, the elevation at the new location is

$$f(6+h) = \frac{2}{3}(6+h) - \frac{1}{36}(6+h)^2 = 3 + \frac{1}{3}h - \frac{1}{36}h^2.$$

The difference in elevation is given by

$$f(6+h) - f(6) = \frac{1}{3}h - \frac{1}{36}h^2,$$

and the average slope of the road on the interval $[6, 6+h]$ is

$$\frac{f(6+h) - f(6)}{h} = \frac{1}{3} - \frac{1}{36}h.$$

This is not the number we were looking for, that is, the slope of the road at $x = 6$, since it depends on h. On the other hand, it is reasonable to assume that, if h is a very small number, then $\frac{1}{3} - \frac{1}{36}h$ gives us a good approximation of the slope. If we let h decrease to 0, we obtain

$$\lim_{h \to 0} \frac{f(6+h) - f(6)}{h} = \frac{1}{3}.$$

We say that the number $\frac{1}{3}$ is the slope of the road at $x = 6$.

The slope at a point can also be interpreted as the slope of the tangent line at that point. In Fig. 4.3 the straight line PQ has slope

$$\frac{f(x_0 + h) - f(x_0)}{h}.$$

If point P is fixed and Q approaches P, then the secant line gradually "approaches" the tangent line at P. The slope of the tangent line can be found by calculating the limit

$$\lim_{h \to 0} \frac{f(x_0 + h) - f(x_0)}{h}.$$

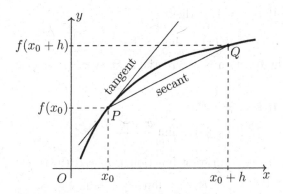

Fig. 4.3   Tangent and secant.

Here is still another interpretation of the above limit. Suppose a train leaves a railway station. Let's denote its distance from that station by y. The distance y depends on time t, so we can write $y = f(t)$. Let t_0 be a fixed instant. Between instances t_0 and $t_0 + h$ the average velocity of the train is

$$\frac{f(t_0 + h) - f(t_0)}{h}.$$

The limit

$$\lim_{h \to 0} \frac{f(t_0 + h) - f(t_0)}{h}$$

represents the instantaneous velocity of the train at the instant t_0.

EXERCISES 4.1

(1) Let $f(x) = x^2 + 1$. Find the slope of the tangent line to the graph of f at $x = 2$.
(2) Let $f(x) = (x + 1)^2$. Find the slope of the tangent line to the graph of f at $x = 2$.

4.2   Derivative

Definition 4.2.1. If the limit

$$\lim_{h \to 0} \frac{f(x + h) - f(x)}{h} \tag{4.1}$$

exists, it is called the *derivative* of f at x. If the derivative of f exists at a point x, we say that f is *differentiable* at x. If f is differentiable at every point of an interval, we simply say that f is differentiable on that interval.

In general the limit (4.1) has different values at different points, it is thus a function of x. This function will be denoted by f'.

Example 4.2.2. If $f(x) = 1/x$, then

$$f'(x) = \lim_{h \to 0} \frac{\frac{1}{x+h} - \frac{1}{x}}{h} = -\frac{1}{x^2}.$$

In this example the function is differentiable at every point of its domain, that is, for $x \neq 0$.

Example 4.2.3. If $f(x) = x$, then

$$f'(x) = \lim_{h \to 0} \frac{x + h - x}{h} = 1.$$

Example 4.2.4. If f is a constant function $f(x) = c$, then

$$f'(x) = \lim_{h \to 0} \frac{c - c}{h} = 0.$$

Examples 4.2.3 and 4.2.4 are very simple but important, for they are the base for calculations of many other, more complicated derivatives. For this reason it is advisable to remember them well.

EXERCISES 4.2

(1) Using the definition find the derivatives of the following functions:
   (a) $f(x) = 2x$,
   (b) $f(x) = x^2$,
   (c) $f(x) = \frac{1}{x^2}$.
(2) Find the equation of the tangent line to the graph of $y = x^3$ at $x = 1$.
(3) Find the point on the parabola $y = x^2$ that is closest to the line $y = x - 1$. (Hint: use the fact that the tangent line at that point is parallel to the line $y = x - 1$.)
(4) Let f be a continuous, differentiable function. Use the definition of the derivative to prove that the function $g(x) = (f(x))^2$ is differentiable and express g' in terms of f and f'.
(5) Give an example of a continuous function that is not differentiable at some point.

4.3   Differentiability and continuity

Theorem 4.3.1. *A function differentiable at a point is continuous at that point.*

Proof. For every $h \neq 0$, we have

$$f(x + h) = \frac{f(x + h) - f(x)}{h} h + f(x).$$

If f is differentiable at x, then

$$\lim_{h \to 0} f(x + h) = \lim_{h \to 0} \frac{f(x + h) - f(x)}{h} \lim_{h \to 0} h + \lim_{h \to 0} f(x)$$

$$= f'(x) \cdot 0 + f(x) = f(x)$$

which proves the continuity of f at x, by Corollary 3.5.3. □

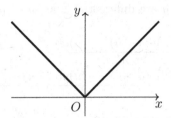

Fig. 4.4   Graph of $y = |x|$.

The converse is not true. For instance, the function $f(x) = |x|$ (see Fig. 4.4) is continuous at 0 (it is continuous everywhere), but it is not differentiable at 0. Indeed,

$$\lim_{h \to 0+} \frac{|0 + h| - 0}{h} = 1 \text{ and } \lim_{h \to 0-} \frac{|0 + h| - 0}{h} = -1,$$

and therefore the limit does not exist. This also has a geometrical meaning: the graph of $f(x) = |x|$ has a corner at 0, so there is no tangent line at that point. Karl Weierstrass (1815–1897) constructed a function which is continuous everywhere, but does not have a derivative at any point.

EXERCISES 4.3

(1) Give an example of a function that is not differentiable at integral points $x = 0, \pm 1, \pm 2 \ldots$, but is differentiable at every other point. Is it possible if the function is required to be continuous everywhere?

(2) Let

$$f(x) = \begin{cases} x^2, & \text{if } x \text{ is a rational number,} \\ 0, & \text{if } x \text{ is an irrational number.} \end{cases}$$

Show that f is differentiable at $x = 0$ but it is not differentiable at any other point.

(3) Let f be a differentiable function such that $f(x) \neq 0$ for all $x \in \mathbb{R}$. Use the definition of the derivative to prove that the function $g(x) = \frac{1}{f(x)}$ is differentiable and express g' in terms of f and f'.

(4) Possible or not: f is not differentiable at 0, but f^2 is?

(5) Possible or not: f^2 is not differentiable at 0, but f is?

(6) Give an example of a function differentiable everywhere except at two distinct points of your choice.

4.4   Derivative of a product

Theorem 4.4.1. *If functions f and g are differentiable at a point, then the product fg is differentiable at that point and its derivative is given by*

$$(fg)' = f'g + fg'. \tag{4.2}$$

Proof. Assume that f and g are differentiable at some point $x = a$. Let $\varphi(x) = f(x)g(x)$. Then, for $h \neq 0$,

$$\frac{\varphi(a+h) - \varphi(a)}{h} = \frac{f(a+h)g(a+h) - f(a)g(a)}{h}$$

$$= \frac{f(a+h)g(a+h) - f(a)g(a+h) + f(a)g(a+h) - f(a)g(a)}{h}$$

$$= \frac{f(a+h) - f(a)}{h} g(a+h) + f(a) \frac{g(a+h) - g(a)}{h}$$

Since g is differentiable, thus continuous, we have $\lim_{h \to 0} g(a+h) = g(a)$. Hence

$$\lim_{h \to 0} \frac{\varphi(a+h) - \varphi(a)}{h} = \lim_{h \to 0} \frac{f(a+h) - f(a)}{h} \lim_{h \to 0} g(a+h)$$

$$+ \lim_{h \to 0} f(a) \lim_{h \to 0} \frac{g(a+h) - g(a)}{h}$$

$$= f'(a)g(a) + f(a)g'(a).$$

$\square$

From (4.2) we get

$$(x^2)' = (xx)' = 1 \cdot x + x \cdot 1 = 2x$$

and

$$(x^3)' = (x^2 x)' = 2xx + x^2 \cdot 1 = 3x^2.$$

In general, for every natural number n, we have

$$(x^n)' = nx^{n-1}. \tag{4.3}$$

To prove this, notice that the formula holds for $n = 1$, since $x^1 = x$ and $x^0 = 1$. Assume now that (4.3) holds for some $n \in \mathbb{N}$. Then

$$\left(x^{n+1}\right)' = (x^n \, x)' = (x^n)'x + x^n = (n+1)x^n.$$

Hence formula (4.3) follows for all natural n by induction.

S: Isn't x^n a number? Do we differentiate numbers now?

T: You are right; x^n is a number. No, we don't differentiate numbers; we only differentiate functions. However we didn't have a name for the function on which to hang the prime.

S: Couldn't you say, *Let $f(x) = x^n$?* We have done that in the past. Then there would be a place to hang the prime.

T: You are right. We could do that every time, but we think it would be a nuisance.

S: Couldn't we give names to these power functions — once for all?

T: Sure, we could do that. In fact some authors have done that, but there are no universally accepted names for them.

S: When we get to the sine function, we'll have a name on which to hang the prime; so we'll write sin$'$ x, won't we?

T: Sorry, that would not be socially acceptable; it just isn't done. So we'll write (sin x)$'$ instead. In fact we abuse the notation again already in the next paragraph, but the reader should be able to tell from the context what is meant.

If $f(x) = c$, a constant function, then $f'(x) = (c)' = 0$ and hence, by (4.2), we obtain $(f(x)g(x))' = (cg(x))' = (c)'g(x) + cg'(x) = cg'(x)$. We thus have the following property of derivatives:

4.4.2. $(cg)' = cg'$ *(where c is a constant).*

EXERCISES 4.4

(1) Prove 4.4.2 using the definition of the derivative.
(2) Let f be a differentiable function on an interval (a, b) such that $f(x) \neq 0$ for all $x \in (a, b)$. Assume that $\frac{1}{f}$ is differentiable and use Theorem 4.4.1 to prove that

$$\left(\frac{1}{f}\right)' = -\frac{f'}{f^2}.$$

(Hint: $1 = f \cdot \frac{1}{f}$)
(3) Use the Induction Principle and Theorem 4.4.1 to prove that if f is differentiable on an interval (a, b), then, for any $n \in \mathbb{N}$, $g(x) = (f(x))^n$ is differentiable on (a, b) and $g'(x) = n (f(x))^{n-1} f'(x)$.
(4) Possible or not: f is differentiable at a, g is not differentiable at a, but the product fg is differentiable at a?
(5) Possible or not: Neither f nor g is differentiable at a, but the product fg is differentiable at a?
(6) Possible or not: Neither f nor g is continuous at a, but the product fg is differentiable at a?
(7) Find a formula for the derivative of the product of three functions.

4.5   Derivative of a sum and a difference

Theorem 4.5.1. *If functions f and g are differentiable at a point, then the sum $f + g$ is differentiable at that point and its derivative is given by*

$$(f + g)' = f' + g'.$$

Proof. Assume that f and g are differentiable at some point $x = a$. If $\varphi(x) = f(x) + g(x)$, then

$$\frac{\varphi(a + h) - \varphi(a)}{h} = \frac{f(a + h) - f(a)}{h} + \frac{g(a + h) - g(a)}{h}.$$

The assertion follows by passing to the limit as $h \to 0$. □

Theorem 4.5.2. *If functions f and g are differentiable at a point, then the difference $f - g$ is differentiable at that point and its derivative is given by*

$$(f - g)' = f' - g'.$$

Proof. Using 4.4.2 and Theorem 4.5.1 we obtain

$$(f - g)' = (f + (-1)g)' = f' + (-1)g' = f' - g'.$$

$\square$

Using the properties of derivatives presented in this and the preceding section we can easily calculate derivatives of some more complicated functions, for example,

$$(3x^2 - x + 2)' = 6x - 1,$$

$$\left(6x^5 - \frac{x^4}{4} - \frac{x}{2} - 2\right)' = 30x^4 - x^3 - \frac{1}{2}.$$

EXERCISES 4.5

(1) Prove that every polynomial function is differentiable everywhere.
(2) Possible or not: Neither f nor g is differentiable at a, but $f + g$ is differentiable at a?

4.6   Derivative of a quotient

Theorem 4.6.1. *If functions f and g are differentiable at a point and g is different from zero at that point, then the quotient f/g is also differentiable at that point and*

$$\left(\frac{f}{g}\right)' = \frac{f'g - fg'}{g^2}. \tag{4.4}$$

Proof. The proof follows from

$$\frac{\frac{f(x+h)}{g(x+h)} - \frac{f(x)}{g(x)}}{h} = \frac{\frac{f(x+h)-f(x)}{h}g(x) - f(x)\frac{g(x+h)-g(x)}{h}}{g(x)g(x+h)}$$

by letting $h \to 0$ (that is, taking the limit as $h \to 0$). $\square$

Example 4.6.2.

$$\left(\frac{x^2+1}{x^2-x+1}\right)' = \frac{2x(x^2 - x + 1) - (x^2 + 1)(2x - 1)}{(x^2 - x + 1)^2} = \frac{1 - x^2}{(x^2 - x + 1)^2}.$$

Example 4.6.3.

$$\left(\frac{1}{x^2+1}\right)' = \frac{0 \cdot (x^2 + 1) - 1 \cdot 2x}{(x^2 + 1)^2} = \frac{-2x}{(x^2 + 1)^2}.$$

If the numerator is a constant, it is more convenient to use the following simplified formula

4.6.4. $\left(\dfrac{c}{g}\right)' = -\dfrac{c\,g'}{g^2}$ *(where c is a constant).*

From 4.6.4 we obtain a formula for the derivative of negative integral powers of x. Indeed,

$$\left(\frac{1}{x^n}\right)' = \frac{-n\,x^{n-1}}{(x^n)^2} = -\frac{n}{x^{n+1}}$$

which agrees with formula (4.3) in Section 4.4.

S: Why do we only use rational functions in all our examples? If we admitted other functions like sine, logarithm, we could greatly extend the stock of examples and make it more interesting.

T: What is the sine function?

S: The ratio of the side opposite to the acute angle of a right triangle to its hypotenuse.

T: What is hypotenuse?

S: A lecture on geometry would be needed to explain all this.

T: Precisely. We do not admit any geometrical definitions or proofs in this book. Definitions of sine, logarithm and other so called elementary functions will be supplied later and based on concepts of analysis.

EXERCISES 4.6

(1) Assuming that $\left(\dfrac{f}{g}\right)'$ exists, use Theorem 4.4.1 to derive the formula given in (4.2).
(2) Show that every rational function is differentiable at every point of its domain.
(3) Show that the derivative of every rational function is a rational function whose domain is the same as that of the original function.

4.7 Derivative of composition of functions

Given two functions f and g the function F defined by

$$F(x) = (f \circ g)(x) = f(g(x)) \tag{4.5}$$

is called the composition of f and g. In Section 3.5 we proved that the composition of two continuous functions is a continuous function. The following theorem shows that the same is true for differentiability. Formula (4.6) is often called the chain rule.

Theorem 4.7.1. *If a function g is differentiable at a point x_0 and a function f is differentiable at the point $g(x_0)$, then the function $F = f \circ g$ is differentiable at x_0 and we have*

$$F'(x_0) = f'(g(x_0))g'(x_0). \tag{4.6}$$

Proof. Let $t_0 = g(x_0)$ and let f be defined in some interval (a, b) containing t_0. We introduce an auxiliary function

$$\rho(t) = \begin{cases} \frac{f(t)-f(t_0)}{t-t_0} - f'(t_0), & \text{for } t \neq t_0, \\ 0, & \text{for } t = t_0. \end{cases} \tag{4.7}$$

Note that the function ρ is defined in the entire interval (a, b) and continuous at t_0. From (4.7) we get

$$f(t) - f(t_0) = (t - t_0)(f'(t_0) + \rho(t)) \tag{4.8}$$

for all $t \neq t_0$. Since this equality holds also for $t = t_0$, it holds in the entire interval (a, b).

The function g, being differentiable at x_0, is continuous at that point and thus there exists a number $\delta > 0$ such that $g(x) \in (a, b)$ whenever $|x - x_0| < \delta$. Hence, by (4.8), we have

$$f(g(x)) - f(g(x_0)) = (g(x) - g(x_0))(f'(g(x_0)) + \rho(g(x)))$$

whenever $|x - x_0| < \delta$. For $x \neq x_0$, we can write

$$\frac{F(x) - F(x_0)}{x - x_0} = \frac{f(g(x)) - f(g(x_0))}{x - x_0} = \frac{g(x) - g(x_0)}{x - x_0}\left(f'(g(x_0)) + \rho(g(x))\right).$$

Consequently,

$$\lim_{x \to x_0} \frac{F(x) - F(x_0)}{x - x_0} = \lim_{x \to x_0} \frac{g(x) - g(x_0)}{x - x_0}\left(f'(g(x_0)) + \rho(g(x))\right) = g'(x_0)f'(g(x_0)),$$

since

$$\lim_{x \to x_0} \rho(g(x)) = \lim_{t \to t_0} \rho(t) = 0.$$

$\square$

EXERCISES 4.7

(1) Assuming that f, g, and h are differentiable functions, find the derivative of $f \circ (g + h)$.

(2) Assuming that f, g, and h are differentiable functions, find the derivative of $f \circ (gh)$.

(3) Assuming that f is a differentiable function, find the derivative of $f \circ f \circ f$.

(4) Possible or not: f is differentiable, g is not differentiable, $f \circ g$ is differentiable?

(5) Possible or not: f is differentiable, g is not differentiable, $g \circ f$ is differentiable?

(6) Possible or not: f and g are not differentiable, $f \circ g$ is differentiable?

4.8 Derivative of the inverse function.

If f and g are inverses of each other, then $f(g(x)) = x$. If both f and g are differentiable, then by applying the chain rule we obtain $f'(g(x))g'(x) = 1$, and hence

$$g'(x) = \frac{1}{f'(g(x))},$$

provided $f'(g(x)) \neq 0$. This formula shows how to find the derivative of an inverse function, given that both functions are differentiable. However, in important cases, it suffices to assume only that one of them is differentiable.

Theorem 4.8.1. *Let f be a continuous and increasing (or decreasing) function defined in an interval (a,b). If f has a derivative different from zero at a point $t_0 \in (a,b)$, then the inverse function g has a derivative at the point $x_0 = f(t_0)$ and we have*

$$g'(x_0) = \frac{1}{f'(g(x_0))}.$$

Proof. The proof is similar to the proof of Theorem 4.7.1. We define an auxiliary function

$$\rho(t) = \begin{cases} \dfrac{f(t) - f(t_0)}{t - t_0} - f'(t_0), & \text{if } t \neq t_0, \\ 0, & \text{if } t = t_0. \end{cases}$$

Then

$$x - x_0 = f(g(x)) - f(g(x_0)) = (g(x) - g(x_0))\left(f'(g(x_0)) + \rho(g(x))\right),$$

and hence

$$\frac{g(x) - g(x_0)}{x - x_0} = \frac{1}{f'(g(x_0)) + \rho(g(x))} \tag{4.9}$$

for $x \neq x_0$. Since ρ and g are continuous and $g(x_0) = t_0$, $\lim_{x \to x_0} \rho(g(x)) = 0$ and thus

$$\lim_{x \to x_0} \frac{g(x) - g(x_0)}{x - x_0} = \lim_{x \to x_0} \frac{1}{f'(g(x_0)) + \rho(g(x))} = \frac{1}{f'(g(x_0))}.$$

$\square$

EXERCISES 4.8

(1) In Theorem 4.8.1 we assume that f is increasing or decreasing. Where is it used in the proof?

(2) Possible or not: f is differentiable and invertible, but f^{-1} is not differentiable?

(3) Let f be a differentiable function and let $g = f^{-1}$. Is it possible that
$$g'(x) = \frac{1}{f'(x)}?$$

4.9   Differentials

S: In some books the derivative is denoted by $\frac{dy}{dx}$. Is it the same derivative?

T: Yes, it is. That notation goes back to the German mathematician Gottfried Wilhelm Leibniz (1646–1716). One advantage of this notation is that, for example, the formula in Theorem 4.8.1 can be written as

$$\frac{dy}{dx} = \frac{1}{\frac{dx}{dy}}.$$

S: I like this notation, because the formula is now easy to remember.

T: That is true. But it still requires a proof.

S: Why? The proof follows from the general rules of division.

T: No, the rules of division have been stated for numbers and dx and dy are not numbers.

S: So what are they?

T: They actually have no separate meaning. An explanation of their use may be the symbol $\frac{\Delta y}{\Delta x}$, where $\Delta y = f(x) - f(x_0)$ and $\Delta x = x - x_0$. If x approaches x_0, then the differences become small, infinitesimally small and they are then called differentials.

S: What is the definition of a differential?

T: Leibniz probably imagined dx and dy as infinitesimally small numbers. Since the times of Leibniz the language of mathematics has changed. It has become much more precise and logical. Expressions like "an infinitesimally small number" cannot be accepted as definitions. There exists a theory of infinitesimals, but this theory is rather difficult and sophisticated and, at least in calculus, introduces only complications.

In calculus textbooks a differential is defined as

$$dy = f'(x)dx,$$

where dx is an arbitrary number. So, for a fixed x, dy is regarded as a function of dx."

S: Dividing both sides of the last equality by dx we obtain

$$\frac{dy}{dx} = f'(x),$$

but I do not see any use of this manipulation.

T: Neither do I.

S: The chain rule in Section 4.7 can also be written as an algebraic identity

$$\frac{df}{dx} = \frac{df}{dy} \cdot \frac{dy}{dx}.$$

T: Yes, but it also needs a proof. In any case it is worthwhile as a mnemonic way to remember the rule.

4.10 Minimum and maximum of a function revisited

In Section 3.4 we proved the existence of the minimum and maximum of a continuous function in a closed interval. The following theorem can often be used to find these values.

Theorem 4.10.1. *If a function f defined on an interval (a, b) has an extremum at $x_0 \in (a, b)$ and is differentiable at x_0, then $f'(x_0) = 0$.*

Proof. Suppose f has a minimum at x_0, then for every $x_0 + h$ in the (a, b) we have $f(x_0 + h) - f(x_0) \geq 0$, and consequently

$$\frac{f(x_0 + h) - f(x_0)}{h} \leq 0 \text{ if } h < 0,$$

$$\frac{f(x_0 + h) - f(x_0)}{h} \geq 0 \text{ if } h > 0.$$

By letting $h \to 0$ we get, in view of Theorem 2.6.1,

$$f'(x_0) \leq 0 \text{ and } f'(x_0) \geq 0,$$

which proves that $f'(x_0) = 0$.

If f has a maximum at x_0, the above proof only requires some obvious modifications. $\square$

From Theorem 4.10.1 it follows that a function defined on a closed interval may have its maximum and minimum only at points

(a) at which the derivative equals 0;
(b) at which the derivative does not exist;
(c) at the endpoints of the interval.

The function can not have an extremum at any other point.

Example 4.10.2. Find the minimum and the maximum of the function $f(x) = x^2 - 2x - 3$ in the interval $[-2, 2]$, see Fig. 4.5.

The function is differentiable everywhere and $f'(x) = 2x - 2$. The only point for which $f'(x) = 0$ is $x = 1$. Thus f can have its extrema at $x = 1$ or the endpoints $x = -2$ and $x = 2$. Since $f(1) = -4$, $f(-2) = 5$, and $f(2) = -3$, we conclude that the minimum is -4 and the maximum is 5.

Example 4.10.3. Find the minimum and the maximum of $f(x) = |2x + 1|$ in the interval $[-1, 1]$, see Fig. 4.6.

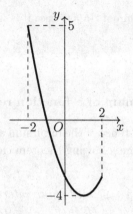

Fig. 4.5　Graph of $y = x^2 - 2x - 3$ for $x \in [-2, 2]$.

Since

$$f(x) = \begin{cases} 2x + 1, & \text{for } x \ge -\frac{1}{2}, \\ -2x - 1, & \text{for } x < -\frac{1}{2}, \end{cases}$$

we have

$$f'(x) = \begin{cases} 2, & \text{for } x > -\frac{1}{2}, \\ -2, & \text{for } x < -\frac{1}{2}, \end{cases}$$

and at $x = -\frac{1}{2}$ there is no derivative. Since the derivative is never equal to 0, the extrema can be reached only at points $x = -\frac{1}{2}$, $x = -1$, or $x = 1$. Since $f(-\frac{1}{2}) = 0$, $f(-1) = 1$, and $f(1) = 3$, the minimum is 0 and the maximum is 3.

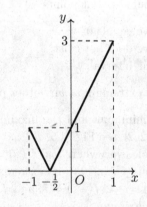

Fig. 4.6　Graph of $y = |2x + 1|$ for $x \in [-1, 1]$.

Example 4.10.4. In the parabola $y = 1-x^2$ there is inscribed a right angle triangle so that one of its sides lies on the x-axis, see Fig. 4.7. Find the triangle whose area is maximal.

It is evident that one of vertices must lie at the point $(1,0)$ or $(-1,0)$. Because of the symmetry we can assume the second case. Then the point $(-1,0)$ is also the end-point of the hypotenuse. The other end of the hypotenuse with coordinates (x,y) lies on the arc of the parabola, we thus have $y = 1 - x^2$. The area of the triangle is then $\frac{1}{2}(1+x)y$. We have to find the maximum of the function

$$P(x) = \frac{1}{2}(1+x)(1-x^2) = \frac{1}{2}(1+x-x^2-x^3)$$

on the interval $(-1,1)$. We have

$$P'(x) = \frac{1}{2}(1 - 2x - 3x^2) = -\frac{1}{2}(3x-1)(x+1).$$

Therefore, $P'(x) = 0$ for $x = \frac{1}{3}$ and $x = -1$. Looking for the extrema we have to take into account points -1, $\frac{1}{3}$, and the endpoint 1. Since $P(-1) = P(1) = 0$ and $P\left(\frac{1}{3}\right) = \frac{16}{27}$, we conclude that the maximal area is $\frac{16}{27}$ and is reached at $x = \frac{1}{3}$.

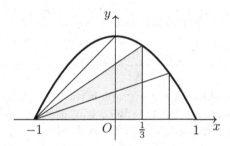

Fig. 4.7   The triangle with the maximal area in Example 4.10.4.

EXERCISES 4.10

(1) Prove Theorem 4.10.1 for the maximum.

(2) Show that $x + \frac{1}{x} \geq 2$ for all $x > 0$.

(3) Find p and q such that the function $f(x) = x^2 + px + q$ attains its minimum value of 3 at $x = 1$.

(4) Suppose P is a polynomial function with an extremum at a point $x = a$ such that $P(a) = 0$. What can be said about the number of times that $(x-a)$ appears as a factor when $P(x)$ is factored into linear and irreducible quadratic factors? Prove your conjecture.

4.11   The Mean Value Theorem

Consider an arc of a curve and the chord joining its ends (see Fig. 4.8). If at each point of the arc there is a tangent, then from among those tangents we can select one which is parallel to the chord. This theorem, which seems obvious geometrically, will be formulated and proved analytically.

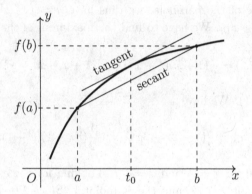

Fig. 4.8   Illustration of the mean value theorem.

Theorem 4.11.1 (Mean Value Theorem). *If a function f is continuous on* $[a, b]$ *and differentiable on* (a, b), *then there exists at least one point* $t_0 \in (a, b)$ *such that*

$$\frac{f(b) - f(a)}{b - a} = f'(t_0). \tag{4.10}$$

Proof. Let g denote the function whose graph is the line joining the points $(a, f(a))$ and $(b, f(b))$ and let $\varphi = f - g$. The function φ is continuous on $[a, b]$ so it must have both a maximum and a minimum on that interval. Since $\varphi(a) = 0 = \varphi(b)$, the only way that both extrema can occur at the endpoints is if the function is identically 0 throughout the interval in which case φ' is also identically zero throughout the interval. If φ is not identically zero, then φ' must have value 0 at that extremum which occurs at an interior point of the interval. In either case, there is a point $t_0 \in (a, b)$ such that $\varphi'(t_0) = 0$. But then we must have $f'(t_0) = g'(t_0)$ which completes the proof because the graph of g is a line whose slope is precisely $\frac{f(b) - f(a)}{b - a}$. □

The Mean Value Theorem is often stated in the following form.

Theorem 4.11.2. *If a function f is continuous on* $[a, b]$ *and differentiable on* (a, b), *then, for any points x and* x_0 *in* $[a, b]$, *we have*

$$f(x) - f(x_0) = (x - x_0)f'(t_0) \quad \textit{for some } t_0 \textit{ between } x \textit{ and } x_0.$$

This form can be used if x is a variable in a fixed interval. One has to remember that in Theorem 4.11.2 the value t_0 depends on x. Overlooking this fact may lead to false results which has happened even to prominent authors.

Corollary 4.11.3. *If f is a function continuous on an interval $[a, b]$ and $f'(x) = 0$ for each $x \in (a, b)$, then f is a constant function on $[a, b]$.*

Proof. If $a \leq x_1 < x_2 \leq b$, then $f(x)$ is continuous on $[x_1, x_2]$ and there exists a point t_0 in (x_1, x_2) such that

$$\frac{f(x_2) - f(x_1)}{x_2 - x_1} = f'(t_0) = 0.$$

Hence $f(x_1) = f(x_2)$.   □

Corollary 4.11.4. *If two functions are continuous on an interval $[a, b]$ and have the same derivative in (a, b), then they differ by an additive constant.*

Proof. The difference of the functions satisfies Corollary 4.11.3 and thus it is a constant.   □

If a function f is differentiable at every point of an interval, then f' need not be continuous in that interval (see Exercise 9 in Section 7.9). However, even if f' is not continuous, it has the intermediate value property.

Theorem 4.11.5. *If a function f is differentiable at every point in the closed interval $[a, b]$, then f' assumes every value between $f'(a)$ and $f'(b)$.*

Proof. Assume f is a function that is differentiable at every point in the closed interval $[a, b]$. Define φ and ψ via

$$\varphi(x) = \begin{cases} \frac{f(x) - f(a)}{x - a}, & \text{if } x \neq a, \\ f'(a), & \text{if } x = a, \end{cases}$$

and

$$\psi(x) = \begin{cases} \frac{f(b) - f(x)}{b - x}, & \text{if } x \neq b, \\ f'(b), & \text{if } x = b. \end{cases}$$

Since φ is continuous in $[a, b]$, it assumes every value between $\varphi(a) = f'(a)$ and $\varphi(b) = \frac{f(b) - f(a)}{b - a}$. Then, by the mean value theorem, f' assumes every value between $f'(a)$ and $\frac{f(b) - f(a)}{b - a}$. Similarly, ψ assumes every value between $\psi(a) = \frac{f(b) - f(a)}{b - a}$ and $\psi(b) = f'(b)$ and thus f' assumes every value between $\frac{f(b) - f(a)}{b - a}$ and $f'(b)$. Consequently, f' assumes every value between $f'(a)$ and $f'(b)$.   □

EXERCISES 4.11

(1) Show that, if f fails to be differentiable at one point of (a, b) then (4.10) need not hold at any $t_0 \in (a, b)$.

(2) Give an example of a function such that $f'(x) = 0$ for all points in its domain but f is not a constant function.

(3) Let f be a differentiable function in an interval $(-\alpha, \alpha)$. Prove that if $f(0) = 0$ and $|f'(x)| \leq \beta$ for all $x \in (-\alpha, \alpha)$ then $|f(x)| \leq \alpha\beta$ for all $x \in (-\alpha, \alpha)$.

(4) Prove: If $f(0) = 0$ and $|f'(x)| \leq M$ for all $x \in \mathbb{R}$, then $|f(x)| \leq M|x|$ for all $x \in \mathbb{R}$.

(5) Let f and g be differentiable on $(0, \infty)$ and continuous on $[0, \infty)$. Prove that if $f(0) = g(0)$ and $f'(x) \leq g'(x)$ for all $x > 0$ then $f(x) \leq g(x)$ for all $x > 0$.

(6) Prove Cauchy's Generalized Mean Value Theorem: If functions f and g are continuous in $[a, b]$ and differentiable in (a, b), then there exists $c \in (a, b)$ such that

$$(f(b) - f(a)) g'(c) = (g(b) - g(a)) f'(c).$$

Hint: Consider the function $\varphi(x) = (f(b) - f(a)) g(x) - (g(b) - g(a)) f(x)$.

(7) Restate the mean value theorem in the context of velocities as illustrated by the following statement: If during the past hour your average velocity was 60 miles per hour, then at least once during the past hour your instantaneous velocity must have been 60 miles per hour.

(8) Give an interpretation for the content of Exercise 6 in terms of the speed of two cars and their relative locations.

4.12 Derivatives of monotone functions

Theorem 4.12.1. *If the derivative of f is positive on an interval, then f is increasing on that interval. If the derivative of f is negative on an interval, then f is decreasing on that interval.*

Proof. Suppose $f'(x) > 0$ for every x in some interval and let x_1 and x_2 be points in that interval with $x_1 < x_2$. By Theorem 4.11.2, there is some $t_0 \in (x_1, x_2)$ such that

$$f(x_2) - f(x_1) = f'(t_0)(x_2 - x_1) > 0,$$

from which it follows that $f(x_1) < f(x_2)$. Thus f is increasing.

The proof of the second part is similar. $\qquad\square$

The above theorem can be used to investigate behavior of functions. For instance the function $f(x) = (1 + x)(1 - x^2)$ (see Fig. 4.9) has the derivative $f'(x) = -(3x - 1)(x + 1)$, which is positive on $(-1, \frac{1}{3})$, and negative on $(-\infty, -1)$ and $(\frac{1}{3}, \infty)$. Hence f is increasing on $(-1, \frac{1}{3})$ and decreasing on $(-\infty, -1)$ and $(\frac{1}{3}, \infty)$.

In Section 3.2 we proved that the equation $x^5 + x - 1 = 0$ has a solution in $(0, 1)$. Since the function $f(x) = x^5 + x - 1$ has the derivative $f'(x) = 5x^4 + 1$ which is positive everywhere, it is increasing. Thus the solution is unique.

The converse of Theorem 4.12.1 is false. For instance, the function $f(x) = x^3$ is increasing everywhere, although its derivative $f'(x) = 3x^2$ equals 0 at $x = 0$; see Fig. 4.10.

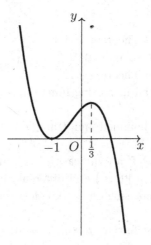

Fig. 4.9 Graph of $y = (1+x)^2(1-x)$.

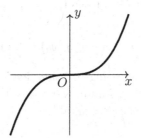

Fig. 4.10 Graph of $y = x^3$.

Theorem 4.12.2. *If $f'(x) \geq 0$ on some interval, then f is non-decreasing on that interval. If $f'(x) \leq 0$ on some interval, then f is non-increasing on that interval.*

The proof of the above theorem is very similar to the proof of Theorem 4.12.1 and is left as an exercise.

In contrast to Theorem 4.12.1, Theorem 4.12.2 can be reversed.

Theorem 4.12.3. *If f is differentiable and non-increasing on an interval, then $f'(x) \leq 0$ on that interval. Similarly, if f is differentiable and non-decreasing on an interval, then $f'(x) \geq 0$ on that interval.*

Proof. In the first case we have
$$\frac{f(x+h) - f(x)}{h} \leq 0$$
for x and $x+h$ in the interval ($h \neq 0$). Hence, letting $h \to 0$, we get $f'(x) \leq 0$. In the second case the proof is similar. $\quad\square$

EXERCISES 4.12

(1) Complete the proof of Theorem 4.12.1.

(2) Provide a proof of Theorem 4.12.2.

(3) Complete the proof of Theorem 4.12.3.

(4) True or false: If f is an increasing function, then f' is an increasing function?

(5) Possible or not: f is bounded and f and f' are increasing on the interval $(0, \infty)$?

(6) Possible or not: f and f' are increasing on $\mathbb{R}$ and the range of f is $\mathbb{R}$?

(7) Let f be continuous on $[0, \infty)$ and differentiable on $(0, \infty)$. Show that if $f(0) = 0$ and f' is increasing on $(0, \infty)$ then the function $g(x) = f(x)/x$ is increasing on $(0, \infty)$.

(8) Let f be a function such that f' is increasing on the closed interval $[a, b]$. Let g be the function whose graph is the line passing through the points $(a, f(a))$ and $(b, f(b))$. Prove that $f(x) < g(x)$ for every $x \in (a, b)$.

(9) Prove: If f is a non-decreasing function defined on an interval (a, b), then at each point $x_0 \in (a, b)$ both $\lim_{x \to x_0-} f(x_0)$ and $\lim_{x \to x_0+} f(x_0)$ exist and, moreover,

$$\lim_{x \to x_0-} f(x_0) \leq f(x_0) \leq \lim_{x \to x_0+} f(x_0).$$

(10) Prove: If f is a non-decreasing function defined on an interval (a, b), then the set consisting of all points in the interval (a, b) at which f fails to be continuous is either finite or countable.

Chapter 5

SEQUENCES AND SERIES

5.1 Convergent sequences

By a sequence we mean a progression of numbers $a_1, a_2, \ldots$. While the word sequence is used to mean both finite and infinite sequences, we will use it exclusively to mean infinite sequences.

Formally, a sequence $a_1, a_2, \ldots$ is a function defined on the set of all natural numbers. The value of that function at n is called the n-th term of the sequence and is denoted by a_n. We will write "a sequence $a_1, a_2, \ldots$" or "a sequence (a_n)". The parentheses are used to distinguish the sequence (a_n) from its n-th term a_n.

Definition 5.1.1. We say that a is the *limit* of a sequence (a_n), if for every $\varepsilon > 0$ there is a number M such that $|a_n - a| < \varepsilon$ for all $n > M$. If a is the limit of (a_n), we say that (a_n) *converges* or *tends* to a and we write

$$\lim_{n \to \infty} a_n = a \quad \text{or} \quad a_n \to a.$$

If a sequence has a limit, it is called a *convergent sequence*. Sequences which are not convergent are called *divergent*.

While we are not requiring the number M in the above definition to be a natural number, it is often convenient to assume that it is, as in the proof of Theorem 5.1.3. Clearly, this does not result in any loss of generality.

Note that the above definition is very similar to the definition of the limit of a function at infinity. As we will see, this similarity can be used to obtain properties of limits of sequences from similar properties of limits of functions.

A sequence whose terms are all equal is called a *constant sequence*. It is easy to see that the constant sequence $a, a, a, \ldots$ converges to a.

Example 5.1.2. We will show that

$$\lim_{n \to \infty} \frac{n}{n+1} = 1.$$

Let ε be an arbitrary positive number. Define $M = \frac{1}{\varepsilon}$. Then, for all $n > M$, we have

$$\left| \frac{n}{n+1} - 1 \right| = \left| \frac{1}{n+1} \right| < \frac{1}{M} < \varepsilon.$$

Since this works for any $\varepsilon > 0$, we have shown that $\lim_{n \to \infty} \frac{n}{n+1} = 1$.

A sequence (a_n) is called *bounded* if there exists a constant B such that $|a_n| \leq B$ for all $n \in \mathbb{N}$.

Theorem 5.1.3. *Convergent sequences are bounded.*

Proof. Let (a_n) be a convergent sequence. If a is the limit of (a_n), then there exists a natural number M such that $|a_n - a| < 1$ for all $n > M$. Hence, for all $n > M$,

$$|a_n| = |(a_n - a) + a| \leq |a_n - a| + |a| < 1 + |a|.$$

If we define

$$B = \max\{|a_1|, |a_2|, \ldots, |a_M|, 1 + |a|\},$$

then we will have $|a_n| \leq B$ for all $n \in \mathbb{N}$. $\quad\square$

Clearly, not every bounded sequence is convergent.

Since the definition of the limit of a convergent sequence is practically identical with the definition of the limit of a function at infinity, it should be possible to obtain properties of limits of sequences as corollaries to theorems already proved for limits of functions. The following simple observation allows us to realize this plan.

Theorem 5.1.4. *Let (a_n) be a sequence and let f be the function defined on $[1, \infty)$ such that $f(x) = a_n$ if $x \in [n, n+1)$. Then*

$$\lim_{n \to \infty} a_n = a \quad \text{if and only if} \quad \lim_{x \to \infty} f(x) = a.$$

Proof. The theorem is an immediate consequence of the definition of the limit of a convergent sequence and the definition of the limit of a function at infinity. $\quad\square$

Theorem 5.1.5. *If $\lim_{n \to \infty} a_n = a$ and $\lim_{n \to \infty} b_n = b$, then*

(a) $\lim_{n \to \infty} (a_n + b_n) = a + b$,

(b) $\lim_{n \to \infty} (a_n - b_n) = a - b$,

(c) $\lim_{n \to \infty} a_n b_n = ab$.

Moreover, if $b_n \neq 0$ and $b \neq 0$, then

(d) $\lim_{n \to \infty} \dfrac{a_n}{b_n} = \dfrac{a}{b}$.

Proof. By Theorem 5.1.4, the above follows immediately from Corollary 2.7.3. $\quad\square$

Theorem 5.1.6. *If $a_n \leq b_n$ for all $n \in \mathbb{N}$, then $\lim_{n \to \infty} a_n \leq \lim_{n \to \infty} b_n$, whenever the limits exist.*

Proof. The property follows from Theorem 2.6.1, by Theorems 5.1.4 and 2.7.2. $\quad\square$

Theorem 5.1.7. *If $a_n \leq b_n \leq c_n$ for all $n \in \mathbb{N}$ and $\lim_{n \to \infty} a_n = \lim_{n \to \infty} c_n = l$, then* $\lim_{n \to \infty} b_n = l$.

Proof. The property follows from Theorem 2.6.2, by Theorems 5.1.4 and 2.7.2. $\square$

Corollary 5.1.8. *If $|a_n| \leq b_n$ for every $n \in \mathbb{N}$ and $\lim_{n \to \infty} b_n = 0$, then $\lim_{n \to \infty} a_n = 0$.*

S: It is nice that we can obtain properties of limits of sequences from theorems on limits of functions. Would it be possible to first prove theorems about sequences and then derive from them similar properties of functions?

T: Actually, this is a fairly common approach. First we define limits of sequences. Then we can define the limit of a function using limits of sequences: $\lim_{x \to x_0} f(x) = l$ if for every sequence (x_n) in the domain of f such that $x_n \to x_0$ and $x_n \neq x_0$ we have $\lim_{n \to \infty} f(x_n) = l$. This definition is attributed to Eduard Heine (1821–1881). The sequences used in the definition are sometimes called *Heine sequences*.

S: Which approach is better: starting from functions or starting from sequences?

T: It is a matter of taste. The definitions are equivalent. The use of Heine's approach is sometimes very helpful, see for example the proof of Theorem 11.1.1. Sequences can also be used to define the limit of a function at infinity. In that case we use sequences tending to infinity.

Definition 5.1.9. We say that a sequence (x_n) *tends to infinity* and write $x_n \to \infty$ or $\lim_{n \to \infty} x_n = \infty$, if for every number K there is a number M such that $x_n > K$ for all $n > M$.

It is important to remember that a sequence tending to infinity does not have a limit in the sense of Definition 5.1.1.

EXERCISES 5.1

(1) Show that $\lim_{n \to \infty} a_n = a$ if and only if for every $\varepsilon > 0$ there exists a natural number M such that $|a_n - a| < \varepsilon$ for all $n > M$.

(2) Show that $\lim_{n \to \infty} a_n = a$ if and only if for every natural number k there exists a natural number M such that $|a_n - a| < \frac{1}{k}$ for all $n > M$.

(3) Prove that the sequence $a, a, a, \ldots$ converges to a.

(4) Provide a detailed proof of Theorem 5.1.4.

(5) Provide a detailed proof of Corollary 5.1.8.

(6) Use Definition 5.1.1 to prove that $\lim_{n \to \infty} \frac{1}{n^2} = 0$.

(7) Use Definition 5.1.1 to prove that $\lim_{n \to \infty} \frac{n+1}{n+2} = 1$.

(8) Use Definition 5.1.1 to prove that $\left(\frac{n^2+1}{n+1} \right)$ diverges.

(9) Use Definition 5.1.1 to prove that $((-1)^n)$ diverges.

(10) Use Definition 5.1.1 to prove that $\lim_{n \to \infty} \frac{1}{n^2} \neq 10$.

(11) Use Definition 5.1.1 to prove that $\lim_{n \to \infty} \frac{2n^2+1}{1+n+n^2} \neq 1$.

(12) Prove that $\lim_{n \to \infty} \frac{1}{n!} = 0$ (where $n! = 1 \cdot 2 \cdot 3 \cdots \cdot n$).

(13) Possible or not: (a_n) converges, (b_n) diverges, and $(a_n + b_n)$ converges?

(14) Possible or not: (a_n) and (b_n) diverge, but $(a_n + b_n)$ converges?

(15) Possible or not: (a_n) converges, (b_n) diverges, and $(a_n b_n)$ converges?

(16) Possible or not: (a_n) and (b_n) diverge, but $(a_n b_n)$ converges?

(17) Is it possible to define a continuous function f satisfying the conditions in Theorem 5.1.4?

(18) True or false: If $a_n < b_n$ for all $n \in \mathbb{N}$, then $\lim_{n \to \infty} a_n < \lim_{n \to \infty} b_n$, whenever the limits exist?

(19) Prove: If $\lim_{n \to \infty} a_n = a$, then $\lim_{n \to \infty} |a_n| = |a|$.

(20) Prove: $\lim_{n \to \infty} a_n = 0$ if and only if $\lim_{n \to \infty} |a_n| = 0$.

(21) Give an example of a bounded sequence which is not convergent.

(22) True or false: If $0 \leq a_n \leq b_n$ for all $n \in \mathbb{N}$ and (b_n) converges, then (a_n) converges?

(23) Prove: If $\lim_{n \to \infty} a_n = a$ and $\lim_{n \to \infty} b_n = a$, then the sequence $a_1, b_1, a_2, b_2, \ldots$ converges to a.

(24) Let (a_n) be an arbitrary sequence and let $b_n = a_{k+n}$ for some $k \in \mathbb{N}$. Prove that the sequence (b_n) converges if and only if the sequence (a_n) converges. Moreover, if both sequences converge, then they have the same limit.

(25) True or false: $\lim_{x \to x_0} f(x) = k$ if and only if $\lim_{n \to \infty} f(x_n) = k$ for every sequence $x_1, x_2, \ldots$ in the domain of f such that $\lim_{n \to \infty} x_n = x_0$? Justify your response with a proof or a counterexample.

(26) Prove that $\lim_{x \to x_0} f(x) = l$ if and only if $\lim_{n \to \infty} f(x_n) = l$ for every sequence (x_n) in the domain of f such that $x_n \to x_0$ and $x_n \neq x_0$.

(27) Prove that $\lim_{x \to \infty} f(x) = k$ if and only if $\lim_{n \to \infty} f(x_n) = k$ for every sequence $x_1, x_2, \ldots$ in the domain of f such that $\lim_{n \to \infty} x_n = \infty$.

(28) Define $\lim_{n \to \infty} x_n = -\infty$.

(29) Show that the order of terms of a sequence does not affect its convergence.

5.2   Monotone sequences and subsequences

A sequence (a_n), is called *non-decreasing* if $a_n \leq a_{n+1}$ for every $n \in \mathbb{N}$. Similarly, (a_n) is called *non-increasing* if $a_n \geq a_{n+1}$ for every $n \in \mathbb{N}$. Both non-decreasing and non-increasing sequences are called *monotone*. A constant sequence is non-increasing and at the same time non-decreasing.

The following important property of sequences is an immediate consequence of Theorems 2.8.4 and 5.1.4.

Theorem 5.2.1. *Every bounded and monotone sequence is convergent.*

If we remove from a sequence certain terms such that infinitely many terms are left, then the remaining terms form a new sequence which is called a subsequence of the initial sequence. This intuitive definition can be precisely formulated as follows.

Definition 5.2.2. Let (a_n) be a sequence and let (p_n) be an increasing sequence of natural numbers. The sequence $a_{p_1}, a_{p_2}, a_{p_3}, \ldots$ is called a *subsequence* of (a_n).

Sequences (a_{2n}), (a_{2n-1}), or (a_{m+n}), where m is a fixed natural number, are subsequences of (a_n). Similarly,

$$\frac{1}{2}, \frac{1}{4}, \frac{1}{8}, \frac{1}{16}, \ldots,$$

$$\frac{1}{2}, \frac{1}{20}, \frac{1}{200}, \frac{1}{2000}, \ldots$$

are subsequences of the sequence $\frac{1}{2}, \frac{1}{3}, \frac{1}{4}, \frac{1}{5}, \ldots$.

Theorem 5.2.3. *A subsequence of a convergent sequence converges to the same limit.*

Proof. Let $\lim_{n \to \infty} a_n = a$ and let (a_{p_n}) be a subsequence of (a_n). For every $\varepsilon > 0$ there is a number M such that $|a_n - a| < \varepsilon$ for all $n > M$. Since $p_n \geq n$, which is easily checked by induction, we have $|a_{p_n} - a| < \varepsilon$ for all $n > M$, proving the theorem. $\square$

The next theorem is usually credited to two mathematicians: Bernard Bolzano (1781–1848) and Karl Weierstrass (1815–1897). It describes one of the fundamental properties of the real numbers.

Theorem 5.2.4. *(Bolzano–Weierstrass Theorem). Every bounded sequence contains a convergent subsequence.*

Proof. Let (a_n) be a bounded sequence. For every $n \in \mathbb{N}$, let x_n be the least upper bound of the set $\{a_n, a_{n+1}, a_{n+2}, \ldots\}$. Since the sequence (x_n) is non-increasing and bounded, it converges to a limit x, by Theorem 5.2.1. Now, let a_{p_1} be the first term in the sequence $a_1, a_2, a_3, \ldots$ such that

$$|a_{p_1} - x_1| < 1.$$

Next, let a_{p_2} be the first term in the sequence $a_{p_1+1}, a_{p_1+2}, a_{p_1+3}, \ldots$ such that

$$|a_{p_2} - x_{p_1+1}| < \frac{1}{2}.$$

In general, let $a_{p_{n+1}}$ be the first term in the sequence $a_{p_n+1}, a_{p_n+2}, a_{p_n+3}, \ldots$ such that

$$|a_{p_{n+1}} - x_{p_n+1}| < \frac{1}{n+1}.$$

Since

$$|a_{p_n} - x| \leq |a_{p_n} - x_{p_n+1}| + |x_{p_n+1} - x| < \frac{1}{n+1} + |x_{p_n+1} - x| \to 0$$

as $n \to \infty$, (a_{p_n}) is a convergent subsequence of (a_n). $\square$

From the Bolzano–Weierstrass theorem we obtain the following useful criterion for convergence of sequences of real numbers.

Corollary 5.2.5. *The following conditions are equivalent:*

(a) *The sequence (a_n) is convergent;*

(b) *For every $\varepsilon > 0$ there is a number M such that $|a_m - a_n| < \varepsilon$ for all $m, n > M$.*

Proof. First assume that (a_n) is a convergent sequence. Let $a = \lim_{n \to \infty} a_n$ and let $\varepsilon > 0$. Then there exists a number M such that $|a_n - a| < \frac{\varepsilon}{2}$ for all $n > M$. Thus, for all $m, n > M$, we have

$$|a_m - a_n| \leq |a_m - a| + |a - a_n| < \varepsilon.$$

Now assume that (b) holds. Let $\varepsilon > 0$. Then there is a number $n_0 \in \mathbb{N}$ such that $|a_m - a_n| < \frac{\varepsilon}{2}$ for all $m, n > n_0$. Since, for all $n > n_0$ we have

$$|a_n| \leq |a_n - a_{n_0+1}| + |a_{n_0+1}| < \frac{\varepsilon}{2} + |a_{n_0+1}|,$$

for all $n \in \mathbb{N}$ we have

$$|a_n| \leq \max\left\{ |a_1|, \ldots, |a_{n_0}|, \frac{\varepsilon}{2} + |a_{n_0+1}| \right\}.$$

Therefore the sequence (a_n) is bounded. By the Bolzano–Weierstrass theorem, (a_n) has a convergent subsequence (a_{p_n}). Let $a = \lim_{n \to \infty} a_{p_n}$. Then there exists a number M such that

$$|a_{p_n} - a| < \frac{\varepsilon}{2}$$

for all $n > M$. Consequently, for all $n > \max\{n_0, M\}$ we have

$$|a_n - a| \leq |a_n - a_{p_n}| + |a_{p_n} - a| < \varepsilon,$$

proving convergence of (a_n). □

A sequence satisfying condition (b) in Corollary 5.2.5 is called a *Cauchy sequence*. According to the above, a sequence of real numbers is convergent if and only if it is a Cauchy sequence. This is of practical importance. In order to prove convergence of a sequence using the definition of convergence we need to know the limit of the sequence. Since the limit does not appear in condition (b), it is possible to prove convergence without knowing the limit.

EXERCISES 5.2

(1) Check whether the following sequences are monotone:

(a) $\left(\frac{1}{n^2}\right)$,   (b) $\left(\frac{n}{n+1}\right)$,   (c) $\left(\frac{n^4}{n!}\right)$.

(2) Define $a_1 = 1$ and $a_{n+1} = \frac{3a_n+4}{2a_n+3}$ for $n \geq 1$. Show that (a_n) converges.

(3) Define $a_1 = 1$ and $a_{n+1} = \frac{a_n+2}{a_n+1}$ for $n \geq 1$. Show that (a_n) converges. (Hint: Find a_{n+2} in terms of a_n and then use the previous exercise.)

(4) Prove: If every subsequence of (a_n) has a subsequence convergent to a, then (a_n) converges to a.

(5) Prove: If (a_n) has two convergent subsequences that converge to different limits, then (a_n) is divergent.

(6) Possible or not: The sequence (a_n) is divergent, but every convergent subsequence of (a_n) converges to a?

(7) Prove that $\left(\left(1+\frac{1}{n}\right)^n\right)$ is a convergent sequence. The following steps will help.

    (a) Let $x, y \in \mathbb{R}$. Use the Induction Principle to prove that, for every $n \in \mathbb{N}$, we have
$$(x+y)^n = x^n + \frac{n}{1}x^{n-1}y + \frac{n(n-1)}{1\cdot 2}x^{n-2}y^2 + \cdots + \frac{n(n-1)\cdots 2}{1\cdot 2\cdots(n-1)}xy^{n-1} + y^n.$$

    (b) Show that $\left(\left(1+\frac{1}{n}\right)^n\right)$ is a monotone sequence by using part (a) to rewrite each of $\left(1+\frac{1}{n}\right)^n$ and $\left(1+\frac{1}{n+1}\right)^{n+1}$ and comparing the results term-by-term.

    (c) Show that $\left(\left(1+\frac{1}{n}\right)^n\right)$ is bounded by using part (a) and comparing term-by-term with the sequence
$$a_n = 2 + 1 + \frac{1}{2} + \frac{1}{4} + \cdots + \frac{1}{2^{n-2}}.$$

(8) Show that the sequence
$$a_n = 1 - \frac{1}{2} + \frac{1}{3} - \frac{1}{4} + \cdots + (-1)^{n+1}\frac{1}{n}$$
is a Cauchy sequence.

(9) A function on $[a, b]$ is called *uniformly continuous* if for every $\varepsilon > 0$ there is a $\delta > 0$ such that $|f(x_1) - f(x_2)| < \varepsilon$ for all $x_1, x_2 \in [a, b]$ such that $|x_1 - x_2| < \delta$. Prove that every continuous function on a bounded closed interval $[a, b]$ is uniformly continuous.

5.3   Geometric sequence and series

The sequence (q^n), where q is an arbitrary real number, is called a *geometric sequence*. An example of such a sequence is
$$\frac{1}{3}, \frac{1}{9}, \frac{1}{27}, \frac{1}{81}, \ldots,$$
which is $\left(\frac{1}{3^n}\right)$. The choice $q = 1$ produces the constant sequence $1, 1, 1, \ldots$ which converges to 1. If $q = -1$, then the resulting geometric sequence $-1, 1, -1, 1, \ldots$ is not convergent. It contains a subsequence convergent to 1 and a subsequence convergent to -1.

Theorem 5.3.1. *If $|q| < 1$, then $\lim\limits_{n\to\infty} q^n = 0$.*

Proof. If $0 < q < 1$, then we have

$$q^{n+1} = q \cdot q^n < q^n \quad \text{and} \quad 0 < q^n < 1,$$

and thus the sequence (q^n) is non-increasing and bounded. Hence, by Theorem 5.2.1, it converges to some limit l. Moreover,

$$l = \lim_{n\to\infty} q^{n+1} = \lim_{n\to\infty} q \cdot q^n = q \cdot \lim_{n\to\infty} q^n = q \cdot l,$$

because (q^{n+1}) is a subsequence of (q^n) and thus $\lim_{n\to\infty} q^{n+1} = \lim_{n\to\infty} q^n$. The only l that satisfies the equality $l = ql$ for $0 < q < 1$ is $l = 0$. This proves the theorem for $0 < q < 1$.

If $-1 < q < 0$, then $0 < |q| < 1$, and thus $\lim_{n\to\infty} |q|^n = 0$, by the first part of the proof. This implies $\lim_{n\to\infty} q^n = 0$, by Theorem 5.1.8, because $|q^n| = |q|^n$.

Since obviously $\lim_{n\to\infty} q^n = 0$ if $q = 0$, the proof is complete. $\qquad\square$

Note that, if $q > 1$, then $\lim_{n\to\infty} q^n = \infty$. If $q < -1$, then the sequence (q^n) oscillates between positive and negative numbers of larger and larger absolute values.

Now we consider a sequence (s_n) defined by induction as follows:

$$s_1 = 1, \quad s_{n+1} = s_n + q^n \tag{5.1}$$

for $n \in \mathbb{N}$ and some $q \in \mathbb{R}$. Thus we have

$$s_2 = 1 + q,$$
$$s_3 = 1 + q + q^2,$$
$$s_4 = 1 + q + q^2 + q^3,$$

and generally

$$s_{n+1} = 1 + q + q^2 + \cdots + q^n.$$

The last expression has a clear intuitive meaning, but its precise definition is (5.1). The sequence (s_n) is called a *geometric series*.

Theorem 5.3.2. *If $|q| < 1$, then the sequence defined by (5.1) is convergent and*

$$\lim_{n\to\infty} s_n = \frac{1}{1-q}. \tag{5.2}$$

Proof. It follows by induction that

$$(1-q)s_n = 1 - q^n$$

for every $n \in \mathbb{N}$. Dividing both sides by $1 - q$ we obtain

$$s_n = \frac{1-q^n}{1-q},$$

and hence, by Theorem 5.3.1,

$$\lim_{n\to\infty} s_n = \lim_{n\to\infty} \frac{1-q^n}{1-q} = \frac{1}{1-q}.$$

$\qquad\square$

The equality in (5.2) is often expressed in the form

$$1 + q + q^2 + q^3 + \cdots = \frac{1}{1-q}.$$

This equality has a simple geometric interpretation. It is easily seen that the triangles AOB and CBD in Fig. 5.1 are similar. Hence $AO/BA = BC/CD$, that is,

$$\frac{1 + q + q^2 + \cdots}{1} = \frac{1}{1-q}.$$

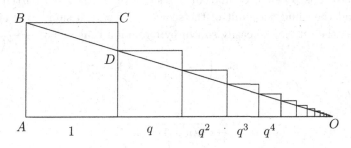

Fig. 5.1 $\quad \frac{1+q+q^2+\cdots}{1} = \frac{1}{1-q}.$

EXERCISES 5.3

(1) Show that $\lim_{n\to\infty} q^{n^2} = 0$ if $|q| < 1$.
(2) Find the limit $\lim_{n\to\infty} \frac{2^n - n^2}{n^2 \cdot 2^n}$.
(3) Find the limit $\lim_{n\to\infty} (2 \cdot 3^{1+n} \cdot 5^{1-n})$.
(4) Find the limit $\lim_{n\to\infty} (-2)^{3n} \cdot 3^{-2n}$.
(5) Pove that $\lim_{n\to\infty} q^n = \infty$ if $q > 1$.
(6) Let $k \in \mathbb{N}$. Show that $1 + q^k + q^{2k} + q^{3k} + \cdots = \frac{1}{1-q^k}$ for every $|q| < 1$.
(7) Show that $1 - q + q^2 - q^3 + \cdots = \frac{1}{1+q}$ for every $|q| < 1$.
(8) Prove: If $q \geq 1$ then, for the sequence defined in (5.1), we have $\lim_{n\to\infty} s_n = \infty$.
(9) Express $0.1 + 0.01 + 0.001 + \ldots$ as a quotient of two natural numbers.

5.4 Convergent series

Starting from an arbitrary sequence (a_n) we can define a new sequence (s_n) by using the formulas

$$s_1 = a_1 \text{ and } s_{n+1} = s_n + a_{n+1} \text{ for all } n \in \mathbb{N}. \tag{5.3}$$

In other words,

$$s_n = a_1 + \cdots + a_n \quad \text{or} \quad s_n = \sum_{k=1}^{n} a_k.$$

The number s_n is called the *n-th partial sums* of the sequence (a_n). If the sequence (s_n) converges to some number s, that is,

$$s = \lim_{n \to \infty} s_n = \lim_{n \to \infty} \sum_{k=1}^{n} a_k,$$

we write

$$\sum_{k=1}^{\infty} a_k = s.$$

Sometimes it is possible to find an explicit form of the n-th partial sum of a sequence and then find the limit of the sequence of partial sums. For example, if $a_n = \frac{1}{n(n+1)}$, then it can be easily shown by induction that $s_n = 1 - \frac{1}{n+1}$. Since

$$\lim_{n \to \infty} \left(1 - \frac{1}{n+1} \right) = 1,$$

we have

$$\sum_{n=1}^{\infty} \frac{1}{n(n+1)} = 1.$$

Instead of saying that the sequence of partial sums of a sequence (a_n) converges, we will say simply that the *series* $\sum_{n=1}^{\infty} a_n$ *converges*. Similarly, if the limit the sequence of partial sums of the sequence (a_n) is s, we say that the *sum of the series* $\sum_{n=1}^{\infty} a_n$ *is s*.

S: What is the definition of a series?

T: It is rather difficult to give a logical definition of a series. If we were to speak of a series we simultaneously think of both: the sequence of terms

$$a_1, a_2, a_3, \ldots$$

and the sequence of partial sums

$$s_1, s_2, s_3, \ldots$$

S: So the theory of series is a part of the theory of sequences — sequences of particular form. Could we say that a series is a sequence of partial sums?

T: This does not solve the problem. Any sequence $a_1, a_2, a_3, \ldots$ can be considered as a sequence of partial sums: it suffices to take

$$b_1 = a_1 \text{ and } b_n = a_n - a_{n-1} \text{ for } n \geq 2.$$

Then $b_1 + \cdots + b_n = a_n$.

S: And how is this problem handled in other books?

T: Their authors usually speak of series without giving a precise definition. The prominent specialist in the theory of series Konrad Knopp (1882–1957) says that a series is a sequence that is defined in a special way: one first gives a sequence $a_1, a_2, a_3, \ldots$, and then constructs a new sequence by letting

$$s_1 = a_1, \quad s_2 = a_1 + a_2, \quad s_3 = a_1 + a_2 + a_3, \quad \ldots .$$

According to Knopp, a series is a symbol $a_1 + a_2 + a_3 + \ldots$ or $\sum_{n=1}^{\infty} a_n$.

S: We never define anything to be a symbol. Is it correct?

T: It may be correct, but such definitions are not used. Knopp wrote his famous book *Theory and Applications of Infinite Series* in 1921. Since that time mathematics has become more rigorous and it is not common to define mathematical objects by symbols.

S: Then what actually is a series?

T: If one necessarily wants to have a definition, then we can say that a series is a pair of sequences a_1, a_2, a_3, and $s_1, s_2, s_3, \ldots$ connected by (5.3).

S: What is the use of such a definition?

T: None, but we have a definition that is adequate and logically correct. In fact, we don't really need a definition of a series. It is possible to express everything about series in terms of sequences and limits of sequences. However, using the name series for sequences obtained as partial sums of sequences has a long tradition and is often convenient. For example, it is more convenient to say that the series $\sum_{n=1}^{\infty} a_n$ converges instead of saying that the sequence of partial sums of the sequence (a_n) converges.

Since $\sum_{n=1}^{\infty} a_n$ denotes the limit of a sequence of numbers, namely the sequence of partial sums of the sequence (a_n), it is a number. From that point of view the symbol $\sum_{n=1}^{\infty} (-1)^n$ is meaningless, because the sequence of partial sums of the sequence $-1, 1, -1, 1, \ldots$ is not convergent. If the sequence of partial sums of a sequence (a_n) is not convergent, we simply say that the *series* $\sum_{n=1}^{\infty} a_n$ *diverges*.

The sequence of partial sums of the sequence $1, 1, 1, \ldots$ is not convergent, because the n-th partial sum of this sequence is n. Since, according to Definition 5.1.9, we have $\lim_{n \to \infty} n = \infty$, we will write $\sum_{n=1}^{\infty} 1 = \infty$. However, this series will still be called divergent because the sequence of partial sums is not convergent.

We would like to point out a certain inconsistency: the symbol $\sum_{n=1}^{\infty} a_n$ is used to denote the sum as well as the "series itself." We follow here a long tradition. This inconsistency should not cause any problems, since it is always clear from the context what is meant.

Theorem 5.4.1. *If the series* $\sum_{n=1}^{\infty} a_n$ *converges, then* $\lim_{n \to \infty} a_n = 0$.

Proof. Let $s_n = \sum_{k=1}^{n} a_k$ for $n \in \mathbb{N}$. Convergence of the series $\sum_{n=1}^{\infty} a_n$ means that, for some $s \in \mathbb{R}$, we have $\lim_{n \to \infty} s_n = s$. Therefore, for every $\varepsilon > 0$ there exists a number M such that for all $n > M$ we have

$$|s_n - s| < \frac{1}{2}\varepsilon.$$

For $n > M + 1$ we have

$$|a_n| = |s_n - s_{n-1}| \leq |s_n - s| + |s - s_{n-1}| \; < \frac{1}{2}\varepsilon + \frac{1}{2}\varepsilon = \varepsilon.$$

Since ε is an arbitrary positive number, we conclude that $\lim_{n \to \infty} a_n = 0$. □

The converse of Theorem 5.4.1 is not true. For example, we have $\lim_{n \to \infty} \frac{1}{n} = 0$, but the series $\sum_{n=1}^{\infty} \frac{1}{n}$ diverges. Divergence of this series follows immediately from the so called integral test (Theorem 11.5.1). It can also be obtained directly by considering partial sums of the form

$$1 + \frac{1}{2},$$

$$1 + \frac{1}{2} + \left(\frac{1}{3} + \frac{1}{4}\right),$$

$$1 + \frac{1}{2} + \left(\frac{1}{3} + \frac{1}{4}\right) + \left(\frac{1}{5} + \frac{1}{6} + \frac{1}{7} + \frac{1}{8}\right),$$

and so on.

Theorem 5.4.2. *If the series $\sum_{n=1}^{\infty} a_n$ converges, then, for any constant $c \in \mathbb{R}$, the series $\sum_{n=1}^{\infty} c a_n$ converges and we have*

$$\sum_{n=1}^{\infty} c a_n = c \sum_{n=1}^{\infty} a_n. \tag{5.4}$$

Proof. For any $m \in \mathbb{N}$, by an easy induction we obtain

$$\sum_{n=1}^{m} c a_n = c \sum_{n=1}^{m} a_n. \tag{5.5}$$

Hence, by letting $m \to \infty$, we get (5.4). □

The phrase "by letting $m \to \infty$" does not mean that we merely replace m by ∞. It is a short way of indicating that we take the limit as $m \to \infty$ on both sides of (5.5). Such an argument is valid only if the limits exist. The detailed proof of (5.4) would require the following steps:

$$\sum_{n=1}^{\infty} c a_n = \lim_{m \to \infty} \sum_{n=1}^{m} c a_n = \lim_{m \to \infty} \left(c \sum_{n=1}^{m} a_n\right) = c \lim_{m \to \infty} \sum_{n=1}^{m} a_n = c \sum_{n=1}^{\infty} a_n.$$

The first and the last equality follow from the definition of the sum of a series, the second equality follows from (5.5), and finally the third equality follows from Theorem 5.1.5. Whenever a phrase like "by letting $m \to \infty$" is used, we should check that the argument is justified.

Theorem 5.4.3. *If the series $\sum_{n=1}^{\infty} a_n$ and $\sum_{n=1}^{\infty} b_n$ are convergent, then the series $\sum_{n=1}^{\infty}(a_n + b_n)$ is convergent and*

$$\sum_{n=1}^{\infty}(a_n + b_n) = \sum_{n=1}^{\infty} a_n + \sum_{n=1}^{\infty} b_n. \tag{5.6}$$

Proof. Since $\sum_{n=1}^{m}(a_n + b_n) = \sum_{n=1}^{m} a_n + \sum_{n=1}^{m} b_n$ for every $m \in \mathbb{N}$, we obtain (5.6) by letting $m \to \infty$. $\qquad\square$

It is natural to ask whether convergence of $\sum_{n=1}^{\infty} a_n$ and $\sum_{n=1}^{\infty} b_n$ implies convergence of $\sum_{n=1}^{\infty} a_n b_n$. In general, the answer is negative. For example, the series $\sum_{n=1}^{\infty} \frac{(-1)^{n+1}}{\sqrt{n}}$ converges (see Theorem 5.5.4), but the series $\sum_{n=1}^{\infty} \left(\frac{(-1)^{n+1}}{\sqrt{n}}\right)^2 = \sum_{n=1}^{\infty} \frac{1}{n}$ diverges. On the other hand, for absolutely convergent series, discussed in the next section, the answer is positive.

EXERCISES 5.4

(1) Show that
$$\sum_{n=1}^{\infty} \frac{1}{n(n+1)} = 1 - \frac{1}{n+1}.$$

(2) Find a sequence for which the sequence of partial sums is $1, \frac{1}{2}, \frac{1}{3}, \frac{1}{4}, \frac{1}{5}, \ldots$

(3) Determine convergence and find the sum if convergent.

(a) $\displaystyle\sum_{n=1}^{\infty} \frac{n^2}{n^2 + 1}$,

(d) $\displaystyle\sum_{n=1}^{\infty} \left(1 + \frac{1}{n}\right)^n$,

(b) $\displaystyle\sum_{n=1}^{\infty} \frac{1}{n^2 + 2n}$,

(e) $\displaystyle\sum_{n=1}^{\infty} (-2)^{3n} \cdot 3^{-2n}$,

(c) $\displaystyle\sum_{n=1}^{\infty} \frac{1}{n^2 + 4n + 3}$,

(f) $\displaystyle\sum_{n=1}^{\infty} 2 \cdot 3^{1+n} \cdot 5^{1-n}$.

(4) Show that $\sum_{n=1}^{\infty} \frac{1}{n}$ diverges by completing the proof started in this section.

(5) Let $\sum_{n=1}^{\infty} a_n$ be a convergent series. Show that, if $\sum_{n=1}^{\infty} a_n = s$, then
$$\sum_{n=m+1}^{\infty} a_n = s - a_1 - \cdots - a_m$$
for any $m \in \mathbb{N}$.

(6) Possible or not: $\sum_{n=1}^{\infty} a_n$ converges, $\sum_{n=1}^{\infty} b_n$ diverges, and $\sum_{n=1}^{\infty}(a_n + b_n)$ converges. Justify your response?

(7) Possible or not: Both $\sum_{n=1}^{\infty} a_n$ and $\sum_{n=1}^{\infty} b_n$ diverge, but $\sum_{n=1}^{\infty}(a_n + b_n)$ converges? Justify your response.

(8) Use Theorem 5.2.1 to prove that $\sum_{n=1}^{\infty} \left(\frac{1}{2}\right)^{n(n+1)/2}$ converges.

5.5 Absolutely convergent series

Proving that a given series is convergent can often be carried out by comparing it with some other series known to be convergent.

Theorem 5.5.1 (Comparison test). *If $|a_n| \leq b_n$ for all $n \in \mathbb{N}$ and the series $\sum_{n=1}^{\infty} b_n$ converges, then the series $\sum_{n=1}^{\infty} a_n$ converges and we have $|\sum_{n=1}^{\infty} a_n| \leq \sum_{n=1}^{\infty} b_n$.*

Proof. Let (p_n) be the sequence obtained from (a_n) by replacing all negative terms by 0, that is, $p_n = \frac{1}{2}(|a_n| + a_n)$. Similarly, let (q_n) be the sequence which is obtained by replacing all positive terms by 0 and changing the sign of the remaining terms: $q_n = p_n - a_n$. Then

$$a_n = p_n - q_n. \tag{5.7}$$

Moreover, since $|a_n| \leq b_n$, we have

$$0 \leq p_n \leq b_n \quad \text{and} \quad 0 \leq q_n \leq b_n. \tag{5.8}$$

For $n = 1, 2, 3, \ldots$ let

$$s_n = \sum_{k=1}^{n} a_k, \quad t_n = \sum_{k=1}^{n} b_k, \quad u_n = \sum_{k=1}^{n} p_k, \quad \text{and} \quad v_n = \sum_{k=1}^{n} q_k.$$

By induction, from (5.7) and (5.8) we obtain

$$s_n = u_n - v_n, \quad 0 \leq u_n \leq t_n, \quad \text{and} \quad 0 \leq v_n \leq t_n. \tag{5.9}$$

By the assumption of the theorem, the sequence (t_n) is convergent and, consequently, it is bounded. Therefore, there is a number M such that $t_n \leq M$ for all $n \in \mathbb{N}$. Hence

$$0 \leq u_n \leq M \quad \text{and} \quad 0 \leq v_n \leq M \tag{5.10}$$

for all $n \in \mathbb{N}$. Moreover, the sequences (u_n) and (v_n) are non-decreasing (as sequences of partial sums of sequences with non-negative terms). Therefore, by Theorem 5.2.1, they are convergent. This, in view of (5.7), implies convergence of the series $\sum_{n=1}^{\infty} a_n$, by Theorems 5.4.2 and 5.4.3.

Moreover, from (5.9) it follows that $-t_n \leq s_n \leq t_n$ and, as $n \to \infty$, we obtain

$$-\sum_{n=1}^{\infty} b_n \leq \sum_{n=1}^{\infty} a_n \leq \sum_{n=1}^{\infty} b_n,$$

which completes the proof. $\qquad\square$

Definition 5.5.2. A series $\sum_{n=1}^{\infty} a_n$ is called *absolutely convergent*, if the series $\sum_{n=1}^{\infty} |a_n|$ is convergent. To indicate that a series $\sum_{n=1}^{\infty} a_n$ is absolutely convergent we can write $\sum_{n=1}^{\infty} |a_n| < \infty$.

Obviously, every convergent series of non-negative terms is absolutely convergent. Note that the series $\sum_{n=1}^{\infty} a_n$ in Theorem 5.5.1 is actually absolutely convergent. Not every convergent series is absolutely convergent (see, for example, Exercise 2). On the other hand, by letting $b_n = |a_n|$ in Theorem 5.5.1, we obtain the following important property.

Corollary 5.5.3. *Every absolutely convergent series converges.*

A convergent series that is not absolutely convergent is called *conditionally convergent*. The following theorem will allow us to easily produce examples of conditionally convergent series. In calculus textbooks this theorem is usually called the Alternating Series Test.

Theorem 5.5.4 (Leibniz's criterion). *If $a_0 > a_1 > a_2 > \ldots$ and $\lim_{n \to \infty} a_n = 0$, then the series*

$$\sum_{n=0}^{\infty} (-1)^n a_n \tag{5.11}$$

converges. Moreover, for each $m \in \mathbb{N}$, we have

$$s_{2m-1} < \sum_{n=0}^{\infty} (-1)^n a_n < s_{2m}, \tag{5.12}$$

where $s_m = \sum_{n=0}^{m} (-1)^n a_n$.

Proof. If $a_0 > a_1 > a_2 > \ldots$ and $\lim_{n \to \infty} a_n = 0$, then we have

$$a_{2m} - a_{2m+1} > 0, \quad a_{2m+2} > 0, \quad \text{and} \quad a_{2m+1} - a_{2m+2} > 0,$$

for all $m \in \mathbb{N}$. Hence

$$s_{2m-1} < s_{2m-1} + (a_{2m} - a_{2m+1}) = s_{2m+1} < s_{2m+1} + a_{2m+2}$$
$$= s_{2m+2} = s_{2m} - (a_{2m+1} - a_{2m+2}) < s_{2m},$$

which proves that

$$s_{2m-1} < s_{2m+1} < s_{2m+2} < s_{2m},$$

for all $m \in \mathbb{N}$. This shows that the sequences (s_{2m}) and (s_{2m-1}) are both monotone and bounded, and thus convergent. Since $s_{2m} - s_{2m-1} = a_{2m}$, we have

$$\lim_{m \to \infty} (s_{2m} - s_{2m-1}) = \lim_{m \to \infty} a_{2m} = 0$$

and, consequently,

$$\lim_{m \to \infty} s_{2m} = \lim_{m \to \infty} s_{2m-1}.$$

This implies the convergence of the sequence (s_m). Let $s = \lim_{m \to \infty} s_m$. Since the sequence (s_{2m}) is decreasing, the sequence (s_{2m-1}) is increasing, and both sequences converge to s, for every $m \in \mathbb{N}$ we have

$$s_{2m-1} < s < s_{2m},$$

proving (5.12). $\square$

Since $1 > \frac{1}{2} > \frac{1}{3} > \ldots$ and $\lim_{n \to \infty} \frac{1}{n} = 0$, the series $\sum_{n=0}^{\infty} \frac{(-1)^n}{n+1}$ converges. Since the series $\sum_{n=0}^{\infty} \frac{1}{n+1}$ diverges, $\sum_{n=0}^{\infty} \frac{(-1)^n}{n+1}$ is a conditionally convergent series.

When proving theorems involving absolutely convergent series the following theorem is often useful.

Theorem 5.5.5. *The folowing conditions are equivalent:*

(a) *The series* $\sum_{n=1}^{\infty} a_n$ *is absolutely convergent.*

(b) *There exists a constant M such that* $\sum_{n=1}^{m} |a_n| < M$ *for all $m \in \mathbb{N}$.*

(c) *For every $\varepsilon > 0$ there exists $n_0 \in \mathbb{N}$ such that* $\sum_{n=n_0}^{\infty} |a_n| < \varepsilon$.

Proof. Let

$$x_m = \sum_{n=1}^{m} |a_n|.$$

If the series $\sum_{n=1}^{\infty} a_n$ is absolutely convergent, then the sequence (x_n) is convergent, and thus bounded. This means that there exists a constant M such that

$$\sum_{n=1}^{m} |a_n| = x_m = |x_m| < M$$

for all $m \in \mathbb{N}$. Therefore (a) implies (b).

If (b) holds, then the sequence (x_m) is bounded. Since it is a non-decreasing sequence, it is convergent. Let $\lim_{m \to \infty} x_m = x$. Then, for any $\varepsilon > 0$ there exists $n_0 \in \mathbb{N}$ such that

$$|x - x_{n_0-1}| = \sum_{n=n_0}^{\infty} |a_n| < \varepsilon.$$

Thus (b) implies (c).

If (c) holds, then there exists an $n_0 \in \mathbb{N}$ such that $\sum_{n=n_0}^{\infty} |a_n| < 1$. Then for every $m \in \mathbb{N}$, we have

$$|x_m| < |a_1| + \cdots + |a_{n_0-1}| + 1,$$

so (x_n) is a bounded sequence. Since (x_n) is also a non-decreasing sequence, it is convergent, which means that the series $\sum_{n=1}^{\infty} a_n$ is absolutely convergent. Therefore (c) implies (a). $\qquad\square$

The following property of absolutely convergent series is frequently used.

Corollary 5.5.6. *If $\sum_{n=1}^{\infty} a_n$ is an absolutely convergent series and (b_n) is a subsequence of (a_n), then the series $\sum_{n=1}^{\infty} b_n$ is absolutely convergent.*

We leave the proof of this corollary as an exercise.

EXERCISES 5.5

(1) Prove convergence of the following series:

(a) $\displaystyle\sum_{n=1}^{\infty} \frac{1}{2^n + \sqrt{n}}$,

(b) $\displaystyle\sum_{n=1}^{\infty} \frac{(-1)^n}{n - 3^n}$.

(2) Let $(a_n) = (1, -1, \frac{1}{2}, -\frac{1}{2}, \frac{1}{3}, -\frac{1}{3}, \dots)$. Prove that the series $\sum_{n=1}^{\infty} a_n$ is conditionally convergent.

(3) In the proof of Theorem 5.5.4 we claim that the sequences (s_{2m}) and (s_{2m-1}) are bounded. Justify that claim.

(4) Prove: If (b_n) is a sequence such that $\lim_{n\to\infty} b_{2n} = \lim_{n\to\infty} b_{2n-1}$, then the sequence (b_n) is convergent.

(5) Prove: If $a_0 > a_1 > a_2 > \dots$ and $\lim_{n\to\infty} a_n = 0$, then for all $m \in \mathbb{N}$

$$\left| \sum_{n=0}^{\infty} (-1)^n a_n - \sum_{n=0}^{m} (-1)^n a_n \right| < a_{m+1}.$$

(6) How many terms of the series $\sum_{n=1}^{\infty} \frac{(-1)^{n+1}}{n}$ should one use to approximate the sum with an error less than 0.001?

(7) How many terms of the series $\sum_{n=1}^{\infty} \frac{(-1)^{n+1}}{n!}$ should one use to approximate the sum with an error less than 0.001?

(8) Explain how Corollary 5.5.6 can be obtained from Theorem 5.5.5.

(9) Prove: If $\sum_{n=1}^{\infty} a_n$ and $\sum_{n=1}^{\infty} b_n$ converge absolutely, then $\sum_{n=1}^{\infty} a_n b_n$ converges absolutely. Does this imply that $\sum_{n=1}^{\infty} a_n b_n = \sum_{n=1}^{\infty} a_n \cdot \sum_{n=1}^{\infty} b_n$?

(10) Prove: If the series $\sum_{n=1}^{\infty} a_n$ converges absolutely and the sequence (b_n) is bounded, then the series $\sum_{n=1}^{\infty} a_n b_n$ converges absolutely.

(11) True or false: If the series $\sum_{n=1}^{\infty} a_n$ converges and the sequence (b_n) is bounded, then the series $\sum_{n=1}^{\infty} a_n b_n$ converges? Justify your response.

(12) True or false: If the series $\sum_{n=1}^{\infty} a_n$ converges and $\lim_{n\to\infty} b_n = 1$, then the series $\sum_{n=1}^{\infty} a_n b_n$ converges? Justify your response.

(13) Let $\sum_{n=1}^{\infty} a_n$ be a conditionally convergent series. Define

$$p_n = \frac{1}{2}(|a_n| + a_n) \quad \text{and} \quad q_n = \frac{1}{2}(|a_n| - a_n).$$

Prove that both series $\sum_{n=1}^{\infty} p_n$ and $\sum_{n=1}^{\infty} q_n$ diverge.

(14) Prove: $\sum_{n=1}^{\infty} a_n$ converges if and only if for every $\varepsilon > 0$ there exists $m \in \mathbb{N}$ such that $|a_m + a_{m+1} + \dots + a_{m+p}| < \varepsilon$ for all $p \in \mathbb{N}$.

(15) Prove: If the limit $\lim_{n\to\infty} \frac{a_n}{b_n}$ exists (is finite) and is not 0, then $\sum_{n=1}^{\infty} a_n$ converges absolutely if and only if $\sum_{n=1}^{\infty} b_n$ converges absolutely.

(16) True or false: If the limit $\lim_{n\to\infty} \frac{a_n}{b_n}$ exists (is finite) and is not 0, then $\sum_{n=1}^{\infty} a_n$ converges if and only if $\sum_{n=1}^{\infty} b_n$ converges?

5.6   Rearrangements and partitions of absolutely convergent series

S: What would happen if we put all the terms of a convergent series $\sum_{n=1}^{\infty} a_n$ into a sack, mixed them thoroughly and picked them at random one after another and formed a new series? Would that new series converge? Would the sum be the same?

T: Various things can happen. For instance, the series $\sum_{n=1}^{\infty} \frac{(-1)^{n+1}}{n}$ is convergent, but it can be rearranged into a divergent series. We can pick from it first only positive numbers so as to reach an arbitrary large number. Taking next only negative terms, we can come back with partial sums down to negative numbers and reach an arbitrary large negative number. Continuing so we can form a series whose partial sums "oscillate between ∞ and $-\infty$," so that it is not convergent. Still more surprising is that by properly rearranging the terms of the above series we can make the new series converge to any given number K. This is possible because the given series is not absolutely convergent.

S: And what happens if the series is absolutely convergent?

T: Then no rearrangement effects the absolute convergence or the sum of the series. Since mixing is not really a mathematical concept, we must make these ideas more precise. What does a "sack" of terms mean? This should mean that we do not care about the order of terms. But in a sack there may exist identical terms. That is why we cannot identify the contents of a sack with a set. The difficulty disappears if we carefully distinguish between the set of all numbers that appear in a series and terms of the series. A term of a series is a number with an index showing the position of that number in the series. If the same number, say 1, appears twice in a series, then it is counted as two different terms of the series. The exact definition of mixing can be based on the concept of rearrangement.

Let (p_n) be a sequence of natural numbers such that every natural number appears in the sequence exactly once. Such a sequence will be called a *rearrangement* of $\mathbb{N}$. If (a_n) is an arbitrary sequence of real numbers and (p_n) is a rearrangement of $\mathbb{N}$, then the sequence (a_{p_n}) will be called a *rearrangement* of the sequence (a_n).

Theorem 5.6.1. *Let $\sum_{n=1}^{\infty} a_n$ be an absolutely convergent series and let (p_n) be a rearrangement of $\mathbb{N}$. Then the series $\sum_{n=1}^{\infty} a_{p_n}$ is absolutely convergent and*

$$\sum_{n=1}^{\infty} a_{p_n} = \sum_{n=1}^{\infty} a_n. \tag{5.13}$$

Proof. First note that, for every $m \in \mathbb{N}$,

$$\sum_{n=1}^{m} |a_{p_n}| \leq \sum_{n=1}^{P_m} |a_n| \leq \sum_{n=1}^{\infty} |a_n|,$$

where $P_m = \max\{p_1, \ldots, p_m\}$. This proves that $\sum_{n=1}^{\infty} a_{p_n}$ is absolutely convergent by Theorem 5.5.5. To prove (5.13) we have to show that

$$\lim_{m \to \infty} \sum_{n=1}^{m} a_{p_n} = \lim_{m \to \infty} \sum_{n=1}^{m} a_n. \tag{5.14}$$

Since both limits exist, (5.14) is equivalent to

$$\lim_{m \to \infty} \left(\sum_{n=1}^{m} a_{p_n} - \sum_{n=1}^{m} a_n \right) = 0.$$

Take an arbitrary $\varepsilon > 0$. Absolute convergence of $\sum_{n=1}^{\infty} a_n$ implies that there exists a $n_0 \in \mathbb{N}$ such that

$$\sum_{n=n_0}^{\infty} |a_n| < \varepsilon.$$

Let $k \in \mathbb{N}$ be such that every $n < n_0$ is among the numbers $p_1, p_2, \ldots, p_k$. Then, for any $m > k$, the difference

$$\sum_{n=1}^{m} a_{p_n} - \sum_{n=1}^{m} a_n,$$

after canceling all terms with opposite signs, will have no term of the sequence (a_n) with index less than n_0. Consequently, for any $m > k$,

$$\left| \sum_{n=1}^{m} a_{p_n} - \sum_{n=1}^{m} a_n \right| \le \sum_{n=n_0}^{\infty} |a_n| < \varepsilon.$$

This proves (5.13).   □

The next theorem is about another very important property of absolutely convergent series. If we use terms of an absolutely convergent series and form two series from them, then they both will be absolutely convergent and the sum of their sums will equal the sum of the original series. Again, conditionally convergent series do not have that property (see Exercise 13 in Section 5.5).

Let (p_n) and (q_n) be sequences of natural numbers. If no natural number appears in both of them and every natural number appears in one of them, then we say that $\{(p_n), (q_n)\}$ is a *partition* of $\mathbb{N}$. In other words, $\{(p_n), (q_n)\}$ is a partition of $\mathbb{N}$, if

$$\{p_n : n \in \mathbb{N}\} \cap \{q_n : n \in \mathbb{N}\} = \emptyset \quad \text{and} \quad \{p_n : n \in \mathbb{N}\} \cup \{q_n : n \in \mathbb{N}\} = \mathbb{N}.$$

For example, $\{(2n-1), (2n)\}$ is a partition of $\mathbb{N}$.

Theorem 5.6.2. *If $\sum_{n=1}^{\infty} a_n$ is an absolutely convergent series and $\{(p_n), (q_n)\}$ is a partition of $\mathbb{N}$, then both series*

$$\sum_{n=1}^{\infty} a_{p_n} \quad and \quad \sum_{n=1}^{\infty} a_{q_n}$$

are absolutely convergent and

$$\sum_{n=1}^{\infty} a_{p_n} + \sum_{n=1}^{\infty} a_{q_n} = \sum_{n=1}^{\infty} a_n. \tag{5.15}$$

Proof. Without loss of generality, we can assume that both sequences (p_n) and (q_n) are increasing, because, if necessary, we can rearrange them and then use Theorem 5.6.1. Under this assumption we have

$$\sum_{n=1}^{m} |a_{p_n}| \le \sum_{n=1}^{p_m} |a_n| \le \sum_{n=1}^{\infty} |a_n| \quad \text{and} \quad \sum_{n=1}^{m} |a_{q_n}| \le \sum_{n=1}^{q_m} |a_n| \le \sum_{n=1}^{\infty} |a_n|$$

for every $m \in \mathbb{N}$. Consequently, by Theorem 5.5.5, both series $\sum_{n=1}^{\infty} a_{p_n}$ and $\sum_{n=1}^{\infty} a_{q_n}$ converge absolutely. Thus, to prove that (5.15), it suffices to show that

$$\lim_{m \to \infty} \left(\sum_{n=1}^{m} a_{p_n} + \sum_{n=1}^{m} a_{q_n} - \sum_{n=1}^{m} a_n \right) = 0. \tag{5.16}$$

Let $\varepsilon > 0$. Absolute convergence of $\sum_{n=1}^{\infty} a_n$ implies that there exists a $n_0 \in \mathbb{N}$ such that $\sum_{n=n_0}^{\infty} |a_n| < \varepsilon$. Since every natural number appears in either (p_n) or (q_n), there must be a number k such that all numbers $1, 2, \dots, n_0 - 1$ appear among the numbers $p_1, p_2, \dots, p_k, q_1, q_2, \dots, q_k$. Then, for any $m > k$, the sum

$$\sum_{n=1}^{m} a_{p_n} + \sum_{n=1}^{m} a_{q_n} - \sum_{n=1}^{m} a_n,$$

after canceling all terms with opposite signs, will have no term with index less than n_0. Consequently, for any $m > k$,

$$\left| \sum_{n=1}^{m} a_{p_n} + \sum_{n=1}^{m} a_{q_n} - \sum_{n=1}^{m} a_n \right| \le \sum_{n=n_0}^{\infty} |a_n| < \varepsilon.$$

This proves (5.16). $\qquad\qquad\qquad\qquad\qquad\qquad\qquad\qquad\qquad\qquad\qquad\quad$ □

We can define a partition on $\mathbb{N}$ into any number of sequences: infinite sequences $(p_{1,n}), \dots, (p_{j,n})$ form a *partition* of $\mathbb{N}$ if every natural number appears in one of the sequences and no natural number appears in two of them. In other words, $\{(p_{1,n}), \dots, (p_{j,n})\}$ is a partition of $\mathbb{N}$ if

$$\bigcup_{k=1}^{j} \{p_{k,n} : n \in \mathbb{N}\} = \mathbb{N}$$

and

$$\{p_{k_1,n} : n \in \mathbb{N}\} \cap \{p_{k_2,n} : n \in \mathbb{N}\} = \emptyset \quad \text{whenever } k_1 \ne k_2.$$

Theorem 5.6.2 can be easily generalized to any finite number of sequences by induction.

Theorem 5.6.3. *Let $\sum_{n=1}^{\infty} a_n$ be an absolutely convergent series and let*

$$\{(p_{1,n}), \dots, (p_{j,n})\}$$

be a partition of $\mathbb{N}$. Then each of the series

$$\sum_{n=1}^{\infty} a_{p_{1,n}}, \dots, \sum_{n=1}^{\infty} a_{p_{j,n}}$$

is absolutely convergent and

$$\sum_{n=1}^{\infty} a_{p_{1,n}} + \cdots + \sum_{n=1}^{\infty} a_{p_{j,n}} = \sum_{n=1}^{\infty} a_n. \tag{5.17}$$

For some applications it is necessary to consider a partition into an infinite sequence of sequences: an infinite sequence $(p_{1,n}), (p_{2,n}), \ldots$ of infinite sequences is called a *partition* of $\mathbb{N}$ if every natural number appears in one of the sequences $(p_{j,n})$ and no natural number appears in two of them. In other words, $\{(p_{1,n}), (p_{2,n}), \ldots\}$ is a partition of $\mathbb{N}$ if

$$\bigcup_{k=1}^{\infty} \{p_{k,n} : n \in \mathbb{N}\} = \mathbb{N}$$

and

$$\{p_{k_1,n} : n \in \mathbb{N}\} \cap \{p_{k_2,n} : n \in \mathbb{N}\} = \emptyset \quad \text{whenever } k_1 \neq k_2.$$

Theorem 5.6.4. *Let $\sum_{n=1}^{\infty} a_n$ be an absolutely convergent series and let*

$$\{(p_{1,n}), (p_{2,n}), \ldots\}$$

be an infinite partition of $\mathbb{N}$. Then, for each $j \in \mathbb{N}$, the series $\sum_{n=1}^{\infty} a_{p_{j,n}}$ converges absolutely and we have

$$\sum_{j=1}^{\infty} \sum_{n=1}^{\infty} a_{p_{j,n}} = \sum_{n=1}^{\infty} a_n. \tag{5.18}$$

Proof. First note that Corollary 5.5.6 and Theorem 5.6.1 imply that, for every $j \in \mathbb{N}$, the series $\sum_{n=1}^{\infty} a_{p_{j,n}}$ is absolutely convergent. Now (5.18) can be written as

$$\lim_{m \to \infty} \left(\sum_{n=1}^{\infty} a_{p_{1,n}} + \sum_{n=1}^{\infty} a_{p_{2,n}} + \cdots + \sum_{n=1}^{\infty} a_{p_{m,n}} \right) = \sum_{n=1}^{\infty} a_n. \tag{5.19}$$

Fix an $\varepsilon > 0$. Then there exists a $n_0 \in \mathbb{N}$ such that $\sum_{n=n_0}^{\infty} |a_n| < \varepsilon$. Let $j \in \mathbb{N}$ be such that every natural number less than n_0 appears in one of the sequences $\{p_{1,n}\}, \{p_{2,n}\}, \ldots, \{p_{j,n}\}$. Now consider $m > j$. Let (q_n) be a sequence of natural numbers such that

$$\{(p_{1,n}), (p_{2,n}), \ldots, (p_{m,n}), (q_n)\}$$

is a partition of $\mathbb{N}$. Then, by Theorem 5.6.3,

$$\sum_{n=1}^{\infty} a_{p_{1,n}} + \sum_{n=1}^{\infty} a_{p_{2,n}} + \cdots + \sum_{n=1}^{\infty} a_{p_{m,n}} + \sum_{n=1}^{\infty} a_{q_n} = \sum_{n=1}^{\infty} a_n.$$

Consequently,

$$\sum_{n=1}^{\infty} a_n - \left(\sum_{n=1}^{\infty} a_{p_{1,n}} + \sum_{n=1}^{\infty} a_{p_{2,n}} + \cdots + \sum_{n=1}^{\infty} a_{p_{m,n}} \right) = \sum_{n=1}^{\infty} a_{q_n}.$$

But since no natural number less than n_0 appears in the sequence (q_n) we must have

$$\left| \sum_{n=1}^{\infty} a_n - \left(\sum_{n=1}^{\infty} a_{p_{1,n}} + \sum_{n=1}^{\infty} a_{p_{2,n}} + \cdots + \sum_{n=1}^{\infty} a_{p_{m,n}} \right) \right|$$

$$= \left| \sum_{n=1}^{\infty} a_{q_n} \right| \leq \sum_{n=1}^{\infty} |a_{q_n}| \leq \sum_{n=n_0}^{\infty} |a_n| < \varepsilon.$$

This proves (5.19) and thus (5.18). $\qquad \square$

Equality (5.18) can be written in a more intuitive but less precise way as

$$(a_{1,1} + a_{1,2} + a_{1,3} + \dots) + (a_{2,1} + a_{2,2} + a_{2,3} + \dots) + \dots$$
$$+ (a_{n,1} + a_{n,2} + a_{n,3} + \dots) + \dots = a_1 + a_2 + a_3 + \dots \,.$$

Since we are adding the same numbers on the left and on the right, although in a different order, the equality may seem obvious. We know that the algebraic operation of addition has the property that the sum does not depend on the order. However the operation of adding infinitely many numbers is not a purely algebraic operation. Therefore the rules of addition do not automatically apply to series. The theory of convergent series is largely devoted to determining which properties of algebraic operations on numbers extend to series and under what conditions these properties extend. For instance, the above theorem together with the previous discussion concerning the series $\sum_{n=1}^{\infty} \frac{(-1)^{n+1}}{n}$ shows that absolute convergence suffices and is crucial for "reordering" (or "regrouping") the terms of an infinite series. A good understanding of the preceding theorem as well as the following theorem will be extremely important in the chapter on the Lebesgue integral.

The following theorem is somewhat converse to Theorem 5.6.4.

Theorem 5.6.5. *Let $\{(p_{1,n}), (p_{2,n}), \dots\}$ be an infinite partition of $\mathbb{N}$ and let (a_n) be a sequence of real numbers such that the series $\sum_{n=1}^{\infty} a_{p_{j,n}}$ converges absolutely for every $j \in \mathbb{N}$. For every $j \in \mathbb{N}$ define*

$$A_j = \sum_{n=1}^{\infty} |a_{p_{j,n}}|.$$

If the series $\sum_{j=1}^{\infty} A_j$ converges absolutely, then so does the series $\sum_{n=1}^{\infty} a_n$ and we have

$$\sum_{n=1}^{\infty} a_n = \sum_{j=1}^{\infty} \sum_{n=1}^{\infty} a_{p_{j,n}}. \tag{5.20}$$

Proof. If we prove that the series $\sum_{n=1}^{\infty} a_n$ converges absolutely, then (5.20) will follow from Theorem 5.6.4. Note that, for every $m \in \mathbb{N}$, there is a $j_0 \in \mathbb{N}$ such that every natural number less than or equal to m appears in one of the sequences $(p_{1,n}), \dots, (p_{j_0,n})$. Then

$$\sum_{n=1}^{m} |a_n| \leq A_1 + \dots + A_{j_0} \leq \sum_{j=1}^{\infty} A_j.$$

This implies absolute convergence of $\sum_{n=1}^{\infty} a_n$ by Theorem 5.5.5.   □

EXERCISES 5.6

(1) Let $(a_n) = (1, -1, \frac{1}{2}, -\frac{1}{2}, \frac{1}{3}, -\frac{1}{3}, \dots)$. Find a rearrangement of $\sum_{n=1}^{\infty} a_n$ which diverges.

(2) Let $\sum_{n=1}^{\infty} a_n$ be a conditionally convergent series and let $K \in \mathbb{R}$. Prove that there exists a rearrangement $\sum_{n=1}^{\infty} b_n$ of $\sum_{n=1}^{\infty} a_n$ such that $\sum_{n=1}^{\infty} b_n = K$.

(3) Prove Theorem 5.6.3.

(4) Prove: Let $\{(p_{1,n}), (p_{2,n}), \ldots, (p_{m,n})\}$ be a finite partition of $\mathbb{N}$ and let (a_n) be a sequence of numbers such that each series $\sum_{n=1}^{\infty} a_{p_{i,n}}$ converges absolutely for $i = 1, 2, \ldots m$. Prove that $\sum_{n=1}^{\infty} a_n$ also converges absolutely and we have

$$\sum_{n=1}^{\infty} a_n = \sum_{i=1}^{m} \sum_{n=1}^{\infty} a_{p_{i,n}}.$$

(5) Let $\sum_{n=1}^{\infty} a_n$ be an absolutely convergent series and let, for all $n \in \mathbb{N}$, $b_n = a_{2n-1} + a_{2n}$. Prove that the series $\sum_{n=1}^{\infty} b_n$ is absolutely convergent and that $\sum_{n=1}^{\infty} a_n = \sum_{n=1}^{\infty} b_n$.

(6) Let $\sum_{n=1}^{\infty} a_n$ be an absolutely convergent series and let, for all $n \in \mathbb{N}$, $b_n = a_{3n-2} + a_{3n-1} + a_{3n}$. Prove that the series $\sum_{n=1}^{\infty} b_n$ is absolutely convergent and that $\sum_{n=1}^{\infty} a_n = \sum_{n=1}^{\infty} b_n$.

(7) Find the sum of the series $\sum_{n=1}^{\infty} \frac{n}{2^n}$.

(8) Let a_i be a real number for each integer i. Let q and m be fixed integers. Prove: If either of the two series

$$\sum_{i=m}^{\infty} a_i \quad \text{or} \quad \sum_{i=m+q+1}^{\infty} a_i$$

is convergent, then the other series is also convergent and we have

$$\sum_{i=m}^{\infty} a_i = \sum_{i=m}^{m+q} a_i + \sum_{i=m+q+1}^{\infty} a_i.$$

(9) For brevity, we refer to the series

$$1 + \frac{1}{2} - \frac{1}{2} + \frac{1}{4} - \frac{1}{4} + \frac{1}{4} - \frac{1}{4} + \frac{1}{8} - \frac{1}{8} + \frac{1}{8} - \frac{1}{8} + \frac{1}{8} - \frac{1}{8} + \frac{1}{8} - \frac{1}{8} + \cdots$$

as $\sum_{n=1}^{\infty} a_n$. The terms of $\sum_{n=1}^{\infty} a_n$ come in strings of length 2^m terms where each of the 2^m terms has denominator 2^m. Let $\{(p_{1,j}), (p_{2,j}), \ldots\}$ be the infinite partition of $\mathbb{N}$ defined as follows: $p_{i,j}$ is chosen so that $a_{p_{i,j}}$ is the i-th term in the j-th string of $\sum_{n=1}^{\infty} a_n$ whose length $2^m \geq i$. For example, $p_{1,3} = 4$, because $a_4 = \frac{1}{4}$ is the 1-st term in the 3-rd string of $\sum_{n=1}^{\infty} a_n$ whose length $2^m \geq 1$; and $p_{4,2} = 11$ because $a_{11} = -\frac{1}{8}$ is the 4-th term in the 2-nd string whose length $2^m \geq 4$.

(a) Verify that each series $\sum_{n=1}^{n} a_{p_{i,n}}$ converges absolutely, but that $\sum_{n=1}^{\infty} a_n$ does not converge absolutely. Explain why this does not violate Theorem 5.6.5.

(b) Find $\sum_{i=1}^{\infty} \sum_{n=1}^{\infty} a_{p_{i,n}}$.

5.7   Decimal expansions

Let (a_n) be an infinite sequence of digits, that is, each $a_n \in \{0,1,2,3,4,5,6,7,8,9\}$. From Theorem 5.5.1 it follows that the series

$$\sum_{n=1}^{\infty} \frac{a_n}{10^n}$$

is always convergent, because

$$0 \le \frac{a_n}{10^n} \le \frac{9}{10^n}$$

and

$$\sum_{n=1}^{\infty} \frac{9}{10^n} = \frac{9}{10}\left(1 + \frac{1}{10} + \frac{1}{10^2} + \cdots\right) = \frac{9}{10}\frac{1}{1-\frac{1}{10}} = 1. \qquad (5.21)$$

Therefore every sequence of digits corresponds to a number $A = \sum_{n=1}^{\infty} \frac{a_n}{10^n}$ such that $0 \le A \le 1$.

Let A_n be the n-th partial sum of the series $\sum_{n=1}^{\infty} \frac{a_n}{10^n}$, that is,

$$A_n = \sum_{k=1}^{n} \frac{a_{k}}{10^k}.$$

In the standard notation we write

$$A_n = 0.a_1 a_2 \ldots a_m.$$

For example

$$\frac{5}{10} + \frac{7}{10^2} + \frac{7}{10^3} + \frac{2}{10^4} = 0.5772.$$

We can prove that for all n we have

$$A_n \le A \le A_n + \frac{1}{10^n}. \qquad (5.22)$$

Indeed, since the sequence (A_n) is non-decreasing, we have $A_n \le A$ for all n. In order to prove the second inequality we write

$$A = A_n + \sum_{k=n+1}^{\infty} \frac{a_k}{10^k}.$$

For the last sum we have

$$\sum_{k=n+1}^{\infty} \frac{a_k}{10^k} = \sum_{k=1}^{\infty} \frac{a_{n+k}}{10^{n+k}} \le \frac{1}{10^n}\sum_{k=1}^{\infty} \frac{9}{10^k} = \frac{1}{10^n},$$

which completes the proof of (5.22).

Let us consider the case when $a_n = 2$ for all n. Then

$$A = \sum_{k=1}^{\infty} \frac{2}{10^k} = \frac{2}{10}\frac{1}{1-\frac{1}{10}} = \frac{2}{9}.$$

For $n = 5$ we have

$$A_5 \leq A \leq A_5 + \frac{1}{10^5},$$

that is,

$$0.22222 \leq \frac{2}{9} \leq 0.22223.$$

We can write $\frac{2}{9} \approx 0.22222$, which is read as "$\frac{2}{9}$ equals approximately 0.22222." We can also say that the number 0.22222 is the decimal approximation of $\frac{2}{9}$ up to 5 decimal places. We write 278.3023 to denote the number $278 + 0.3023$. The expression $A \approx 278.3023$ is read "A equals approximately 278.3023". In this book we adopt the convention that $A \approx 278.3023$ means

$$278.3023 \leq A < 278.3024.$$

In general, if we write $A \approx b_k \ldots b_0.a_1 \ldots a_n$, where $b_0, \ldots, b_k, a_1, \ldots, a_n$ are digits, then

$$b_k \ldots b_0.a_1 \ldots a_n \leq A < b_k \ldots b_0.a_1 \ldots a_n + \frac{1}{10^n}.$$

At the beginning of this section we show that every sequence of digits $0.a_1a_2a_3 \ldots$ represents a number $0 \leq x \leq 1$. Now we will show the converse, that is, that every real number $0 \leq x \leq 1$ has such a representation. Since $0 = 0.000 \ldots$ and $1 = 0.999 \ldots$, we can assume that $0 < x < 1$. Let a_1 be the greatest integer such that

$$\frac{a_1}{10} \leq x.$$

Obviously, $a_1 \in \{0, 1, \ldots, 9\}$. Next, define a_2 to be the greatest integer such that

$$\frac{a_1}{10} + \frac{a_2}{10^2} \leq x.$$

In general, define a_n to be the greatest integer such that

$$\frac{a_1}{10} + \frac{a_2}{10^2} + \cdots + \frac{a_n}{10^n} \leq x. \tag{5.23}$$

Clearly, $a_n \in \{0, 1, \ldots, 9\}$ for all $n \in \mathbb{N}$. We will prove that $0.a_1a_2a_3 \ldots$ is a decimal representation of x, that is,

$$x = \sum_{k=1}^{\infty} \frac{a_k}{10^k}. \tag{5.24}$$

By (5.23), we have $\sum_{k=1}^{\infty} \frac{a_k}{10^k} \leq x$. Suppose $\sum_{k=1}^{\infty} \frac{a_k}{10^k} < x$. Let n be the least integer such that

$$\frac{1}{10^n} \leq x - \sum_{k=1}^{\infty} \frac{a_k}{10^k}$$

or equivalently

$$\sum_{k=1}^{\infty} \frac{a_k}{10^k} + \frac{1}{10^n} \leq x.$$

This implies

$$\frac{a_1}{10} + \frac{a_2}{10^2} + \cdots + \frac{a_{n-1}}{10^{n-1}} + \frac{a_n + 1}{10^n} \leq x,$$

contradicting the definition of a_n. This proves (5.24).

We have proved that every $x \in [0, 1]$ has a decimal expansion, but it is not unique. For example $0.5 = 0.4999\ldots$. However, if we exclude sequences $a_1, a_2, a_3, \ldots$ for which there is an index n_0 such that $a_n = 9$ for all $n \geq n_0$, then every number can be represented by exactly one such a sequence of digits. To prove this suppose that there are two different representations of a number x:

$$x = 0.a_1 a_2 a_3 \ldots, \quad \text{and} \quad x = 0.b_1 b_2 b_3 \ldots.$$

Let k be the least natural number for which $a_k \neq b_k$. We may assume that $a_k < b_k$. Let n be the least natural number greater than k such that $a_n \neq 9$ and let $c = a_n + 1$. Then

$$x = 0.a_1 a_2 a_3 \cdots < 0.a_1 a_2 \ldots a_{n-1} c \leq 0.b_1 b_2 \ldots b_n \leq 0.b_1 b_2 b_3 \cdots = x.$$

This contradiction shows that x cannot have two such representations. But does our exclusion mean that some numbers in $[0, 1]$ will no longer have a decimal representation of the form $0.a_1 a_2 a_3 \ldots$? Yes, it does mean that; 1 is an example. Are there others? No, there are not. To see this, suppose $x = 0.a_1 a_2 a_3 \ldots$ where $a_n = 9$ for $n > n_0$, but $a_{n_0} < 9$, then,

$$
\begin{aligned}
x &= \frac{a_1}{10} + \frac{a_2}{10^2} + \cdots + \frac{a_{n_0}}{10^{n_0}} + \frac{9}{10^{n_0+1}} + \frac{9}{10^{n_0+2}} + \frac{9}{10^{n_0+3}} + \cdots \\
&= \frac{a_1}{10} + \frac{a_2}{10^2} + \cdots + \frac{a_{n_0}}{10^{n_0}} + \frac{9}{10^{n_0+1}} \left[1 + \frac{1}{10} + \frac{1}{10^2} + \cdots \right] \\
&= \frac{a_1}{10} + \frac{a_2}{10^2} + \cdots + \frac{a_{n_0}}{10^{n_0}} + \frac{9}{10^{n_0+1}} \cdot \frac{1}{1 - \frac{1}{10}} \\
&= \frac{a_1}{10} + \frac{a_2}{10^2} + \cdots + \frac{a_{n_0} + 1}{10^{n_0}}.
\end{aligned}
$$

This proves that each $x \in [0, 1)$ has exactly one decimal expansion provided we observe our suggested exclusion.

Notice that, since any real number x can be written as $x = a + b$ where a is an integer and $b \in [0, 1)$, x has a decimal expansion, which is unique under the assumption discussed above.

EXERCISES 5.7

(1) In the proof given above that every $x \in [0, 1]$ has a decimal representation, the followihg is written, "Suppose $\sum_{k=1}^{\infty} \frac{a_k}{10^k} < x$. Let n be the least integer such that $\frac{1}{10^n} \leq x - \sum_{k=1}^{\infty} \frac{a_k}{10^k}$." Prove that such a natural number does exist.

(2) Let $0.a_1 a_2 a_3 \ldots$ be the decimal representation for $\frac{2}{13}$. Tell what a_n is for each $n \in \mathbb{N}$.

(3) Without writing out all the details, explain why each rational number in $[0, 1]$ must have a "repeating" decimal representation.

(4) Is the sum of the series $\sum_{n=1}^{\infty} \left(\frac{1}{10}\right)^{n(n+1)/2}$ a rational number, an irrational number, or doesn't the series have a sum?

Chapter 6

SEQUENCES AND SERIES OF FUNCTIONS

6.1 Power series

In Section 5.3 we proved that the geometric series

$$\sum_{n=0}^{\infty} x^n \tag{6.1}$$

converges for every $x \in (-1, 1)$. The sum depends on x and thus it represents a function defined on the interval $(-1, 1)$. We have shown that the function is $\frac{1}{1-x}$. The geometric series is an example of a power series.

Here is another example of a power series

$$\sum_{n=1}^{\infty} \frac{(-1)^{n+1}}{n} x^n. \tag{6.2}$$

This series is also convergent for every $x \in (-1, 1)$ because

$$\left| (-1)^{n+1} \frac{x^n}{n} \right| \leq |x|^n$$

for all $n \in \mathbb{N}$ and the series $\sum_{n=1}^{\infty} |x|^n$ converges for every $x \in (-1, 1)$.

Any series of the form

$$\sum_{n=0}^{\infty} a_n x^n, \tag{6.3}$$

where the coefficients $a_0, a_1, a_2, \ldots$ are constants, is called a *power series*.

Power series play a very important role in mathematical analysis. They allow us to introduce several new functions of fundamental significance. For instance, series (6.2) is connected with the logarithm function.

One of the basic problems concerning power series is determining the set of all points x for which a given power series converges. Clearly, every power series converges for $x = 0$ and the sum is a_0.

S: The first term of the series $\sum_{n=0}^{\infty} a_n x^n$ is $a_0 x^0$. In particular, if $x = 0$, we get $a_0 0^0$. What is the value of 0^0?

T: 1.

S: Is this something obvious?

T: Not at all. On the one hand, we have $\lim_{x \to 0+} x^0 = 1$ and thus it is reasonable to define $0^0 = 1$. On the other hand, however, $\lim_{x \to 0+} 0^x = 0$. Moreover, using methods discussed in Chapter 8, one can show that $\lim_{x \to 0+} x^x = 1$."

T: Oh! I get it. Since we get 1 more often that 0, we define $0^0 = 1$.

S: Not really. Defining $0^0 = 1$ is a matter of convenience. We want the first term to be the constant a_0. Note that in the series $\sum_{n=0}^{\infty} a_n x^n$ all the terms have the same form because $0^0 = 1$. Also for the sake of convenience we define $0! = 1$. In this case too, of course, there are other reasons for this choice.

Theorem 6.1.1. *If a power series*

$$\sum_{n=0}^{\infty} a_n x^n \tag{6.4}$$

is convergent at some point $x_0 \neq 0$, then there exists a number $r > 0$ such that the series is absolutely convergent for all $x \in (-r, r)$ and divergent (that is, not convergent) whenever $|x| > r$. The case $r = \infty$ is also admitted, which means that the series converges absolutely for all $x \in (-\infty, \infty)$.

Recall that absolute convergence of (6.4) means that the series

$$\sum_{n=0}^{\infty} |a_n||x^n| \tag{6.5}$$

is convergent. By Theorem 5.5.3 every absolutely convergent series is convergent. Thus, if we erase the word "absolutely" in Theorem 6.1.1, the theorem remains true, but becomes weaker.

Proof of Theorem 6.1.1. Assume that the series (6.4) is convergent for some x_0 and set $z = |x_0|$. Then $\lim_{n \to \infty} |a_n z^n| = 0$, by Theorem 5.4.1. Hence, by Theorem 5.1.3, there is some M such that $|a_n z^n| < M$ for all $n \in \mathbb{N}$. If $|x| < z$, then

$$|a_n||x|^n = |a_n z^n| \frac{|x|^n}{z^n} \leq M \frac{|x|^n}{z^n}$$

for all $n \in \mathbb{N}$. Since $\frac{|x|}{z} < 1$, the series $\sum_{n=1}^{\infty} \frac{M|x|^n}{z^n}$ is convergent, and hence the series (6.5) is convergent. This means the series (6.4) is absolutely convergent for all $|x| < z$.

If the set of all x for which the series (6.5) converges is unbounded, then for any x there is a z such that $|x| < z$ and such that $\sum_{n=0}^{\infty} a_n z^n$ converges. By the argument above, the series (6.4) converges. Conseqently, the series (6.4) converges for all $x \in (-\infty, \infty)$. (In this case we write $r = \infty$.)

If the set of all x for which the series (6.5) converges is bounded, let the least upper bound of that set be r. If $x \in (-r, r)$, then there exists a number z with $|x| < z < r$, such that the series $\sum_{n=0}^{\infty} a_n z^n$ converges. This implies, by the comparison test, convergence of the series (6.5). We have thus proved absolute convergence of the series (6.4) in $(-r, r)$.

If $|x| > r$, divergence of the series (6.5) follows from the definition of r and the first part of this proof. $\qquad\qquad\qquad\qquad\qquad\qquad\qquad\qquad\qquad\qquad\qquad\qquad\qquad$ $\square$

The number r in the above theorem is called the *radius of convergence* of the series (6.4). Although ∞ is not a number, it is convenient to say "the radius is infinity" and write $r = \infty$ when the series converges absolutely for all x. The set of all numbers x for which the series (6.4) converges is called the *interval of convergence* of that series. Theorem 6.1.1 implies the following:

(1) If $r = 0$, then the interval of convergence is reduced to a single point $\{0\}$.
(2) If $r = \infty$, then the interval of convergence is the entire real line.
(3) If $0 < r < \infty$, then the interval of convergence is one of the following intervals:

$$(-r, r), \ [-r, r), \ (-r, r], \ [-r, r].$$

Note that Theorem 6.1.1 says nothing about convergence of the series at the endpoints. Those have to be checked separately. For example, the interval of convergence of the series $\sum_{n=0}^{\infty} x^n$ is $(-1, 1)$ since both series

$$\sum_{n=0}^{\infty} 1^n \quad \text{and} \quad \sum_{n=0}^{\infty} (-1)^n$$

diverge.

The interval of convergence of series $\sum_{n=1}^{\infty} \frac{(-1)^{n+1}}{n} x^n$ is $(-1, 1]$. Convergence of the series

$$\sum_{n=1}^{\infty} \frac{(-1)^{n+1}}{n} 1^n = \sum_{n=1}^{\infty} \frac{(-1)^{n+1}}{n}$$

follows from Theorem 5.5.4. Divergence of the series

$$\sum_{n=1}^{\infty} \frac{(-1)^{n+1}}{n} (-1)^n = -\sum_{n=1}^{\infty} \frac{1}{n}$$

was discussed in Section 5.4. Since the series converges at $x = 1$, it converges for every $|x| < 1$. On the other hand, since it diverges at $x = -1$, it diverges for every $|x| > 1$.

In applications of power series discussed in the following sections convergence at the endpoints of the interval of convergence will not play any role. Therefore, for the sake of simplicity, by the interval of convergence we will always mean the open interval $(-r, r)$.

EXERCISES 6.1

(1) Complete the proof of Theorem 6.1.1 by showing that series (4) diverges when $|x| > r$.

(2) Prove: If the radius of convergence of $\sum_{n=0}^{\infty} a_n x^n$ is r, then the radius of convergence of $\sum_{n=0}^{\infty} a_n x^{kn}$ is $r^{1/k}$ for every $k \in \mathbb{N}$.

(3) If the radius of convergence of $\sum_{n=0}^{\infty} a_n x^n$ is r and $\lim_{n \to \infty} b_n = 1$, what can you say about the radius of convergence of $\sum_{n=0}^{\infty} a_n b_n x^n$.

(4) If the radius of convergence of $\sum_{n=0}^{\infty} a_n x^n$ is r and the sequence (b_n) is convergent, what can you say about the radius of convergence of $\sum_{n=0}^{\infty} a_n b_n x^n$.

(5) If the radius of convergence of $\sum_{n=0}^{\infty} a_n x^n$ is r and the sequence (b_n) is bounded, what can you say about the radius of convergence of $\sum_{n=0}^{\infty} a_n b_n x^n$.

(6) Find the radius of convergence of the series $\sum_{n=0}^{\infty} n x^n$. (Hint: use Theorem 5.6.5.)

6.2   Sequences of functions

Partial sums $s_n(x)$ of a power series form a sequence of functions of x. Generally, by a sequence of functions we mean a sequence (f_n) whose terms are functions defined on a common domain D. A sequence of functions (f_n) becomes a sequence of numbers, if all its terms are evaluated at an arbitrary number from D. Thus a sequence of functions (f_n) associates with each point $x \in D$ a numerical sequence $(f_n(x))$.

Let (f_n) be a sequence of functions defined on an interval I. Suppose the sequence $(f_n(x))$ is convergent for every $x \in I$. Since the limits depend on x, they define a function in I:

$$f(x) = \lim_{n \to \infty} f_n(x) \text{ for } x \in I.$$

In this case we say that the sequence (f_n) converges to f and write $\lim_{n \to \infty} f_n = f$ on I or simply $f_n \to f$ on I. This type of convergence is called *pointwise convergence* and f is called the *pointwise limit* of the sequence (f_n).

For example, in Theorem 5.3.1 we show that $\lim_{n \to \infty} q^n = 0$ for all $|q| < 1$. Equivalently, we can say that the pointwise limit of the sequence of functions $f_n(x) = x^n$ on the interval $I = (-1, 1)$ is the zero function.

Pointwise convergence is a weak convergence. The great French mathematician Augustin-Louis Cauchy (1789–1857) believed that the pointwise limit of a sequence of continuous functions would also be continuous. That he was wrong is easily seen from the following example.

For $n = 1, 2, 3, \ldots$, define, as shown on Fig. 6.1,

$$f_n(x) = \begin{cases} 1, & \text{for } x > 1/n, \\ nx, & \text{for } 0 \le x \le 1/n, \\ 0, & \text{for } x < 0. \end{cases}$$

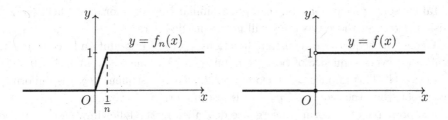

Fig. 6.1 A sequence of continuous functions convergent to a function that is not continuous.

The limit of this sequence of functions is the discontinuous function:

$$f(x) = \begin{cases} 1, & \text{for } x > 0, \\ 0, & \text{for } x \leq 0. \end{cases}$$

To ensure continuity of the limit of a sequence of continuous functions a stronger type of convergence is needed.

Definition 6.2.1. We say that a sequence of functions f_n is *converges uniformly* to a function f on an interval I and write $f_n \rightrightarrows f$ in I, if there exists a sequence of positive numbers (ε_n) such that $\lim_{n \to \infty} \varepsilon_n = 0$ and the inequality

$$|f_n(x) - f(x)| < \varepsilon_n \tag{6.6}$$

holds for all sufficiently large $n \in \mathbb{N}$ and all $x \in I$, that is, there is some $n_0 \in \mathbb{N}$ such that (6.6) holds for all $n > n_0$ and for all $x \in I$.

Note that $f_n \rightrightarrows f$ in I implies $\lim_{n \to \infty} f_n(x) = f(x)$ for every $x \in I$. In other words, *uniform convergence implies pointwise convergence to the same limit*.

The sequence of functions $f_n(x) = x^n$ converges uniformly to 0 in the interval $[0, \alpha]$ for any $0 < \alpha < 1$, because then $|x^n - 0| < \alpha^n$ in $[0, \alpha]$ and $\lim_{n \to \infty} \alpha^n = 0$. Although $\lim_{n \to \infty} x^n = 0$ for every $x \in [0, 1)$, the sequence of functions $f_n(x) = x^n$ is not uniformly convergent in $[0, 1)$. Indeed, for every $n \in \mathbb{N}$, if $|x^n - 0| < \varepsilon_n$ holds for every $x \in [0, 1)$, then $\varepsilon_n \geq 1$.

Theorem 6.2.2. *If $f_n \rightrightarrows f$ and $g_n \rightrightarrows g$ in I, then $f_n + g_n \rightrightarrows f + g$ and $f_n - g_n \rightrightarrows f - g$ in I.*

Proof. If $f_n \rightrightarrows f$ and $g_n \rightrightarrows g$ in I, then there exist two sequences of positive numbers (ε_n) and (δ_n) convergent to zero such that $|f_n(x) - f(x)| < \varepsilon_n$ and $|g_n(x) - g(x)| < \delta_n$ for all sufficiently large $n \in \mathbb{N}$ and all $x \in I$. Hence, using the triangle inequality, we obtain

$$|(f_n + g_n)(x) - (f + g)(x)| < \varepsilon_n + \delta_n \quad \text{and} \quad |(f_n - g_n)(x) - (f - g)(x)| < \varepsilon_n + \delta_n,$$

for all sufficiently large $n \in \mathbb{N}$ and for all $x \in I$. This proves the theorem, because the sequence $(\varepsilon_n + \delta_n)$ converges to zero. $\square$

Taking some precautions, one can prove similar theorems for multiplication and division. However those theorems will not be needed here.

A constant sequence of constant functions $f_n(x) = c$ is uniformly convergent, because for every sequence of positive numbers (ε_n) we have $|f_n(x) - c| = 0 < \varepsilon_n$ for all $n \in \mathbb{N}$. This implies, by Theorem 6.2.2, that if a sequence (g_n) is uniformly convergent, then the sequence $(g_n + c)$ is also uniformly convergent.

The property of uniform convergence described in the following theorem is the main reason this type of convergence is important.

Theorem 6.2.3. *The limit of a uniformly convergent sequence of continuous functions is continuous. More precisely, if $f_n \rightrightarrows f$ in an interval I and all functions f_n are continuous in I, then f is continuous in I.*

Proof. Assume that $f_n \rightrightarrows f$ in an interval I and all functions f_n are continuous in I. Let $f(x) = \lim_{n\to\infty} f_n(x)$ for $x \in I$. Let x_0 be an arbitrary point in I. From the equality

$$f(x) - f(x_0) = f(x) - f_n(x) + f_n(x) - f_n(x_0) + f_n(x_0) - f(x_0)$$

and the triangle inequality, we get

$$|f(x) - f(x_0)| \le |f(x) - f_n(x)| + |f_n(x) - f_n(x_0)| + |f_n(x_0) - f(x_0)|.$$

Since $f_n \rightrightarrows f$ in I, there is a sequence of positive numbers (ε_n) convergent to zero such that

$$|f(x) - f(x_0)| < \varepsilon_n + |f_n(x) - f_n(x_0)| + \varepsilon_n$$

for all $x \in I$ and all sufficiently large $n \in \mathbb{N}$.

Let $\varepsilon > 0$. Since $\lim_{n\to\infty} \varepsilon_n = 0$, there exists an index m such that $\varepsilon_m < \frac{1}{3}\varepsilon$. Hence

$$|f(x) - f(x_0)| \le \frac{2}{3}\varepsilon + |f_m(x) - f_m(x_0)|$$

for all $x \in I$. Continuity of f_m at x_0 implies that there exists a number $\delta > 0$ such that

$$|f_m(x) - f_m(x_0)| < \frac{1}{3}\varepsilon$$

whenever $|x - x_0| < \delta$. Hence

$$|f(x) - f(x_0)| < \varepsilon \quad \text{whenever} \quad |x - x_0| < \delta.$$

This proves that f is continuous at x_0. Since x_0 is an arbitrary point of the interval I, this proves the continuity of f in I. $\qquad\square$

EXERCISES 6.2

(1) Find the following limits:

(a) $\lim\limits_{n\to\infty} \dfrac{1}{1+x^{2n}}$,

(b) $\lim\limits_{n\to\infty} \dfrac{2x+nx^2}{1+nx^2}$,

(c) $\lim\limits_{n\to\infty} \dfrac{nx+x^2+nx^3}{1+nx^2}$,

(d) $\lim\limits_{n\to\infty} \dfrac{(n+1)x+n^2x^3}{1+n^2x^2}$.

(2) Prove that the sequence of functions

$$f_1(x)=\sqrt{x}, \quad f_2(x)=\sqrt{x+\sqrt{x}}, \quad f_3(x)=\sqrt{x+\sqrt{x+\sqrt{x}}},\ldots$$

converges in $[0,\infty)$. Find the limit.

(3) Let f_n, g_n, f, and g be functions defined on an interval I. If $\lim_{n\to\infty} f_n = f$ and $\lim_{n\to\infty} g_n = g$ on I, show that

(a) $\lim\limits_{n\to\infty} (f_n+g_n) = f+g$ on I,

(b) $\lim\limits_{n\to\infty} (f_n-g_n) = f-g$ on I,

(c) $\lim\limits_{n\to\infty} f_n g_n = fg$ on I.

Moreover, if $g_n(x) \neq 0$ and $g(x) \neq 0$ for all $x \in I$, then

(d) $\lim\limits_{n\to\infty} \dfrac{f_n}{g_n} = \dfrac{f}{g}$ on I.

(4) Let f_n and f be functions defined on an interval I. If $\lim_{n\to\infty} f_n = f$ on I, show that $\lim_{n\to\infty} |f_n| = |f|$ on I.

(5) Let f_n be functions defined on an interval I. Show that $\lim_{n\to\infty} f_n = 0$ on I if and only if $\lim_{n\to\infty} |f_n| = 0$ on I.

(6) Show that $f_n \rightrightarrows f$ in I implies $\lim_{n\to\infty} f_n(x) = f(x)$ for every $x \in I$.

(7) Discuss uniform convergence of the following sequences of functions:

(a) $f_n(x) = \dfrac{1}{1+x^{2n}}$,

(b) $f_n(x) = \dfrac{2x+nx^2}{1+nx^2}$,

(c) $f_n(x) = \dfrac{nx+x^2+nx^3}{1+nx^2}$,

(d) $f_n(x) = \dfrac{(n+1)x+n^2x^3}{1+n^2x^2}$.

(8) Prove: $f_n \rightrightarrows f$ on an interval I if and only if for every $\varepsilon > 0$ there exists $n_0 \in \mathbb{N}$ such that for all $n \geq n_0$ and all $x \in I$ we have $|f_n(x) - f(x)| < \varepsilon$.

(9) Let $f, f_1, f_2, \ldots$ be continuous functions on a bounded closed interval $[a,b]$. Prove that

$$f_n \rightrightarrows f \text{ in } [a,b] \quad \text{if and only if} \quad \lim_{n\to\infty} \max_{x\in[a,b]} |f_n(x) - f(x)| = 0.$$

(10) True or false: If $f_n \rightrightarrows f$ and $g_n \rightrightarrows g$ on $\mathbb{R}$, then $f_n g_n \rightrightarrows fg$?

(11) True or false: Let f, f_n, g, g_n be continuous functions on a bounded closed interval $[a,b]$. If $f_n \rightrightarrows f$ and $g_n \rightrightarrows g$ in $[a,b]$, then $f_n g_n \rightrightarrows fg$ on $[a,b]$?

(12) Formulate and prove a theorem on uniform convergence of sequences of the form $\left(\dfrac{f_n}{g_n}\right)$.

(13) Let f be a function with bounded derivative on $\mathbb{R}$ and let $f_n(x) = f(x + \frac{1}{n})$ for $n \in \mathbb{N}$. Prove that $f_n \rightrightarrows f$ on $\mathbb{R}$.

(14) Let (f_n) be a uniformly convergent sequence of continuous functions on $[a, b]$ and let $c \in [a, b]$. Prove that

$$\lim_{n \to \infty} \lim_{x \to c} f_n(x) = \lim_{x \to c} \lim_{n \to \infty} f_n(x).$$

(15) Let $f, f_1, f_2, \ldots$ be continuous functions on a closed bounded interval $[a, b]$. Prove that $f_n \rightrightarrows f$ if and only if $f_n(x_n) \to f(x)$ whenever $x_n \to x$ in $[a, b]$.

6.3   Uniform convergence of power series

A power series is said to be uniformly convergent on an interval, if the sequence of partial sums is uniformly convergent on that interval.

Theorem 6.3.1. *If the radius of convergence of a power series is r, then the series is uniformly convergent on $[-b, b]$ for every $0 < b < r$.*

Proof. Let r be the radius of convergence of a power series

$$\sum_{n=0}^{\infty} a_n x^n. \tag{6.7}$$

If $0 < b < c < r$, then the series

$$\sum_{n=0}^{\infty} |a_n| c^n \tag{6.8}$$

converges and, by comparison with (6.8), the series (6.7) converges for every $x \in [-b, b]$. Let $s(x) = \sum_{n=0}^{\infty} a_n x^n$ for $x \in [-b, b]$. Then

$$\left| s(x) - \sum_{k=0}^{n} a_k x^k \right| = \left| \sum_{k=n+1}^{\infty} a_k x^k \right| \le \sum_{k=n+1}^{\infty} |a_k x^k| \le \sum_{k=n+1}^{\infty} |a_k| c^k \tag{6.9}$$

for every $x \in [-b, b]$. Since

$$\lim_{n \to \infty} \sum_{k=n+1}^{\infty} |a_k| c^k = 0,$$

(6.9) proves uniform convergence of (6.7). $\square$

Corollary 6.3.2. *The sum of a power series is continuous in its interval of convergence.*

Proof. If x is an arbitrary point in the interval of convergence $(-r, r)$ of a power series, then $x \in (-b, b)$ for some $0 < b < r$. By Theorem 6.3.1, the series converges uniformly in $[-b, b]$ and, by Theorem 6.2.3, its sum is a continuous function on $[-b, b]$. This proves that the sum is continuous at x, which is an arbitrary point of $(-r, r)$. $\square$

EXERCISES 6.3

(1) Show that the series $\sum_{n=0}^{\infty} x^n$ is not uniformly convergent in $(-1, 1)$.

(2) Let f, f_n $(n = 1, 2, 3, \dots)$ be continuous functions on the interval $[0, 1]$. If $f_n \rightrightarrows f$ in $[0, r]$ for every $0 < r < 1$, does it necessarily mean that $f_n \rightrightarrows f$ in $[0, 1]$?

(3) A series of functions $\sum_{n=1}^{\infty} f_n(x)$ is said to converge uniformly in an interval I if the sequence of partial sums $s_n(x) = f_1(x) + \cdots + f_n(x)$ converges uniformly in I. Prove the following theorem known as the Weierstrass M-test:

If $|f_n(x)| \leq M_n$ for all $n \in \mathbb{N}$ and all $x \in I$ and the series $\sum_{n=1}^{\infty} M_n$ converges, then the series $\sum_{n=0}^{\infty} f_n(x)$ converges uniformly in I.

(4) Find a sequence of continuous functions f_n on $[0, 1]$ such that the series $\sum_{n=0}^{\infty} f_n(x)$ converges uniformly on $[0, 1]$ but the series $\sum_{n=0}^{\infty} \max_{[0,1]} |f_n(x)|$ diverges. Is it going to be a counterexample for the Weierstrass M-test?

(5) Prove: If $|f_n(x)| \leq g_n(x)$ for all $n \in \mathbb{N}$ and all $x \in [a, b]$ and the series $\sum_{n=1}^{\infty} g_n(x)$ converges uniformly in $[a, b]$, then the series $\sum_{n=1}^{\infty} f_n(x)$ converges uniformly in $[a, b]$.

6.4 Differentiation of sequences of functions

Now we want to address the following question: Under what conditions will (f_n') converge to f' given that (f_n) converges to f? The theorem presented here will be used to prove a theorem on differentiation of power series and therefore it is stated in a rather unpretentious form. Its formulation can be strengthened in various directions, but then proofs become more difficult.

Theorem 6.4.1. If $f_n \rightrightarrows f$ and $f_n' \rightrightarrows g$ on an interval (a, b) and functions f_n' are continuous on (a, b), then f is differentiable on (a, b) and $f' = g$.

Proof. Uniform convergence of (f_n') means that there exists a sequence of positive numbers ε_n, convergent to 0, such that

$$|f_n'(x) - g(x)| < \varepsilon_n$$

for all $x \in (a, b)$ and for all sufficiently large $n \in \mathbb{N}$. Since the functions f_n' are continuous, so is the limit g, by Theorem 6.2.3.

Let $x_0 \in (a, b)$ and $\varepsilon > 0$. Then there exists a $\delta > 0$ such that

$$|g(x) - g(x_0)| < \varepsilon \quad \text{for} \quad |x - x_0| < \delta.$$

By the mean value theorem (Theorem 4.11.1) and the triangle inequality, we have

$$\left| \frac{f_n(x) - f_n(x_0)}{x - x_0} - g(x_0) \right| = |f_n'(\xi) - g(x_0)|$$

$$\leq |f_n'(\xi) - g(\xi)| + |g(\xi) - g(x_0)| \leq \varepsilon_n + \varepsilon \qquad (6.10)$$

for $|x - x_0| < \delta$, where ξ is a number between x and x_0 and hence it satisfies the inequality $|\xi - x_0| < \delta$. By letting $n \to \infty$ in (6.10) we obtain

$$\left| \frac{f(x) - f(x_0)}{x - x_0} - g(x_0) \right| \leq \varepsilon \quad \text{for} \quad |x - x_0| < \delta.$$

This proves that

$$\lim_{x \to x_0} \frac{f(x) - f(x_0)}{x - x_0} = g(x_0),$$

that is, $f'(x_0) = g(x_0)$. Since x_0 is an arbitrary point of in (a, b), the proof is complete. $\qquad \square$

In the proof of Theorem 6.4.3 we will use the following lemma.

Lemma 6.4.2. *If $0 < q < 1$, then $\lim_{n \to \infty} nq^n = 0$.*

Proof. Since $\lim_{n \to \infty} \frac{(n+1)q}{n} = q < 1$, there exists a number n_0 such that $\frac{(n+1)q}{n} < 1$ for all $n > n_0$. Hence $(n+1)q^{n+1} < nq^n$ for $n > n_0$, which means that the sequence (nq^n) is decreasing for $n > n_0$. Since it is also bounded, it converges to some limit L. From

$$(n + 1)q^{n+1} = qnq^n + q^{n+1},$$

by letting $n \to \infty$, we obtain $L = qL$. Since $q \neq 1$, we conclude that $L = 0$. $\qquad \square$

The following theorem is of critical importance in mathematical analysis.

Theorem 6.4.3. *The power series*

$$\sum_{n=0}^{\infty} a_n x^n \qquad (6.11)$$

and

$$\sum_{n=1}^{\infty} na_n x^{n-1} \qquad (6.12)$$

have the same radius of convergence. Moreover, the function $f(x) = \sum_{n=0}^{\infty} a_n x^n$ is differentiable on the interval of convergence and

$$f'(x) = \left(\sum_{n=0}^{\infty} a_n x^n \right)' = \sum_{n=1}^{\infty} na_n x^{n-1}.$$

Proof. Let r be the radius of convergence of (6.11) and let x be an arbitrary number such that $0 < |x| < r$. Choose b so that $|x| < b < r$. Since, according to Lemma 6.4.2, the sequence $\left(n \left| \frac{x}{b} \right|^n \right)$ converges to 0, it is bounded. Consequently,

$$|na_n x^{n-1}| = \left| \frac{1}{x} n \left(\frac{x}{b} \right)^n a_n b^n \right| \leq \frac{1}{|x|} \left(n \left| \frac{x}{b} \right|^n \right) |a_n| b^n \leq M |a_n| b^n,$$

for some constant M. Since the series $\sum_{n=1}^{\infty} |a_n| b^n$ converges, the series (6.12) converges at x. Thus (6.12) converges at every point at which (6.11) converges.

Conversely, the series (6.11) is convergent at every point at which (6.12) is convergent, because

$$|a_n x^n| \le |x| \left(n|a_n| \left|x^{n-1}\right|\right).$$

Therefore series (6.11) and (6.12) have the same radius of convergence.

Now let

$$f_n(x) = \sum_{k=0}^n a_k x^k, \quad g_n(x) = \sum_{k=1}^n k a_k x^{k-1}$$

and

$$f(x) = \sum_{k=0}^\infty a_k x^k, \quad g(x) = \sum_{k=1}^\infty k a_k x^{k-1}.$$

Note that, for all $n \in \mathbb{N}$, f_n and g_n are polynomials and we have $f_n' = g_n$. For every $0 < b < r$ the sequences (f_n) and (g_n) converge uniformly on the interval $[-b, b]$ and hence $f' = g$ in $[-b, b]$. This implies the equality in the entire interval of convergence $(-r, r)$. $\qquad\square$

Note that Theorem 6.4.3 says that a power series can be differentiated term by term.

EXERCISES 6.4

(1) Give an example of a sequence (f_n) of differentiable functions on an interval I such that $f_n \rightrightarrows 0$ on I, but the sequence (f_n') is not convergent on I.

(2) Let $f_n(x) = \frac{n^2 x |x|}{1+n^2 x^2}$. Find $f_n'(x)$ and then find pointwise limits of both sequences (f_n) and (f_n') and call them f and g, respectively. Is f differentiable on $\mathbb{R}$? Does $f' = g$?

(3) Prove: If both series $\sum_{n=1}^\infty f_n(x)$ and $\sum_{n=1}^\infty f_n'(x)$ converge uniformly on I and f_n' is continuous for every $n \in \mathbb{N}$, then the function $f(x) = \sum_{n=1}^\infty f_n(x)$ is differentiable on I and we have $f'(x) = \sum_{n=1}^\infty f_n'(x)$.

(4) Prove: If $0 < q < 1$ and $k \in \mathbb{N}$, then $\lim_{n\to\infty} n^k q^n = 0$.

(5) Find the sums of the following series:

(a) $\sum_{n=1}^\infty n x^n$ for $|x| < 1$,

(b) $\sum_{n=1}^\infty n^2 x^n$ for $|x| < 1$,

(c) $\sum_{n=1}^\infty \frac{n}{3^n}$,

(d) $\sum_{n=1}^\infty \frac{n^2}{7^n}$.

(6) Prove: If $r > 0$ and $\sum_{n=0}^\infty a_n x^n = \sum_{n=0}^\infty b_n x^n$ for $|x| < r$ then $a_n = b_n$ for each $n \in \mathbb{N}$.

Chapter 7

ELEMENTARY FUNCTIONS

In this chapter we give rigorous definitions of elementary functions and establish their properties. By elementary functions we mean here exponential functions, logarithms, trigonometric and inverse trigonometric functions. In elementary calculus these functions are often considered without rigorous definitions. On the other hand, if we want to carefully define these function, we have to choose between different approaches. Our goal was to choose definitions that are natural and mathematically convenient.

7.1 The exponential function e^x

We now approach the didactic problem of introducing the exponential function whose value at x is a^x. This function, perhaps the most important function in calculus, can be introduced in a great many ways. For instance it can be introduced as a generalization of the power a^n to nonnatural number values of the exponent n. First we introduce a^x for negative exponents by declaring that $a^{-n} = 1/a^n$. Next we define $a^{1/n}$ as the only positive solution of the equation $x^n = a$.

S: But there are values of a for which the equation has no solution, for example $x^2 = -1$.

T: Yes, and for that reason we assume that $a > 0$. We have to remember that the exponential function is an extension of the power with positive base.

S: How do we know that a solution exists for $a > 0$ and that it is unique?

T: This will be shown in the next section.

S: And how do we define a^x for other values of x?

T: The exponents m/n are introduced by $a^{m/n} = \left(a^{1/n}\right)^m$. Finally, if x is an irrational number we approximate it by a sequence of rational numbers x_n and then we prove that the sequence (a^{x_n}) is convergent. The value of a^x is defined as the limit of (a^{x_n}).

S: Since there are infinitely many sequences of rational numbers convergent to x, we may obtain infinitely many values for a^x.

T: No. One can prove that the limit does not depend on the choice of the sequence converging to x.

S: This way of introducing the exponential function seems to be long and complicated.

T: So it is. For this reason authors often introduce the exponential function e^x as the sum of a series: $e^x = 1 + \frac{x}{1!} + \frac{x^2}{2!} + \frac{x^3}{3!} + \ldots$

S: What is the meaning of e?

T: This letter has, at the beginning, no separate meaning. The symbol e^x has to be regarded as a whole, like $f(x)$.

S: And what is the connection with a^x?

T: Using the function $y = e^x$ one can define $y = a^x$ by a proper construction.

S: This way looks artificial.

T: Because of this there were mathematicians who were looking for another, possibly simpler definition. For example, the English mathematician Godfrey Harold Hardy (1877–1947) used the integral. One can also use the functional equation $f(x + y) = f(x)f(y)$ as the starting point.

We see that there are many possibilities for introducing the exponential function. It is difficult to decide which way is the best. In this book we will start from the fact that the derivative of $y = e^x$ is the same function. This approach seems to be the most useful one.

Definition 7.1.1. The *exponential function* e^x is the only solution of the differential equation

$$y' = y \tag{7.1}$$

satisfying the initial condition $y(0) = 1$.

In other words, the exponential function $y = e^x$ is a function which is invariant under differentiation and has the value 1 at 0. In order to be able to use this definition we have to be able to answer two questions affirmatively: Does a solution exist? Is it unique?

Existence of a solution of

$$y' = y, \quad y(0) = 1 \tag{7.2}$$

follows immediately as a very particular case of a general theorem on existence of solutions of differential equations. However, we can omit this general theory by using a power series argument.

Let's assume for the moment that there exists a solution of (7.2) which can be represented by a power series:

$$y(x) = \sum_{n=0}^{\infty} a_n x^n. \tag{7.3}$$

Then by Theorem 6.4.3 we have

$$y'(x) = \sum_{n=0}^{\infty} nax^{n-1} = \sum_{n=0}^{\infty} (n+1)a_{n+1}x^n.$$

Looking at the above equation and (7.3), we see that in order for (7.2) to be satisfied we must have

$$a_{n+1} = \frac{1}{n+1}a_n, \tag{7.4}$$

for every n. In particular from (7.4) we get

$$a_1 = \frac{a_0}{1}, a_2 = \frac{a_0}{1 \cdot 2}, a_3 = \frac{a_0}{1 \cdot 2 \cdot 3},$$

and, in general,

$$a_n = \frac{a_0}{n!},$$

where $n! = 1 \cdot 2 \cdots n$. Since we wish to have $y(0) = 1$, we must take $a_0 = 1$. We have thus proved that if there is a function defined by a power series that satisfies (7.2), then it must be the function given by

$$e^x = 1 + \frac{x}{1!} + \frac{x^2}{2!} + \cdots. \tag{7.5}$$

It is straightforward to show that (7.5) satisfies (7.2) in the interval of convergence of the power series. We will prove that the series converges for every x, which means the interval of convergence is the entire real line.

Let x be an arbitrary number and let m be a natural number greater than $|x|$. We have

$$\frac{|x|^{m+1}}{(m+1)!} = \frac{|x|^m}{m!} \cdot \frac{|x|}{m+1} < \frac{|x|^m}{m!} \cdot \frac{|x|}{m}$$

and by induction

$$\frac{|x|^{m+k}}{(m+k)!} < \frac{|x|^m}{m!} \cdot \left(\frac{|x|}{m}\right)^k. \tag{7.6}$$

Since, for every $n \in \mathbb{N}$, we have

$$\sum_{k=1}^{n} \frac{|x|^m}{m!} \left(\frac{|x|}{m}\right)^k = \frac{|x|^m}{m!} \sum_{k=1}^{n} \left(\frac{|x|}{m}\right)^k,$$

and since the series

$$\sum_{k=1}^{\infty} \left(\frac{|x|}{m}\right)^k$$

is a convergent geometric series, convergence of the series

$$\sum_{k=1}^{\infty} \frac{|x|^{m+k}}{(m+k)!}$$

follows by (7.6). This implies convergence of the series in (7.5) for every $x \in \mathbb{R}$.

S: We assumed that there was a power series satisfying (7.2). How can we assume this fact when our aim was just to prove existence of a solution?

T: There is no vicious circle in this argument because the assumption on existence served only to find the coefficients. Having found those coefficients, we disregard our assumption and we prove that the series is convergent and represents a function with the desired properties.

S: Would it not be simpler to just start from series (7.5) since we have to use it anyway?

T: Many authors do so. However, such a definition is a little artificial because there is no explanation for the choice of coefficients. The actual explanation is the differential equation (7.2). We could say that e^x is the only function expandable into a power series and satisfying (7.2).

S: But in Definition 7.1.1 there is no assumption that the function is expandable. Is that definition correct?

T: Yes, it is. One can prove that each function satisfying (7.2) is expandable and this ought to be proved in order to show that the definition is correct.

Now we show that our solution of the differential equation $y' = y$ with $y(0) = 1$ is unique. Indeed, assume that there are two solutions f and g and consider the auxiliary function

$$h(x) = f(x)g(x_0 - x),$$

where x_0 is an arbitrary point. Then

$$h'(x) = f'(x)g(x_0 - x) - f(x)g'(x_0 - x).$$

(To understand the calculations one has to clearly distinguish the function $g'(x_0 - x)$ from the function $(g(x_0 - x))' = -g'(x_0 - x)$.) Since f and g are solutions of the equation $y' = y$, we have

$$h'(x) = f(x)g(x_0 - x) - f(x)g(x_0 - x) = 0.$$

Thus h is a constant function. Consequently, we have $h(x_0) = h(0)$, which means that

$$f(x_0)g(0) = f(0)g(x_0),$$

and, in view of the condition $y(0) = 1$,

$$f(x_0) = g(x_0).$$

Since x_0 is an arbitrary number, f and g are identical and the uniqueness is proved.

The function $y = e^x$ is the only function equal to its derivative which has the value 1 at 0. Since it is differentiable it is continuous (which also follows from the fact that it expands into a power series).

Since the function

$$h(x) = f(x)g(x_0 - x) = e^x e^{x_0 - x}$$

is constant, as shown above, we have

$$e^0 e^{x_0} = e^x e^{x_0 - x}.$$

Recalling that $e^0 = 1$, and replacing x by s and x_0 by $s + t$ we get,

$$e^{s+t} = e^s e^t \tag{7.7}$$

for arbitrary $s, t \in \mathbb{R}$.

From the expansion

$$e^x = 1 + \frac{x}{1!} + \frac{x^2}{2!} + \frac{x^3}{3!} + \ldots, \tag{7.8}$$

it is easily seen that the function is positive for $x \geq 0$. By (7.7) one can easily show that it is positive also for $x < 0$. In fact, for $s = x$ and $t = -x$ we have $e^x e^{-x} = e^0 = 1$, and hence

$$e^{-x} = \frac{1}{e^x}.$$

Thus, the exponential function is positive for all $x \in \mathbb{R}$. It is also an increasing function, since its derivative is everywhere positive.

Since the function $y = e^x$ is increasing and always positive, we know that it assumes each positive value at most once. We are going to prove that it assumes each positive value exactly once. Since we may write

$$e^x = 1 + x + \sum_{k=2}^{\infty} \frac{x^k}{k!},$$

for every positive x we have

$$1 + x < e^x, \tag{7.9}$$

(see Fig. 7.1). Let a be any positive number. Taking in turn a and $\frac{1}{a}$ we get

$$a < 1 + a < e^a \quad \text{and} \quad \frac{1}{a} < 1 + \frac{1}{a} < e^{1/a},$$

and hence

$$e^{-1/a} < a < e^a.$$

Since the function $y = e^x$ is continuous it assumes each value between $e^{-1/a}$ and e^a, and in particular it assumes the value a. Moreover $y = e^x$ assumes a only once, because it is an increasing function.

Instead of e^x we sometimes write $\exp(x)$. This is convenient when we describe a composition of the exponential function with another function. For example, instead of $e^{\frac{x}{1+x^2}}$ we can write $\exp\left(\frac{x}{1+x^2}\right)$.

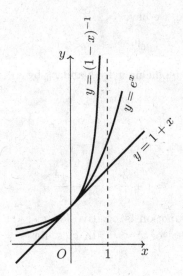

Fig. 7.1   Graphs of $y = 1 + x$, $y = e^x$, and $y = (1 - x)^{-1}$.

EXERCISES 7.1

(1) Prove that the function (7.5) satisfies the differential equation(7.2).
(2) Use the method presented in this section to prove that the differential equation $y' = -y$, $y(0) = 1$, has a solution defined for all $x \in \mathbb{R}$.
(3) Prove uniqueness of solution of the differential equation $y' = -y$, $y(0) = 1$.
(4) Prove uniqueness of solution of the differential equation $y' = y$, $y(0) = 0$.
(5) Find all differentiable functions f such that $f(x + y) = f(x)f(y)$ for all $x \in \mathbb{R}$ and such that $f(0) = 1$. Justify that all of them have been found.
(6) Prove that inequality (7.9) holds also for $x < 0$.
(7) Prove that for $x < 1$ we have $e^x \leq \frac{1}{1-x}$.
(8) Find $\lim_{x \to -\infty} e^x$ and $\lim_{x \to \infty} e^x$.

7.2   General exponential functions

We have proved that for every $a > 0$ there is a unique number c such that

$$a = e^c.$$

That number is called the *natural logarithm* of a and is denoted by

$$c = \ln a.$$

Thus the two equalities are equivalent and we have

$$a = e^{\ln a} \quad \text{for} \quad a > 0. \tag{7.10}$$

In view of

$$e^s e^t = e^{s+t} \tag{7.11}$$

we also have

$$a^2 = e^{\ln a} e^{\ln a} = e^{2 \ln a},$$

$$a^3 = e^{2 \ln a} e^{\ln a} = e^{3 \ln a},$$

and generally

$$a^n = e^{n \ln a} \quad \text{for} \quad n \in \mathbb{N}. \tag{7.12}$$

Moreover we have

$$a^0 = e^0 = 1$$

and

$$a^{-n} = \frac{1}{a^n} = \frac{1}{e^{n \ln a}} = e^{-n \ln a} \quad \text{for} \quad n \in \mathbb{N}.$$

Thus equation (7.12) holds for all integers n. This allows us to denote $e^{x \ln a}$ by the symbol a^x, $a > 0$. We thus have

$$a^x = e^{x \ln a} \tag{7.13}$$

for all $x \in \mathbb{R}$. We use (7.13) as the definition of the *exponential function with base a*. Replacing s in (7.11) by $s \ln a$ and t by $t \ln a$ we get, in view of the last equality,

$$a^s a^t = a^{s+t}. \tag{7.14}$$

From (7.10) and (7.13) we obtain

$$\left(e^{\ln a}\right)^x = e^{x \ln a}.$$

This equality holds for any real values $\ln a$ and x. If we replace x by t and $\ln a$ by $s \ln a$, we get

$$\left(e^{s \ln a}\right)^t = e^{st \ln a},$$

that is,

$$(a^s)^t = a^{st}. \tag{7.15}$$

Equalities (7.14) and (7.15) hold for all s and t and they represent fundamental properties of exponential functions.

If in (7.14) we substitute for t a natural number n and for s the number $-n$, then we have $a^{-n} a^n = 1$, whence

$$a^{-n} = \frac{1}{a^n}$$

This formula reduces the calculation of a^{-n} to multiplication and division, for a^n can be regarded as a product. Similarly, substituting in (7.15) a natural number n for t and the number $1/n$ for s we obtain

$$\left(a^{1/n}\right)^n = a.$$

Hence we see that if a is a positive number and n is a natural number, then there always exists a positive number b such that $b^n = a$. That number is called the *principal n-th root* of a and is denoted by $\sqrt[n]{a}$. Thus

$$a^{1/n} = \sqrt[n]{a}.$$

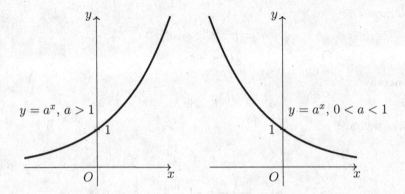

Fig. 7.2　Examples of graphs of $y = a^x$ for $a > 1$ and $0 < a < 1$.

EXERCISES 7.2

(1) Prove that the exponential function a^x is increasing if $a > 1$, and decreasing if $0 < a < 1$, (see Fig. 7.2).

(2) Prove that $(2^x)' = 2^x \cdot \ln 2$.

(3) Find $\lim_{x \to 0} \frac{(2^x - 1)}{x}$.

7.3   The number e

In the symbol a^x the letter a may denote an arbitrary positive number, and x any arbitrary number. On the other hand, when introducing the function $y = e^x$ we emphasized the fact that the letter e had no separate meaning. Now we can say the e represents a number, namely $e = e^1$. The function e^x is thus a special case of the function a^x.

Since e is the value of the function $y = e^x$ at $x = 1$, it is the sum of the series

$$e = 1 + \frac{1}{1!} + \frac{1}{2!} + \frac{1}{3!} + \ldots \tag{7.16}$$

The value of e can be found with arbitrary accuracy from (7.16). To estimate the accuracy of an approximation of e by a partial sum of the series (7.16) we will use the inequalities

$$\sum_{k=0}^{n} \frac{1}{k!} < e < \sum_{k=0}^{n} \frac{1}{k!} + \frac{1}{n! \, n}.$$

(Recall that $0! = 1$.) The first inequality is obvious. To prove the second inequality, we compare the tail of the series in (7.16) with a constant multiple of a geometric series whose sum we can easily find. To this end notice that, for $k > n + 1$, we have

$$\frac{1}{k!} < \frac{1}{(n+1)!(n+1)^{k-n-1}}$$

and consequently

$$\sum_{k=n+1}^{\infty} \frac{1}{k!} < \frac{1}{(n+1)!} \sum_{k=n+1}^{\infty} \frac{1}{(n+1)^{k-n-1}} = \frac{1}{(n+1)!} \cdot \frac{1}{1 - \frac{1}{n+1}} = \frac{1}{n! \, n}.$$

Taking $n = 13$ we have

$$\sum_{k=0}^{13} \frac{1}{k!} = 1 + \frac{1}{1!} + \frac{1}{2!} + \frac{1}{3!} + \frac{1}{4!} + \frac{1}{5!} + \frac{1}{6!} + \frac{1}{7!} + \frac{1}{8!} + \frac{1}{9!} + \frac{1}{10!} + \frac{1}{11!} + \frac{1}{12!} + \frac{1}{13!}. \quad (7.17)$$

The fractions on the right side of (7.17) can be easily found one after the other with arbitrary accuracy. For instance to calculate $1/7!$ it suffices to divide the already found value $1/6!$ by 7. In this way in few minutes we will have all the values:

$$1 = 1,$$
$$1/1! = 1,$$
$$1/2! = 0.5,$$
$$1/3! \approx 0.1666666666666,$$
$$1/4! \approx 0.0416666666666,$$
$$1/5! \approx 0.0083333333333,$$
$$1/6! \approx 0.0013888888888,$$
$$1/7! \approx 0.0001984126984,$$
$$1/8! \approx 0.0000248015873,$$
$$1/9! \approx 0.0000027557319,$$
$$1/10! \approx 0.0000002755731,$$
$$1/11! \approx 0.0000000250521,$$
$$1/12! \approx 0.0000000020876,$$
$$1/13! \approx 0.0000000001605.$$

The sum of the numbers on the right is $S = 2.71828182839$. The numbers $1/3!, 1/4!, \ldots, 1/13!$ were each calculated with an error less than $1/10^{11}$, and each approximation is less than the number approximated. Thus the sum S of those approximations differs from (7.17) by at most $11/10^{11}$ (because we have added 11 approximate numbers). Consequently,

$$S < e < S + \frac{11}{10^{11}} + \frac{1}{13! \cdot 13} < S + \frac{11}{10^{11}} + \frac{2}{10^{11}} = S + \frac{13}{10^{11}},$$

from which we see that

$$2.71828182839 < e < 2.71828182852$$

and thus we may write

$$e \approx 2.718281828.$$

This number is easy to remember, because the sequence 1828 appears twice in it.

EXERCISES 7.3

(1) Using the above method calculate $\sqrt[5]{e}$ with an error less than $1/10^6$.

(2) Using the above method calculate $1/\sqrt{e}$ with an error less than $1/10^6$.

7.4   Natural logarithm

In Section 7.2 we used the fact that for every $a > 0$ there exists exactly one number c such that $a = e^c$ to introduce the natural logarithm $\ln a$ and then define the exponential function a^x in terms of $\ln a$. Now we consider $\ln x$ as a function, called the *natural logarithm function*. Since

$$e^{\ln x} = x \quad \text{for all } x \in (0, \infty)$$

and

$$\ln e^x = x \quad \text{for all } x \in (-\infty, \infty),$$

the functions $y = e^x$ and $y = \ln x$ are inverses of each other. The graph of $y = \ln x$ can be obtained by reflecting the graph of the function $y = e^x$ symmetrically with respect to the line $y = x$ (see Fig. 7.3).

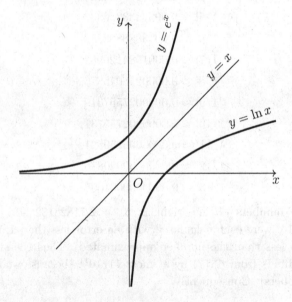

Fig. 7.3   Graphs of $y = e^x$, $y = x$, and $y = \ln x$.

The natural logarithm function $y = \ln x$ is defined in the interval $(0, \infty)$ and is continuous and increasing in this interval. For $0 < x < 1$ it is negative and for $x > 1$ it is positive. Moreover $\ln 1 = 0$. The derivative of the natural logarithm function can be found by the formula given in Theorem 4.8.1. Since $(e^x)' = e^x$, we have

$$(\ln x)' = \frac{1}{e^{\ln x}} = \frac{1}{x}.$$

The function $\ln(x+1)$ is defined in the interval $(-1, \infty)$. In the interval $(-1, 1)$ it expands in a power series which can be found in the following way. First note that

$$\frac{1}{1+x} = \frac{1}{1-(-x)} = 1 - x + x^2 - x^3 + \cdots = \sum_{k=0}^{\infty} (-1)^k x^k, \qquad (7.18)$$

where the series is convergent in the interval $(-1, 1)$. Now, for all $x \in (-1, 1)$, we have

$$(\ln(x+1))' = \frac{1}{1+x} = \sum_{k=0}^{\infty} (-1)^k x^k.$$

Since, by Theorem 6.4.3,

$$\left(\sum_{k=1}^{\infty} \frac{(-1)^{k+1}}{k} x^k\right)' = \sum_{k=0}^{\infty} (-1)^k x^k,$$

we must have

$$\ln(x+1) = C + \sum_{k=1}^{\infty} \frac{(-1)^{k+1}}{k} x^k, \qquad (7.19)$$

for some constant C and all $x \in (-1, 1)$. It remains to find the constant C. We can do that by substituting $x = 0$. This shows that $C = 0$. Consequently

$$\ln(x+1) = \sum_{k=1}^{\infty} \frac{(-1)^{k+1}}{k} x^k = \frac{x}{1} - \frac{x^2}{2} + \frac{x^3}{3} - \frac{x^4}{4} + \cdots, \quad \text{for} \quad -1 < x < 1. \ (7.20)$$

In Section 8.7 we will prove that the equality holds also for $x = 1$. However the series in (7.20) is divergent for $x = -1$ and all $|x| > 1$.

EXERCISES 7.4

(1) Find derivatives of the following functions:

(a) $f(x) = x^2 \ln(x^3)$,

(b) $f(x) = \ln(1 + x^2)$,

(c) $f(x) = \ln\dfrac{1+x}{1-x}$,

(d) $f(x) = e^x \ln x + \ln 3$,

(e) $f(x) = \ln(\ln x)$,

(f) $f(x) = \ln(1 + e^{-x})$.

(2) Find a decimal approximation for $\ln\frac{1}{2}$ with a three decimal place accuracy. Justify your response using (7.20) and Theorem 5.5.4.

(3) Prove that the series in (7.20) diverges for all x such that $|x| > 1$.

(4) Find all continuous functions f defined on $(0, \infty)$ such that

$$f(st) = f(s) + f(t)$$

for all $s, t > 0$.

7.5   Arbitrary logarithms

By differentiating the equality $a^x = e^{x \ln a}$ we get $(a^x)' = e^{x \ln a} \ln a = a^x \ln a$. We thus have

$$(a^x)' = a^x \ln a.$$

Hence it follows that the derivative of $y = a^x$ is positive if $a > 1$ and negative if $0 < a < 1$. Consequently, the function $y = a^x$ is increasing if $a > 1$ and decreasing if $0 < a < 1$. If $a = 1$, then $a^x = 1$ for every x. Therefore, the function $y = a^x$ is invertible if $a \neq 1$. The inverse function of $y = a^x$ is called the *logarithm with base a* and is denoted by $\log_a x$. We always assume that a is positive and different from 1. The domain of the function $y = \log_a x$ is $(0, \infty)$ and the range is all real numbers. The function $y = \log_a x$ is increasing if $a > 1$ and decreasing if $0 < a < 1$, (see Fig. 7.4). In both cases we have $\log_a a = 1$.

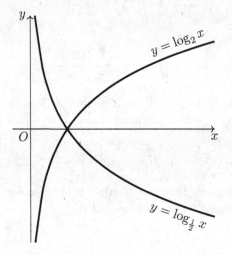

Fig. 7.4   Examples of graphs of $y = \log_a$ for $a > 1$ and $0 < a < 1$.

For all positive numbers s and t the following equalities hold:

$$\log_a st = \log_a s + \log_a t,$$

$$\log_a \frac{s}{t} = \log_a s - \log_a t,$$

$$\log_a s^p = p \log_a s, \text{ for any number } p,$$

$$\log_a s = \frac{\ln s}{\ln a}.$$

The proofs run as follows:

$$\log_a st = \log_a \left(a^{\log_a s} a^{\log_a t}\right) = \log_a \left(a^{\log_a s + \log_a t}\right) = \log_a s + \log_a t,$$

$$\log_a \frac{s}{t} = \log_a \left(\frac{a^{\log_a s}}{a^{\log_a t}} \right) = \log_a \left(a^{\log_a s - \log_a t} \right) = \log_a s - \log_a t,$$

$$\log_a s^p = \log_a \left(\left(a^{\log_a s} \right)^p \right) = \log_a \left(a^{p \log_a s} \right) = p \log_a s,$$

$$\log_a s = \frac{\log_a s \ln a}{\ln a} = \frac{\ln \left(a^{\log_a s} \right)}{\ln a} = \frac{\ln s}{\ln a}.$$

Note that

$$\log_e x = \frac{\ln x}{\ln e} = \ln x$$

and thus all the above formulas hold for the natural logarithm $\ln x$.

Since

$$(\log_a x)' = \left(\frac{\ln x}{\ln a} \right)' = \frac{(\ln x)'}{\ln a} = \frac{\frac{1}{x}}{\ln a},$$

we have

$$(\log_a x)' = \frac{1}{x \ln a}.$$

EXERCISES 7.5

(1) Find derivatives of the following functions:

   (a) $f(x) = 2^x$,
   (b) $f(x) = 2^{(2^x)}$,
   (c) $f(x) = \log_2(1 + x)$,
   (d) $f(x) = \ln \frac{1}{1+3^x}$.

(2) Prove the equality $(\log_x a)' = -\frac{(\log_x a)^2}{x \ln a}$ where a is positive and different from 1.

(3) Show that $\log_a b \log_b a = 1$ for all positive a and b different from 1.

(4) Show that $\log_a x = \frac{\log_b x}{\log_b a}$ for all positive a and b different from 1.

(5) Show that $(\log_x a)' = -\frac{(\log_x a)^2}{x \ln a}$ where a is positive and different from 1.

7.6   Derivatives of power functions

By a *power function* we mean the function

$$f(x) = x^a,$$

where the exponent a is an arbitrary number. Graphs of some power functions are shown in Fig. 7.5. The case when a is an integer was already discussed in Sections 3.1, 4.4, and 4.6. When a is not an integer, we restrict the domain of the function to positive values of x.

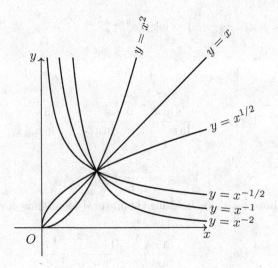

Fig. 7.5   Examples of graphs of $y = x^\alpha$ for different α.

By representing the power in the form

$$x^a = e^{a \ln x},$$

we easily find its derivative

$$(x^a)' = e^{a \ln x} \frac{a}{x} = x^a \frac{a}{x} = ax^{a-1}.$$

Hence the formula

$$(x^a)' = ax^{a-1}, \ x > 0.$$

In particular we have

$$\left(x^{1/n}\right)' = \frac{1}{n} x^{-\frac{n-1}{n}} \quad \text{and} \quad \left(x^{-1/n}\right)' = -\frac{1}{n} x^{-\frac{n+1}{n}},$$

which can be equivalently written as

$$\left(\sqrt[n]{x}\right)' = \frac{1}{n \sqrt[n]{x^{n-1}}} \quad \text{and} \quad \left(\frac{1}{\sqrt[n]{x}}\right)' = -\frac{1}{n \sqrt[n]{x^{n+1}}}.$$

In the case of the square root, which is most used in applications, we have

$$\left(\sqrt{x}\right)' = \frac{1}{2\sqrt{x}}.$$

It is important to remember that the power function $y = x^a$ is different from the exponential function $y = a^x$. However, the derivatives of both functions were found in a similar way by representing them as $x^a = e^{a \ln x}$ and $a^x = e^{x \ln a}$. A similar method can be used to find the derivative of a function $F(x) = g(x)^{f(x)}$ where both f and g are differentiable functions and $g > 0$. Since we have

$$F = e^{f \ln g},$$

we find

$$F' = e^{f \ln g}(f \ln g)' = g^f \left(\frac{f}{g} g' + f' \ln g \right).$$

This formula need not be remembered, because in practice it is more convenient to apply the described method in every case separately.

For example, for $x > 0$, we have

$$(x^x)' = \left(e^{x \ln x} \right)' = e^{x \ln x}(x \ln x)' = x^x(\ln x + 1)$$

and

$$\left((\ln x)^{x^2} \right)' = \left(e^{x^2 \ln(\ln x)} \right)' = e^{x^2 \ln(\ln x)} \left(x^2 \ln(\ln x) \right)'$$

$$= (\ln x)^{x^2} \left(2x \ln(\ln x) + \frac{1}{x \ln x} \right).$$

EXERCISES 7.6

(1) Find derivatives of the following functions:

(a) $f(x) = \ln \frac{\sqrt{1+x} - \sqrt{1-x}}{\sqrt{1+x} + \sqrt{1-x}}$,

(e) $f(x) = \frac{2 + \sqrt{x}}{2 - \sqrt{x}}$,

(b) $f(x) = \frac{(8x^2 + 4x + 3)\sqrt{x^2 - 1}}{15x^2}$,

(f) $f(x) = 2^{\sqrt{1+2x}}$,

(c) $f(x) = 3^x + x^3 + x^x + 3^3$,

(g) $f(x) = (\log_2 x)^{e^x}$.

(d) $f(x) = x^{-2x}$,

(2) Show that

$$\left(\ln(x + \sqrt{x^2 + a}) \right)' = \frac{1}{\sqrt{x^2 + a}}$$

whenever $x^2 + a > 0$.

(3) Find each interval in which $a(x) = xe^{-x}$ is increasing (decreasing).

(4) Find each interval in which $b(x) = (\ln x)^2 - 2 \ln x$ is increasing (decreasing).

(5) Find each interval in which $c(x) = x \cdot \sqrt{\frac{2-x}{2+x}}$ is increasing (decreasing).

7.7 Calculating natural logarithms of natural numbers

Formula (7.20) in Section 7.4, that is,

$$\ln(1 + x) = \frac{x}{1} - \frac{x^2}{2} + \frac{x^3}{3} - \frac{x^4}{4} + \cdots = \sum_{n=0}^{\infty} \frac{(-1)^n}{n} x^n, \quad \text{for } -1 < x < 1,$$

can be used to calculate logarithms, but only for values which are close to 0. If we want to calculate logarithm of natural numbers, we have to use a different formula, which can be obtained from the one above. First we replace x by $-x$ to obtain

$$\ln(1 - x) = -\frac{x}{1} - \frac{x^2}{2} - \frac{x^3}{3} - \frac{x^4}{4} + \cdots = -\sum_{n=0}^{\infty} \frac{x^n}{n}, \quad \text{for } -1 < x < 1.$$

Subtracting the last equality from the preceding one we get

$$\ln\frac{1+x}{1-x} = 2\left[\frac{x}{1} + \frac{x^3}{3} + \frac{x^5}{5} + \frac{x^7}{7} + \dots\right] = 2\sum_{n=1}^{\infty}\frac{x^{2n-1}}{2n-1}, \qquad (7.21)$$

for $-1 < x < 1$. Now, for a given natural number p we select x such that

$$\frac{1+x}{1-x} = \frac{p+1}{p};$$

thus $x = 1/(2p+1)$. Substituting this value in (7.21) we get

$$\ln\frac{p+1}{p} = 2\left[\frac{1}{1(2p+1)} + \frac{1}{3(2p+1)^3} + \frac{1}{5(2p+1)^5} + \dots\right]$$

$$= 2\sum_{n=1}^{\infty}\frac{1}{(2n-1)(2p+1)^{2n-1}}.$$

Finally, adding $\ln p$ to both sides of the above we obtain

$$\ln(p+1) = \ln p + 2\left[\frac{1}{1(2p+1)} + \frac{1}{3(2p+1)^3} + \frac{1}{5(2p+1)^5} + \dots\right]$$

$$= \ln p + 2\sum_{n=1}^{\infty}\frac{1}{(2n-1)(2p+1)^{2n-1}}. \qquad (7.22)$$

Thus we can find the logarithm of $p+1$ if we already know the logarithm of p. In this way we can successively calculate logarithms of natural numbers starting from the already known equality $\ln 1 = 0$. For $p = 1$ we get from (7.22)

$$\ln 2 = 2\left[\frac{1}{1\cdot 3} + \frac{1}{3\cdot 3^3} + \frac{1}{5\cdot 3^5} + \dots\right] = 2\sum_{n=1}^{\infty}\frac{1}{(2n-1)3^{2n-1}}. \qquad (7.23)$$

We have

$$\begin{array}{ll}
1/3 \approx 0.33333333333, & 1/(1\cdot 3) \approx 0.33333333333, \\
1/3^3 \approx 0.03703703703, & 1/(3\cdot 3^3) \approx 0.01234567901, \\
1/3^5 \approx 0.00411522633, & 1/(5\cdot 3^5) \approx 0.00082304526, \\
1/3^7 \approx 0.00045724737, & 1/(7\cdot 3^7) \approx 0.00006532105, \\
1/3^9 \approx 0.00005080526, & 1/(9\cdot 3^9) \approx 0.00000564502, \\
1/3^{11} \approx 0.00000564502, & 1/(11\cdot 3^{11}) \approx 0.00000051318, \\
1/3^{13} \approx 0.00000062722, & 1/(13\cdot 3^{13}) \approx 0.00000004824, \\
1/3^{15} \approx 0.00000006966, & 1/(15\cdot 3^{15}) \approx 0.00000000464, \\
1/3^{17} \approx 0.00000000774, & 1/(17\cdot 3^{17}) \approx 0.00000000045.
\end{array}$$

The numbers of the first column are successively found by dividing the preceding number by 9. The numbers of the second column are found by dividing the numbers of the first column by $1, 3, 5, \dots$, respectively. The sum of the right column is $S = 0.34657359018$. It is an approximate value of the sum of the nine initial terms of the series in (7.23). The possible error is less than $9/10^{11}$. Because all numbers are rounded down, we have

$$2S \leq \ln 2 \leq 2S + \frac{18}{10^{11}} + 2T,$$

where T is the sum of the neglected part of the series in (7.23). Since for $n \geq 9$ we have

$$\frac{1}{(2n+1) \cdot 3^{2n+1}} \leq \frac{1}{19 \cdot 3^{19}} \frac{1}{9^{n-9}},$$

we get

$$T \leq \frac{1}{19 \cdot 3^{19}} \sum_{n=9}^{\infty} \frac{1}{9^{n-9}} = \frac{1}{19 \cdot 3^{19}} \frac{1}{1 - \frac{1}{9}} = \frac{1}{19 \cdot 3^{17} \cdot 8} < \frac{51}{10^{11}}.$$

Hence

$$2S \leq \ln 2 < 2S + \frac{18}{10^{11}} + \frac{102}{10^{11}} = 2S + \frac{120}{10^{11}},$$

which gives us

$$0.69314718036 \leq \ln 2 < 0.69314718156. \tag{7.24}$$

The above calculations show that 8 obtained digits are reliable, so that

$$\ln 2 \approx 0.69314718.$$

In practice the order of calculations is slightly different. We first establish the accuracy with which we want to find the logarithm. Then we calculate the terms of the series with a little better accuracy, usually by two or three digits more, because when summing approximate numbers the error increases. We evaluate consecutive terms of the series as long as we get non-zero numbers (with the chosen accuracy). Such an empiric method allows us to establish more or less the number of terms we have to take into account. This method does not guarantee the degree of accuracy and therefore it is still necessary to estimate the error as we did in the above example.

To calculate $\ln 4$ we need not use the formula (7.22) because $\ln 4 = 2 \ln 2$. Using (7.24) we see that

$$1.38629436079 \leq \ln 4 < 1.38629436312.$$

We thus may write at once

$$\ln 4 \approx 1.38629436.$$

The number $\ln 5$ can be found from (7.22) when $p = 4$:

$$\ln 5 = \ln 4 + 2 \left[\frac{1}{1 \cdot 9} + \frac{1}{3 \cdot 9^3} + \frac{1}{5 \cdot 9^5} + \ldots \right] = \ln 4 + 2 \sum_{n=0}^{\infty} \frac{1}{(2n-1)9^{2n-1}}. \tag{7.25}$$

The approximate sum of the first four terms

$$1/(1 \cdot 9) \approx 0.1111111111,$$
$$1/(3 \cdot 9^3) \approx 0.0004572473,$$
$$1/(5 \cdot 9^5) \approx 0.0000033870,$$
$$1/(7 \cdot 9^7) \approx 0.0000000295,$$

is $U = 0.1115717749$. We thus have

$$\frac{1}{1 \cdot 9} + \frac{1}{3 \cdot 9^3} + \frac{1}{5 \cdot 9^5} + \frac{1}{7 \cdot 9^7} < U + \frac{4}{10^{10}}.$$

The reader can verify that the sum of the neglected part of the series

$$\frac{1}{9 \cdot 9^9} + \frac{1}{11 \cdot 9^{11}} + \frac{1}{13 \cdot 9^{13}} + \cdots$$

is less than $4/10^{10}$. Hence

$$1.60943791059 = 4S + 2U < \ln 5 < 2 \left(2S + \frac{12}{10^{10}} \right) + 2 \left(U + \frac{8}{10^{10}} \right)$$

$$= 4S + 2U + \frac{40}{10^{10}} = 1.60943791459.$$

This leads to the result

$$\ln 5 \approx 1.60943791.$$

The logarithm of 10 can be found from the equality $\ln 10 = \ln 2 + \ln 5$, whence we get

$$\ln 10 \approx 2.30258509,$$

where the correctness of only the last two right-most digits has not been fully justified by our method.

EXERCISES 7.7

(1) Use the method described in this section to find the value of $\ln 3$ to 8 decimal place accuracy.
(2) Find the value of $\ln 6$ to 8 decimal place accuracy.

7.8   Common logarithms

Logarithms change multiplication into addition according to the general formula

$$\log ab = \log a + \log b.$$

Since addition is much easier than multiplication, one can use logarithms in practical calculations. To this aim one needs extended tables which give logarithms of as many numbers as possible. But the range of such tables is always limited so that multiplication of large numbers would be cumbersome. This difficulty can be evaded by introducing *common logarithms*. These are logarithms with base 10. They are also called *decadic logarithms* and can be calculated using the formula

$$\log_{10} x = \frac{\ln x}{\ln 10}.$$

Since

$$\frac{1}{\ln 10} \approx 0.4342944819,$$

the common logarithms can be obtained from natural logarithms just by multiplying them by the above constant factor.

S: What is the advantage of introducing a new kind of logarithms which differ only by a constant factor?

T: Tables of common logarithms need only contain numbers between 1 and 10, because each positive number can be represented in the form

$$y = 10^n x,$$

where $1 < x < 10$. We evidently have

$$\log_{10} y = n + \log_{10} x.$$

The number n is called the *characteristic* of the logarithm and the number $\log_{10} x$ the *mantissa* of the logarithm. The values of the mantissa always lie between 0 and 1 so their digits begin just after the decimal point. The characteristic, when it is positive, stands before the decimal point. For instance, $301 = 10^2 \cdot 3.01$ and hence

$$\log_{10} 301 \approx 2.47784$$

since $\log_{10} 3.01 \approx 0.47784$.

Tables of mantissa were first calculated by Henry Briggs (1561–1631). There were great enthusiasts of common logarithms. The famous French mathematician Pierre Simon Laplace (1749–1827) said: "The invention of logarithms has shortened calculations extending over months to just a few weeks." Other mathematicians regarded the logarithmic tables as the very essence of mathematics.

Nowadays logarithmic tables have completely lost their importance, because they have been replaced by calculators which are much smaller than tables of logarithms, and so much faster, and easier to use. However, sometimes even with a calculator in hand logarithms can prove useful. For example, try to calculate

$$\frac{999^{998}}{998^{999}}.$$

You cannot find 999^{998} and 998^{999} because these numbers are too large for a calculator. The answer can be easily obtained by using the following equality

$$\frac{999^{998}}{998^{999}} = e^{998 \ln 999 - 999 \ln 998}.$$

EXERCISES 7.8

(1) Which number is bigger: 999^{998} or 998^{999}?

(2) Solve the preceding problem without using a calculator at all.

7.9 Sine and cosine

Our aim now is to introduce trigonometric functions and establish connections between them.

S: In most textbooks trigonometric functions are introduced in a geometric way. Those books appear in several editions and have been studied by generations of students.

T: This is a bad situation. Analysis is based on a few axioms of numbers and each step is thoroughly justified. It is very inconsistent to use geometrical definitions there. This is a complete departure from precision of exposition and deviation from the analytic methods which make the core of calculus.

S: Do you mean that the geometrical approach is not precise?

T: It is not. However, it stimulates intuition, which is the first step before introducing precise methods.

S: I see. If we wanted to use geometry for proofs in analysis, we should first state the axioms of geometry and establish their relationship to the axioms of numbers.

T: Right. But even if we did so, the exposition would lose its neatness and simplicity. It would also suggest that axioms of geometry are necessary for introducing trigonometric functions.

S: But in this book there is also a lot of geometry.

T: Well, geometry is treated here as an illustration only and by no means as a proving instrument. A geometric point of view often helps in finding a logically correct, precise, analytical proof.

In physical applications the functions sine and cosine appear as solutions of the differential equation

$$f''(x) = -f(x). \tag{7.26}$$

Assume, for the moment, that there exists a function f satisfying this equation which has a power series representation:

$$f(x) = \sum_{k=0}^{\infty} a_k x^k.$$

Then, in the interval of convergence of the series, we have

$$f'(x) = \sum_{k=0}^{\infty} (k+1)a_{k+1}x^k$$

and

$$f''(x) = \sum_{k=0}^{\infty} (k+1)(k+2)a_{k+2}x^k.$$

For (7.26) to be satisfied, we require that

$$a_{k+2} = -\frac{a_k}{(k+1)(k+2)}$$

for all integers $k \geq 0$. In particular, we have

$$a_2 = -\frac{a_0}{2!}, \quad a_4 = \frac{a_0}{4!}, \quad a_6 = -\frac{a_0}{6!}, \quad \ldots$$

and

$$a_3 = -\frac{a_1}{3!}, \quad a_5 = \frac{a_1}{5!}, \quad a_7 = -\frac{a_1}{7!}, \quad \ldots.$$

Let a_0 and a_1 be arbitrary fixed numbers and let A be the greater of $|a_0|$ and $|a_1|$. It is easy to show that

$$|a_k| \leq \frac{A}{k!} \tag{7.27}$$

for all integers $k \geq 0$. As we know, the series

$$\sum_{k=0}^{\infty} \frac{A}{k!} |x|^k$$

converges for every $x \in \mathbb{R}$. Introducing the functions

$$\cos x = 1 - \frac{x^2}{2!} + \frac{x^4}{4!} - \cdots = \sum_{k=0}^{\infty} (-1)^k \frac{x^{2k}}{(2k)!}, \tag{7.28}$$

$$\sin x = \frac{x}{1!} - \frac{x^3}{3!} + \frac{x^5}{5!} - \cdots = \sum_{k=0}^{\infty} (-1)^k \frac{x^{2k+1}}{(2k+1)!}, \tag{7.29}$$

we see that the defining series also converge for every $x \in \mathbb{R}$. From (7.28) and (7.29) we can easily obtain the following equalities:

$$(\cos x)' = -\sin x \quad \text{and} \quad (\sin x)' = \cos x; \tag{7.30}$$

$$\cos 0 = 1 \quad \text{and} \quad \sin 0 = 0; \tag{7.31}$$

$$\cos(-x) = \cos x \quad \text{and} \quad \sin(-x) = -\sin x. \tag{7.32}$$

Now we give the definition of sine and cosine functions in terms of the equation (7.26).

Definition 7.9.1. The function f which satisfies the equation
$$f'' = -f \quad \text{with} \quad f(0) = 0 \quad \text{and} \quad f'(0) = 1$$
is called the *sine function* and denoted by $f(x) = \sin x$. Similarly, the function g which satisfies the equation
$$g'' = -g \quad \text{with} \quad g(0) = 1 \quad \text{and} \quad g'(0) = 0$$
is called the *cosine function* and denoted by $g(x) = \cos x$.

This definition requires a proof of consistency which would show that the functions sine and cosine not only exist but are unique. It is easy to check that the functions defined in (7.28) and (7.29) have the desired properties, so we have the existence. The uniqueness follows from general theorems in the theory of differential equations. We will present here a direct proof.

In order to prove the uniqueness we first show that, if the functions f and g satisfy
$$f'' = -f \quad \text{and} \quad g'' = -g, \tag{7.33}$$
then
$$f(0)g'(x+y) + f'(0)g(x+y) = f(x)g'(y) + f'(x)g(y) \tag{7.34}$$
for all $x, y \in \mathbb{R}$. In the proof we need not assume at all that the functions f and g have power series representations.

We introduce an auxiliary function
$$h(x) = f(x)g'(x_0 - x) + f'(x)g(x_0 - x), \tag{7.35}$$
where x_0 is an arbitrary fixed number. Then
$$h'(x) = -f(x)g''(x_0 - x) + f''(x)g(x_0 - x) = 0$$
because of (7.33). This shows that h is a constant function. Hence $h(0) = h(x)$, that is, in view of (7.35),
$$f(0)g'(x_0) + f'(0)g(x_0) = f(x)g'(x_0 - x) + f'(x)g(x_0 - x).$$
The above equality holds for all $x, x_0 \in \mathbb{R}$. Thus given any x and y we get the desired formula (7.34), by substituting $x_0 = x + y$.

Assume now that there are two solutions f_1 and f_2 of the equation
$$f'' = -f \tag{7.36}$$
satisfying the same initial conditions, that is, such that
$$f_1(0) = f_2(0) \quad \text{and} \quad f_1'(0) = f_2'(0).$$
Then the difference
$$h = f_1 - f_2$$
also satisfies (7.36) and we have $h(0) = h'(0) = 0$. Letting $g(x) = \sin x$ and replacing f by h in (7.34), we get
$$0 = h(x)\cos y + h'(x)\sin y.$$
If we let $y = 0$, we get $0 = h(x)$, that is, $f_1(x) = f_2(x)$. Thus the solution of the equation (7.33) is unique whenever the initial conditions are given. This completes the proof of cosistency of Definition 7.9.1.

S: The functions sine and cosine used in geometry are different. They are functions of an angle whereas the arguments of the functions defined here are numbers.

T: Right. But there is a simple connections between them. If we denote by $\mathrm{Sin}\,\alpha$ and $\mathrm{Cos}\,\alpha$ the functions of an angle α, then

$$\sin x = \mathrm{Sin}\,\alpha \quad \text{and} \quad \cos x = \mathrm{Cos}\,\alpha$$

whenever

$$\frac{x}{\pi} = \frac{\alpha}{180°}.$$

This relation is so simple that almost all formulas for both kinds of functions are the same.

S: I know that π is the ratio of the length of the circumference of a circle to its diameter. But can we give a precise analytical definition?

T: The number $\pi/2$ is the least positive number x such that $\cos x = 0$.

S: How do we know that such a number exists?

T: We have $\cos 0 = 1$ and

$$\cos 2 = 1 - \frac{2^2}{2!} + \frac{2^4}{4!} - \left(\frac{2^6}{6!} - \frac{2^8}{8!}\right) - \left(\frac{2^{10}}{10!} - \frac{2^{12}}{12!}\right) - \cdots$$

Since the sum of the three initial terms is $-\frac{1}{3}$ and all the remaining terms are negative, we have $\cos 2 < -\frac{1}{3}$. This implies, by Theorem 3.2.4, that $\cos x$ has a root in the interval $(0,2)$. This root will be denoted by $\pi/2$.

S: How do we know that there is only one root in $(0,2)$?

T: We have

$$\sin x = x\left(1 - \frac{x^2}{2\cdot 3}\right) + \frac{x^5}{5!}\left(1 - \frac{x^2}{6\cdot 7}\right) + \cdots$$

which shows that $\sin x > 0$ on the interval $(0,2)$. Since $-\sin x$ is the derivative of $\cos x$, it follows that $\cos x$ is decreasing in $(0,2)$ and therefore it assumes value 0 only once. So $\cos x$ has only one root in the interval $(0,2)$ and consequently the number $\pi/2$ is well-defined.

As we said $\sin x$ and $\mathrm{Sin}\,\alpha$ are different functions: for instance, $\sin \pi = 0$ and $\mathrm{Sin}\,\pi \approx 0.0548$. Many authors consider both as the same function but defined on two different domains and they say "x radians" when the number x is the argument of sin or cos. Such an approach is illogical. If sine is a function of radians, then expressions like $x + \sin x$ or $\sin(\sin x)$ make no sense. The concept of radians is completely unnecessary in mathematics.

By letting $f(x) = \sin x$ and $g(x) = \cos x$ in (7.34) we get the following identity

$$\cos(x + y) = \cos x \cos y - \sin x \sin y, \tag{7.37}$$

and by letting $f(x) = \sin x$ and $g(x) = \sin x$ we get

$$\sin(x + y) = \sin x \cos y + \cos x \sin y. \tag{7.38}$$

Substituting $-x$ for y in (7.37) yields another trigonometric identity:

$$1 = (\cos x)^2 + (\sin x)^2. \tag{7.39}$$

Since $\cos \frac{\pi}{2} = 0$ and $\sin \frac{\pi}{2} > 0$, the above implies $\sin \frac{\pi}{2} = 1$.

From (7.37) and (7.38) we obtain successively

$$\cos(x + \frac{\pi}{2}) = -\sin x, \qquad\qquad \sin(x + \frac{\pi}{2}) = \cos x,$$

$$\cos(x + \pi) = -\cos x, \qquad\qquad \sin(x + \pi) = -\sin x,$$

$$\cos(\pi - x) = -\cos x, \qquad\qquad \sin(\pi - x) = \sin x,$$

$$\cos(x + 2\pi) = \cos x, \qquad\qquad \sin(x + 2\pi) = \sin x.$$

The identities in the last line say that the functions cosine and sine are periodic with period 2π. Graphs of cosine and sine functions look like those in Fig. 7.6.

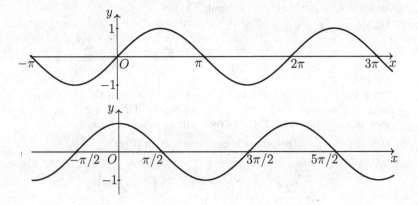

Fig. 7.6   Graphs of $y = \sin x$ and $y = \cos x$.

Equalities (7.37) and (7.38) are of crucial importance in trigonometry.

EXERCISES 7.9

(1) Prove that the function $f(x) = \dfrac{x}{1!} - \dfrac{x^3}{3!} + \dfrac{x^5}{5!} - \cdots = \displaystyle\sum_{k=0}^{\infty} (-1)^k \dfrac{x^{2k+1}}{(2k+1)!}$

satisfies the equation $f'' = -f$ and the conditions $f(0) = 0$, $f'(0) = 1$.

(2) Prove that the function $g(x) = 1 - \dfrac{x^2}{2!} + \dfrac{x^4}{4!} - \cdots = \displaystyle\sum_{k=0}^{\infty} (-1)^k \dfrac{x^{2k}}{(2k)!}$ satisfies

the equation $g'' = -g$ and the conditions $g(0) = 1$, $g'(0) = 0$.

(3) Prove (7.27).

(4) Prove: $\cos x \cos y = \frac{1}{2}(\cos(x + y) + \cos(x - y))$.

(5) Prove: $\cos x \sin y = \frac{1}{2}(\sin(x + y) - \sin(x - y))$.

(6) Prove: $\sin x \sin y = \frac{1}{2}(\cos(x - y) - \cos(x + y))$.

(7) Prove: $\sin\frac{\pi}{4} = \cos\frac{\pi}{4}$.

(8) Prove: $\cos x = 0$ if and only if $x = \frac{\pi}{2} + k\pi$, where k is an integer.

(9) Show that the function

$$f(x) = \begin{cases} x^2 \sin\frac{1}{x^3} & \text{if } x \neq 0, \\ 0 & \text{if } x = 0, \end{cases}$$

is differentiable at every point of the interval $(-1,1)$, but the derivative f' is not continuous in $(-1,1)$.

7.10 Tangent

The function *tangent* is defined by the formula

$$\tan x = \frac{\sin x}{\cos x}.$$

This function is defined for all x for which $\cos x \neq 0$, that is, everywhere except $x = \frac{\pi}{2} + k\pi$, where k is an arbitrary integer. We have

$$\tan 0 = 0 \quad \text{and} \quad \tan\frac{\pi}{4} = 1.$$

Moreover,

$$\tan x > 0 \quad \text{for} \quad 0 < x < \frac{\pi}{2},$$

$$\tan x < 0 \quad \text{for} \quad -\frac{\pi}{2} < x < 0,$$

and

$$\tan(-x) = -\tan x,$$

$$\tan(x + \pi) = \tan x,$$

because

$$\frac{\sin(x + \pi)}{\cos(x + \pi)} = \frac{-\sin x}{-\cos x}.$$

The graph $y = \tan x$ on the interval $(-\pi, \pi)$ is represented in Fig. 7.7.
Since

$$\left(\frac{\sin x}{\cos x}\right)' = \frac{\cos x(\sin x)' - \sin x(\cos x)'}{(\cos x)^2} = \frac{(\cos x)^2 + (\sin x)^2}{(\cos x)^2} = \frac{1}{(\cos x)^2},$$

we have

$$(\tan x)' = \frac{1}{(\cos x)^2}.$$

The derivative is positive everywhere in the domain and the function is increasing on every interval not containing points $\frac{\pi}{2} + k\pi$, where k is an arbitrary integer.

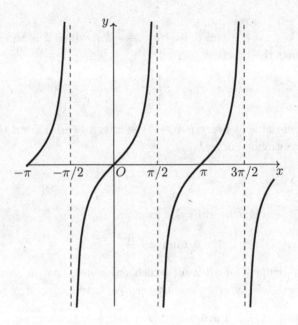

Fig. 7.7   Graph of $y = \tan x$.

Every real number is the value of the tangent function at some point of the interval $(-\frac{\pi}{2}, \frac{\pi}{2})$. Indeed, if a is an arbitrary real number, then there exist numbers $x_1, x_2 \in (-\frac{\pi}{2}, \frac{\pi}{2})$ such that

$$\tan x_1 < a < \tan x_2,$$

because

$$\lim_{x \to -\frac{\pi}{2}+} \tan x = -\infty \quad \text{and} \quad \lim_{x \to \frac{\pi}{2}-} \tan x = \infty.$$

Thus the assertion follows by Theorem 3.2.4.

We also note that

$$\tan(x + y) = \frac{\tan x + \tan y}{1 - \tan x \tan y}, \tag{7.40}$$

which holds for all x and y for which x, y and $x + y$ are different from $\frac{\pi}{2} + k\pi$ for any integer k.

EXERCISES 7.10

(1) Prove that $\lim_{x \to -\frac{\pi}{2}+} \tan x = -\infty$.
(2) Prove (7.40).

7.11 Arctangent

The tangent function is not invertible because, for example, $\tan 0 = \tan \pi$. To construct a function that has some of the properties of an inverse function for the tangent function, we have to restrict the domain of tangent to an interval on which it is a one-to-one function. To this end, we choose the interval $(-\frac{\pi}{2}, \frac{\pi}{2})$. The inverse function for tangent restricted to this interval is called the *arctangent* function and denoted by arctan. The function arctan is defined for all real numbers and its range is the interval $(-\frac{\pi}{2}, \frac{\pi}{2})$. The graph of arctangent is presented in Fig. 7.8.

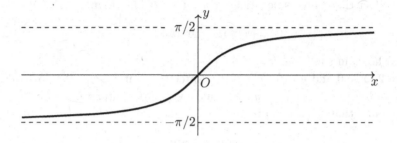

Fig. 7.8 Graph of $y = \arctan x$.

From properties of tangent we easily obtain the following properties of arctangent:

$$\arctan 0 = 0, \quad \arctan 1 = \frac{\pi}{4}, \tag{7.41}$$

$$0 < \arctan x < \frac{\pi}{2} \quad \text{for} \quad x > 0, \tag{7.42}$$

$$-\frac{\pi}{2} < \arctan x < 0 \quad \text{for} \quad x < 0, \tag{7.43}$$

$$\arctan(-x) = -\arctan x \quad \text{for all } x. \tag{7.44}$$

According to the definition of arctangent the equality $\arctan(\tan x) = x$ holds only for $-\frac{\pi}{2} < x < \frac{\pi}{2}$. On the other hand, the equality $\tan(\arctan x) = x$ holds for all $x \in \mathbb{R}$.

The following property of arctangent is useful:

$$\arctan x + \arctan y = \arctan \frac{x+y}{1-xy} \quad \text{whenever} \quad 1 - xy > 0. \tag{7.45}$$

It can be derived from (7.40) as follows. First note that

$$\tan(\arctan x + \arctan y) = \frac{\tan(\arctan x) + \tan(\arctan y)}{1 - \tan(\arctan x)\tan(\arctan y)} = \frac{x+y}{1-xy} \tag{7.46}$$

holds for all x, y. Hence

$$\arctan(\tan(\arctan x + \arctan y)) = \arctan \frac{x+y}{1-xy}. \tag{7.47}$$

The left-hand side can be replaced by $\arctan x + \arctan y$ provided

$$-\frac{\pi}{2} < \arctan x + \arctan y < \frac{\pi}{2}. \tag{7.48}$$

We will show that these inequalities hold whenever $1 - xy > 0$, that is, whenever $xy < 1$. To this end, suppose $xy < 1$. Since $0 < \arctan x < \frac{\pi}{2}$ for $x > 0$ and $-\frac{\pi}{2} < \arctan x < 0$ for $x < 0$, the inequalities in (7.48) hold if $xy \le 0$. If $xy > 0$, we consider two cases:

Case 1: $x > 0$ and $y > 0$. Then $0 < \arctan x + \arctan y < \pi$. Moreover, since $xy < 1$ we have $(x + y)/(1 - xy) > 0$, and by (7.46), $\tan(\arctan x + \arctan y) > 0$. From the fact that $\tan u$ is negative for $u \in (\frac{\pi}{2}, \pi)$, we obtain

$$0 < \arctan x + \arctan y < \frac{\pi}{2},$$

so (7.48) holds in this case.

Case 2: $x < 0$ and $y < 0$. Then $-\pi < \arctan x + \arctan y < 0$. Moreover, since $xy < 1$ we have $(x + y)/(1 - xy) < 0$, and by (7.46), $\tan(\arctan x + \arctan y) < 0$. From the fact that $\tan u$ is positive for $u \in (-\pi, -\frac{\pi}{2})$, we obtain

$$-\frac{\pi}{2} < \arctan x + \arctan y < 0,$$

so (7.48) also holds in this case.

We have proved that $1 - xy > 0$ implies (7.48). Consequently, (7.47) reduces to (7.45).

For all $x \in \mathbb{R}$ we have

$$(\arctan x)' = \frac{1}{1 + x^2}. \tag{7.49}$$

Indeed, using the general formula for the derivative of an inverse function we get

$$(\arctan x)' = (\cos(\arctan x))^2.$$

It thus suffices to prove that

$$(\cos(\arctan x))^2 = \frac{1}{1 + x^2}. \tag{7.50}$$

Denoting $\arctan x$ by y we have $x = \tan y$ and consequently

$$1 + x^2 = 1 + (\tan y)^2 = 1 + \left(\frac{\sin y}{\cos y}\right)^2 = \frac{(\cos y)^2 + (\sin y)^2}{(\cos y)^2}$$

$$= \frac{1}{(\cos y)^2} = \frac{1}{(\cos(\arctan x))^2}.$$

This gives us (7.50) and thus (7.49).

We are now going to show that

$$\arctan x = \frac{x}{1} - \frac{x^3}{3} + \frac{x^5}{5} - \frac{x^7}{7} + \cdots = \sum_{k=0}^{\infty} (-1)^k \frac{x^{2k+1}}{2k+1} \quad \text{for } |x| < 1. \tag{7.51}$$

The series is obviously convergent on $(-1, 1)$ because the absolute values of its terms are bounded by terms of the convergent series $\sum_{n=0}^{\infty} |x|^n$. Moreover

$$\left(\sum_{k=0}^{\infty} (-1)^k \frac{x^{2k+1}}{2k+1} \right)' = \sum_{k=0}^{\infty} (-1)^k x^{2k} = 1 - x^2 + x^4 - x^6 + \cdots = \frac{1}{1+x^2}$$

for $|x| < 1$. Thus, in view of (7.49), for all $x \in (-1, 1)$ we must have

$$\arctan x = C + \frac{x}{1} - \frac{x^3}{3} + \frac{x^5}{5} - \frac{x^7}{7} + \cdots = C + \sum_{k=0}^{\infty} (-1)^k \frac{x^{2k+1}}{2k+1},$$

where C is a constant. This number can be found by substituting $x = 0$ in the above equality which gives $C = 0$. This proves (7.51).

In Section 8.7 we will prove a theorem that will imply that (7.51) holds also for $x = 1$.

EXERCISES 7.11.

(1) Find derivatives of the following functions:

(a) $f(x) = (\tan \sqrt{x})^4$,

(b) $f(x) = \ln \left(\tan \left(\frac{1}{4}\pi + \frac{1}{2}x \right) \right)$,

(c) $f(x) = (\tan x)^{\sin x}$,

(d) $f(x) = x^4 \arctan x$,

(e) $f(x) = \arctan \sqrt{1 - x^2}$,

(f) $f(x) = \sin(\arctan x)$.

(2) Prove (7.41).

(3) Prove (7.42).

(4) Prove (7.43).

(5) Prove (7.44).

(6) Prove that $\arctan x + \arctan \dfrac{1}{x} = \dfrac{\pi}{2}$ for all $x > 0$.

(7) Is there a property similar to the one given in the preceding exercise valid for all $x < 0$? Justify your response.

(8) Prove that $\arctan x + \arctan \dfrac{1}{x} = \dfrac{\pi}{2}$ for all $x > 0$ using derivatives.

7.12 Machin's formula

The power series expansion for arctangent given in (7.51) and the formula (7.45) enable us to evaluate π with arbitrary accuracy. Indeed, we have

$$2 \arctan \frac{1}{5} = \arctan \frac{\frac{1}{5} + \frac{1}{5}}{1 - \frac{1}{25}} = \arctan \frac{5}{12},$$

$$4 \arctan \frac{1}{5} = \arctan \frac{\frac{5}{12} + \frac{5}{12}}{1 - \frac{25}{144}} = \arctan \left(1 + \frac{1}{119} \right),$$

$$4 \arctan \frac{1}{5} - \arctan 1 = \arctan \frac{\left(1 + \frac{1}{119} \right) - 1}{1 + \left(1 + \frac{1}{119} \right)} = \arctan \frac{1}{239}.$$

Since $\arctan 1 = \dfrac{\pi}{4}$, the above equality implies

$$\frac{\pi}{4} = 4\arctan\frac{1}{5} - \arctan\frac{1}{239}.$$

Hence

$$\pi = 16\left(\frac{1}{5} - \frac{1}{3\cdot 5^3} + \frac{1}{5\cdot 5^5} - \cdots\right) - 4\left(\frac{1}{239} - \frac{1}{3\cdot 239^3} + \cdots\right)$$

$$= 16\sum_{n=0}^{\infty}\frac{(-1)^n}{(2n+1)5^{2n+1}} - 4\sum_{n=0}^{\infty}\frac{(-1)^n}{(2n+1)239^{2n+1}}.$$

(7.52)

In order to find an approximate value of π let us take six terms of the first series and only two terms of the second series. Then, according to Theorem 5.5.4, the error will be less than

$$16\frac{1}{13\cdot 5^{13}} + 4\frac{1}{5\cdot 239^5}.$$

(7.53)

Let

$$A = 16\left(\frac{1}{5} - \frac{1}{3\cdot 5^3} + \frac{1}{5\cdot 5^5} - \frac{1}{7\cdot 5^7} + \frac{1}{9\cdot 5^9} - \frac{1}{11\cdot 5^{11}}\right),$$

and

$$B = 4\left(\frac{1}{239} - \frac{1}{3\cdot 239^3}\right).$$

From (7.53) and Theorem 5.5.4 we obtain

$$3.1415926526 \approx A - B < \pi < A - B + 16\frac{1}{13\cdot 5^{13}} + 4\frac{1}{5\cdot 239^5} \approx 3.1415926536.$$

Thus,

$$\pi \approx 3.14159265.$$

Formula (7.52) was given by John Machin (1680-1751) who found π with a one hundred place accuracy when he was 21 years old.

7.13 Newton's binomial

The function $f(x) = (1+x)^\alpha$, where α is an arbitrary real number, is defined for all $x > -1$. Assume that this function expands in a power series:

$$(1+x)^\alpha = c_0 + c_1 x + c_2 x^2 + \cdots = \sum_{k=0}^{\infty} c_k x^k.$$

(7.54)

In order to find the coefficients in the above series we differentiate the above equality and obtain

$$\alpha(1+x)^{\alpha-1} = c_1 + 2c_2 x + 3c_3 x^2 + \cdots = \sum_{k=1}^{\infty} k c_k x^{k-1}.$$

Hence

$$(1+x)(c_1 + 2c_2x + 3c_3x^2 + \ldots) = \alpha(c_0 + c_1x + c_2x^2 + \ldots),$$

that is,

$$c_1 + 2c_2x + 3c_3x^2 + \cdots + c_1x + 2c_2x^2 + 3c_3x^3 + \ldots$$
$$= \alpha c_0 + \alpha c_1x + \alpha c_2x^2 + \alpha c_3x^3 + \ldots$$

From (7.54) we know that $c_0 = 1$. Then, by comparing coefficients of terms with the same powers of x we find

$$c_1 = \frac{\alpha}{1}, \quad c_2 = \frac{\alpha(\alpha - 1)}{1 \cdot 2}, \quad c_3 = \frac{\alpha(\alpha - 1)(\alpha - 2)}{1 \cdot 2 \cdot 3},$$

and in general

$$c_k = \frac{\alpha(\alpha - 1) \ldots (\alpha - k + 1)}{k!}. \tag{7.55}$$

To find the interval of convergence we will use the following general theorem.

Theorem 7.13.1. *If the limit*

$$\lim_{k \to \infty} \left| \frac{c_k}{c_{k+1}} \right| = r$$

exists, then r is the radius of convergence of the series

$$\sum_{k=0}^{\infty} c_k x^k. \tag{7.56}$$

Proof. Let x be an arbitrary number such that $|x| < r$. For $|x| < q < r$ there exists a natural number p such that $|c_k/c_{k+1}| > q$ for all $k \geq p$, that is,

$$|c_{k+1}| < \frac{|c_k|}{q} \quad \text{for } k = p, p+1, p+2, \ldots.$$

Hence

$$|c_{p+1}| < \frac{|c_p|}{q}, \quad |c_{p+2}| < \frac{|c_{p+1}|}{q} < \frac{|c_p|}{q^2},$$

and in general, by induction,

$$|c_{p+k}| < \frac{|c_p|}{q^k} \quad \text{for} \quad k = 1, 2, 3, \ldots.$$

Thus

$$|c_{p+k}x^{p+k}| \leq |c_p x^p| \left| \frac{x}{q} \right|^k \quad \text{for} \quad k = 1, 2, 3, \ldots$$

which implies the convergence of the series (7.56) by comparison with the series $\sum_{k=0}^{\infty} |c_p x^p| \left| \dfrac{x}{q} \right|^k$ which is a constant multiple of a geometric series.

Now, let $|x| > r$. Then there exists a number q such that $r < q < |x|$. By a reasoning similar to the one above we can show that

$$|c_{p+k}x^{p+k}| \geq |c_p x^p| \left|\frac{x}{q}\right|^k \quad \text{for} \quad k = 1, 2, 3, \ldots. \tag{7.57}$$

If the series (7.56) were convergent at x, then also $\sum_{k=0}^{\infty} |c_p x^p| \left|\frac{x}{q}\right|^k$ would also be convergent, which is not true.

We can conclude that the radius of convergence of series (7.56) is r. $\qquad \square$

Theorem 7.13.1 will now be applied to the series (7.54). In this case we have

$$\lim_{k \to \infty} \left|\frac{c_k}{c_{k+1}}\right| = \lim_{k \to \infty} \left|\frac{k+1}{\alpha - k}\right| = 1.$$

So far we have shown that, if (7.54) is true, then the c_k's must be chosen so that $c_0 = 1$ and, for $k \geq 1$, c_k is given by (7.55), and that, if the c_k's are, in fact chosen this way, then the radius of convergence of the series in (7.54) is $r = 1$. To complete the task we must show that, with this choice for the c_k's, (7.54) really is true. This will be the purpose of Exercises 1 and 2 at the end of this section.

Using the notation

$$\frac{\alpha(\alpha-1)\ldots(\alpha-k+1)}{1 \cdot 2 \ldots k} = \binom{\alpha}{k}$$

the *binomial formula* can be written in the form

$$(1+x)^\alpha = \sum_{k=0}^{\infty} \binom{\alpha}{k} x^k \quad \text{for} \ |x| < 1,$$

where we let $\binom{\alpha}{0} = 1$.

If $\alpha = m$ is a natural number, then $\binom{m}{k} = 0$ for $k > m$ and consequently the last formula reduces to a polynomial, namely

$$(1+x)^m = \sum_{k=0}^{m} \binom{m}{k} x^k,$$

and is valid for all x, because if two polynomials are equal for $|x| < 1$, they are equal for every x.

If a and b are arbitrary real numbers and $a \neq 0$, we may write

$$(a+b)^m = a^m \left(1 + \frac{b}{a}\right)^m = a^m \sum_{k=0}^{m} \binom{m}{k} \left(\frac{b}{a}\right)^k$$

and hence

$$(a+b)^m = \sum_{k=0}^{m} \binom{m}{k} a^{m-k} b^k$$

or

$$(a+b)^m = \binom{m}{0}a^m + \binom{m}{1}a^{m-1}b + \binom{m}{2}a^{m-2}b^2 + \cdots + \binom{m}{m}b^m. \qquad (7.58)$$

Since the formula also holds for $a = 0$, it holds for any real numbers a and b.

For initial values of m we get

$$(a+b)^1 = a + b,$$
$$(a+b)^2 = a^2 + 2ab + b^2,$$
$$(a+b)^3 = a^3 + 3a^2b + 3ab^2 + b^3,$$
$$(a+b)^4 = a^4 + 4a^3b + 6a^2b^2 + 4ab^3 + b^4.$$

The coefficients can be arranged into the so-called *Pascal's triangle* (Blaise Pascal 1623–1662):

$$
\begin{array}{ccccccccccc}
 & & & & & 1 & & & & & \\
 & & & & 1 & & 1 & & & & \\
 & & & 1 & & 2 & & 1 & & & \\
 & & 1 & & 3 & & 3 & & 1 & & \\
 & 1 & & 4 & & 6 & & 4 & & 1 & \\
1 & & 5 & & 10 & & 10 & & 5 & & 1
\end{array}
$$

in which every entry is the sum of the two numbers standing just above.

From (7.58) we can easily get various interesting properties of the symbol $\binom{m}{k}$. Taking, for instance, $a = b = 1$ we get

$$2^m = \binom{m}{0} + \binom{m}{1} + \cdots + \binom{m}{m}$$

which would be difficult to deduce directly from the definition of the symbol $\binom{m}{k}$.

Since the radius of convergence of the series in (7.54) is 1, the series converges for every $x \in (-1, 1)$. In particular for $\alpha = -\frac{1}{2}$ we have

$$\frac{1}{\sqrt{1+x}} = 1 - \frac{1}{2}x + \frac{1 \cdot 3}{2 \cdot 4}x^2 - \frac{1 \cdot 3 \cdot 5}{2 \cdot 4 \cdot 6}x^3 + \ldots.$$

This formula can serve for approximate evaluation of square roots. For instance, we may write

$$\sqrt{2} = \frac{7}{5}\left(1 - \frac{1}{50}\right)^{-\frac{1}{2}} = \frac{7}{5}\left(1 + \frac{1}{2} \cdot \frac{1}{50} + \frac{1 \cdot 3}{2 \cdot 4} \cdot \frac{1}{50^2} + \frac{1 \cdot 3 \cdot 5}{2 \cdot 4 \cdot 6} \cdot \frac{1}{50^3} + \ldots\right).$$

Faster convergence would be obtained by using

$$\sqrt{2} = \frac{141}{100}\left(1 - \frac{119}{20000}\right)^{-\frac{1}{2}}.$$

or any other equality which we obtain by taking a rough approximation α of $\sqrt{2}$ (in the above example 1.41) and writing

$$\sqrt{2} = \alpha\left(1 + \frac{\alpha^2 - 2}{2}\right)^{-\frac{1}{2}}.$$

Similarly, using the fact $\sqrt{3} \approx 1.732$ we are lead to the equality

$$\sqrt{3} = 1.732 \left(1 - \frac{176}{3000000}\right)^{-\frac{1}{2}}$$

which allows us to calculate 50 decimal places of $\sqrt{3}$ without particular pain. Such an accuracy is not available on ordinary calculators.

The same method can also be applied to roots of higher order. For example, we can use equalities

$$\sqrt[3]{3} = \frac{10}{7}\left(1 + \frac{29}{1000}\right)^{-\frac{1}{3}} \quad \text{or} \quad \sqrt[5]{1000} = 4\left(1 + \frac{24}{1000}\right)^{-\frac{1}{5}}.$$

EXERCISES 7.13

(1) Prove that $y = \sum_{k=0}^{\infty} \binom{\alpha}{k} x^k$ satisfies both the differential equation

$$(1+x)y' = \alpha y$$

and the condition $y(0) = 1$.

(2) Prove: If $y = f(x)$ satisfies both the differential equation and the condition of Exercise 1, then $f(x) = (1+x)^\alpha$. (Hint: Show that $f(x)/(1+x)^\alpha$ is a constant function.) Refer to Exercise 1 to prove that $(1+x)^\alpha = \sum_{k=0}^{\infty} \binom{\alpha}{k} x^k$ for $|x| < 1$.

(3) Calculate $\sqrt{2}$ up to 6 decimal places and check the results by a calculator.

(4) Calculate $\sqrt{3}$ up to 6 decimal places and check the results by a calculator.

(5) Calculate $\sqrt[3]{3}$ up to 6 decimal places and check the results by a calculator.

(6) Prove: $\binom{m}{0} - \binom{m}{1} + \binom{m}{2} - \cdots + (-1)^m \binom{m}{m} = 0$ for $m \in \mathbb{N}$.

(7) Prove: $\binom{m}{0} + \binom{m}{2} + \binom{m}{4} + \cdots + \binom{m}{m} = 2^{m-1}$ for even $m \in \mathbb{N}$.

(8) Find the radius of convergence for the following series:

(a) $\displaystyle\sum_{k=1}^{\infty} \frac{k^2 x^k}{2^k}$,

(c) $\displaystyle\sum_{k=1}^{\infty} (-3)^k x^{2k}$,

(b) $\displaystyle\sum_{k=1}^{\infty} \frac{(-1)^k}{\sqrt{k}} x^k$,

(d) $\displaystyle\sum_{k=1}^{\infty} \frac{e^k}{k!} x^k$.

7.14　Arccosine and arcsine

The functions cosine and sine are invertible on any interval on which they are monotone. For the cosine function we choose the interval $[0, \pi]$ and for the sine function the interval $[-\frac{\pi}{2}, \frac{\pi}{2}]$. The inverse functions of these restrictions are called *arccosine* and *arcsine* and are denoted by

$$\arccos x \quad \text{and} \quad \arcsin x$$

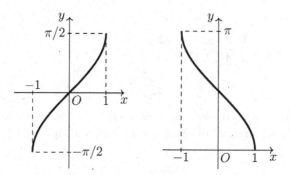

Fig. 7.9   Graphs of $y = \arcsin x$ and $y = \arccos x$.

respectively (see Fig. 7.9). Both are defined on the interval $[-1, 1]$ and we have

$$0 \le \arccos x \le \pi \quad \text{and} \quad -\frac{\pi}{2} \le \arcsin x \le \frac{\pi}{2}.$$

Since, for $-1 \le x \le 1$, we have

$$\sin(\arcsin x) = x = \cos(\arccos x) = \sin(\frac{\pi}{2} - \arccos x),$$

it follows that $\arcsin x = \frac{\pi}{2} - \arccos x$, that is,

$$\arccos x + \arcsin x = \frac{\pi}{2} \quad \text{for} \ -1 \le x \le 1. \tag{7.59}$$

Now we will prove the formulas for derivatives of arccosine and arcsine:

$$(\arccos x)' = -\frac{1}{\sqrt{1 - x^2}} \quad \text{and} \quad (\arcsin x)' = \frac{1}{\sqrt{1 - x^2}}. \tag{7.60}$$

In view of (7.59) it suffices to prove one of them. By the general theorem on differentiation of inverse functions, we have

$$(\arcsin x)' = \frac{1}{\cos(\arcsin x)}. \tag{7.61}$$

Denoting $\arcsin x$ by y we have $x = \sin y$ and hence

$$\sqrt{1 - x^2} = \sqrt{1 - (\sin y)^2} = \sqrt{(\cos y)^2} = \cos y,$$

because $\cos y \ge 0$ for $-\frac{\pi}{2} \le y \le \frac{\pi}{2}$. Thus

$$\sqrt{1 - x^2} = \cos(\arcsin x),$$

which proves (7.60).

To expand the arcsine function into a power series we use the binomial formula. Since

$$\frac{1}{\sqrt{1 - x^2}} = 1 + \frac{1}{2}x^2 + \frac{1 \cdot 3}{2 \cdot 4}x^4 + \frac{1 \cdot 3 \cdot 5}{2 \cdot 4 \cdot 6}x^6 + \dots \tag{7.62}$$

for $|x| < 1$, we have

$$\left(\frac{x}{1} + \frac{1}{2}\cdot\frac{x^3}{3} + \frac{1\cdot3}{2\cdot4}\cdot\frac{x^5}{5} + \frac{1\cdot3\cdot5}{2\cdot4\cdot6}\cdot\frac{x^7}{7} + \ldots\right)'$$

$$= 1 + \frac{1}{2}x^2 + \frac{1\cdot3}{2\cdot4}x^4 + \frac{1\cdot3\cdot5}{2\cdot4\cdot6}x^6 + \cdots = \frac{1}{\sqrt{1-x^2}}.$$

(Note that the differentiated series converges for $|x| < 1$ because the absolute values of its terms are less than those of the series in (7.62) multiplied by $|x|$.) This implies that

$$\arcsin x = C + \frac{x}{1} + \frac{1}{2}\frac{x^3}{3} + \frac{1\cdot3}{2\cdot4}\frac{x^5}{5} + \frac{1\cdot3\cdot5}{2\cdot4\cdot6}\frac{x^7}{7} + \ldots \quad \text{for} \quad |x| < 1.$$

The constant C can be easily established to be 0, because for $x = 0$ both functions vanish. Thus we have

$$\arcsin x = \frac{x}{1} + \frac{1}{2}\frac{x^3}{3} + \frac{1\cdot3}{2\cdot4}\cdot\frac{x^5}{5} + \frac{1\cdot3\cdot5}{2\cdot4\cdot6}\cdot\frac{x^7}{7} + \ldots \quad \text{for} \quad |x| < 1.$$

EXERCISES 7.14

(1) Find the power series for $\arccos x$?

(2) Expand the function $\ln\left(x + \sqrt{1 + x^2}\right)$ into a power series.

(3) Find derivatives of the following functions:

    (a) $f(x) = \arcsin\sqrt{x^3}$,   (c) $f(x) = x^3\arcsin\frac{1}{x}$,

    (b) $f(x) = \arccos\sqrt{1 - x}$,   (d) $f(x) = \sin(\arccos x)$.

(4) Find the derivative of the function

$$f(x) = \sin(\arccos x) - \cos(\arcsin x).$$

Chapter 8

LIMITS REVISITED

8.1 Improper limits

By an *improper limit* we mean a limit, the value of which is not a number but, ∞ or $-\infty$.

Definition 8.1.1. We say that a function f has at a the right-hand limit equal to ∞, and we write

$$f(a+) = \infty \quad or \quad \lim_{x \to a+} f(x) = \infty,$$

if for every number A there exists a number $\varepsilon > 0$ such that $f(x) > A$ for every x satisfying $a < x < a + \varepsilon$. Similarly, we say that a function f has at a the right-hand limit equal $-\infty$, and we write

$$f(a+) = -\infty \quad or \quad \lim_{x \to a+} f(x) = -\infty,$$

if for every number A there exists a number $\varepsilon > 0$ such that $f(x) < A$ for every x satisfying $a < x < a + \varepsilon$.

The left-hand limits are defined in a similar manner, we only need to replace inequalities $a < x < a + \varepsilon$ by $a - \varepsilon < x < a$ and the symbol $a+$ by $a-$.

For instance,

$$\lim_{x \to a+} \frac{1}{x - a} = \infty, \qquad\qquad \lim_{x \to a+} \frac{1}{a - x} = -\infty,$$

$$\lim_{x \to a-} \frac{1}{x - a} = -\infty, \qquad\qquad \lim_{x \to a-} \frac{1}{a - x} = \infty.$$

We can also speak of two-sided improper limits and write $\lim_{x \to a} f(x) = \infty$ whenever $f(a+) = f(a-) = \infty$, and similarly for $-\infty$.

Assume that $f(x) = \frac{1}{g(x)} > 0$ holds for $x \in (a, b)$. Then $g(a+) = 0$ is equivalent to $f(a+) = \infty$. This equivalence follows directly from the definition of improper limits. Similarly, $g(b-) = 0$ is equivalent to $f(b-) = \infty$. On the other hand, if $f(x) = \frac{1}{g(x)} < 0$ in $x \in (a, b)$, then $g(a+) = 0$ is equivalent to $f(a+) = -\infty$, and $g(b-) = 0$ is equivalent to $f(b-) = -\infty$.

We can also define improper limits at infinity.

Definition 8.1.2. We say that a function f has the limit ∞ at infinity if for every number A there exists a number δ such that $f(x) > A$ for all $x > \delta$. We then write

$$f(\infty) = \infty \quad or \quad \lim_{x \to \infty} f(x) = \infty. \tag{8.1}$$

Similarly, we say that a function f has the limit $-\infty$ at infinity if for every number A there exists a number δ such that $f(x) < A$ for all $x > \delta$. We then write

$$f(\infty) = -\infty \quad or \quad \lim_{x \to \infty} f(x) = -\infty. \tag{8.2}$$

For example,

$$e^{\infty} = \infty,$$

$$\ln \infty = \infty, \quad \log_{\frac{1}{2}} \infty = -\infty,$$

$$\infty^{\alpha} = \infty \quad \text{for } \alpha > 0,$$

$$\infty^{\alpha} = 0 \quad \text{for } \alpha < 0.$$

Note that $\lim_{x \to \infty} f(x) = \infty$ is equivalent to $\lim_{x \to 0+} f(1/x) = \infty$, and $\lim_{x \to \infty} f(x) = -\infty$ is equivalent to $\lim_{x \to 0+} f(1/x) = -\infty$.

EXERCISES 8.1

(1) Use Definition 8.1.1 to prove the following:

(a) $\displaystyle \lim_{x \to 0+} \frac{1}{x} = \infty,$

(b) $\displaystyle \lim_{x \to 0-} \frac{1}{x} = -\infty,$

(c) $\displaystyle \lim_{x \to 0+} \ln x = -\infty,$

(d) $\displaystyle \lim_{x \to \frac{\pi}{2}-} \tan x = \infty,$

(e) $\displaystyle \lim_{x \to \frac{\pi}{2}+} \tan x = -\infty.$

(2) Use Definition 8.1.2 to prove the following:

(a) $e^{\infty} = \infty,$

(b) $\ln \infty = \infty,$

(c) $\log_{\frac{1}{2}} \infty = -\infty,$

(d) $\infty^{\alpha} = \infty$ for $\alpha > 0,$

(e) $\infty^{\alpha} = 0$ for $\alpha < 0.$

(3) Prove: $\displaystyle \lim_{x \to \infty} f(x) = \infty$ if and only if $\displaystyle \lim_{x \to 0+} f(1/x) = \infty.$

8.2   The easy l'Hospital's theorem

When evaluating limits the so called *l'Hospital's rule* (Guillaume Francois Antoine de l'Hospital, 1661–1704) can often be used successfully. It has two forms, one with a relatively easy proof and one with a quite difficult proof.

Theorem 8.2.1. *Let f and g be differentiable on some interval $(a - \varepsilon, a + \varepsilon)$ where $\varepsilon > 0$. Then*

$$\lim_{x \to a} \frac{f(x)}{g(x)} = \lim_{x \to a} \frac{f'(x)}{g'(x)}$$

whenever $\lim\limits_{x\to a} f(x) = \lim\limits_{x\to a} g(x) = 0$ *and the limit* $\lim\limits_{x\to a} \dfrac{f'(x)}{g'(x)}$ *(proper or improper) exists.*

This theorem is true for the left-hand limit, the right-hand limit, and the limit at infinity, that is, when a is replaced by $a-$, $a+$, or ∞, respectively. In the last case the interval is of the form (c, ∞).

Proof. The proof will be carried out first for the right-hand limit. To this end, assume that the functions $\dfrac{f}{g}$ and $\dfrac{f'}{g'}$ are defined on an interval $(a, t) \subseteq (a, a + \varepsilon)$. We introduce an auxiliary function

$$\varphi(x) = \begin{cases} f(t)g(x) - f(x)g(t), & \text{if } x \in (a, t], \\ 0, & \text{if } x = a. \end{cases}$$

Note that φ is continuous on $[a, t]$ and $\varphi(a) = \varphi(t) = 0$. Since φ is also differentiable on $(a, a + \varepsilon)$, by the mean value theorem, we have $\varphi'(\xi) = 0$ for some $\xi \in (a, t)$. This means that

$$f(t)g'(\xi) - f'(\xi)g(t) = 0$$

or equivalently

$$\frac{f(t)}{g(t)} = \frac{f'(\xi)}{g'(\xi)}.$$

Hence

$$\lim_{t\to a+} \frac{f(t)}{g(t)} = \lim_{\xi\to a+} \frac{f'(\xi)}{g'(\xi)},$$

whenever the second limit exists.

Replacing, in the above proof, the intervals $[a, t]$, $(a, t]$, and $(a, a + \varepsilon)$ by $[t, a]$, (t, a), and $(a - \varepsilon, a)$, respectively, and the symbol $a+$ by $a-$ we obtain the proof for the left-hand limit. Hence, the theorem also holds for the two-sided limits.

Finally, in the case $a = \infty$, using the first part of the proof, we have

$$\lim_{x\to\infty} \frac{f(x)}{g(x)} = \lim_{x\to 0+} \frac{f(\frac{1}{x})}{g(\frac{1}{x})} = \lim_{x\to 0+} \frac{f'(\frac{1}{x})(-\frac{1}{x^2})}{g'(\frac{1}{x})(-\frac{1}{x^2})} = \lim_{x\to 0+} \frac{f'(\frac{1}{x})}{g'(\frac{1}{x})} = \lim_{x\to\infty} \frac{f'(x)}{g'(x)}.$$

$\square$

For example,

$$\lim_{x\to 0} \frac{\sin x}{x} = \lim_{x\to 0} \frac{\cos x}{1} = 1,$$

$$\lim_{x\to 0} \frac{a^x - 1}{x} = \lim_{x\to 0} \frac{a^x \ln a}{1} = \ln a, \text{ for } a > 0,$$

$$\lim_{x\to 0+} \frac{1 - \cos x}{x} = \lim_{x\to 0+} \frac{\sin x}{1} = 0.$$

EXERCISES 8.2

(1) Find the limits:

(a) $\lim_{x\to 1} \frac{x^3-1}{x^3-2x^2+2x-1}$,

(e) $\lim_{x\to 0} \frac{\sin x - x\cos x}{(\sin x)^3}$,

(b) $\lim_{x\to a+} \frac{\sqrt{x^3-a^3}}{\sqrt{x-a}}$,

(f) $\lim_{x\to 0} \frac{e^x - e^{\sin x}}{x-\sin x}$,

(c) $\lim_{x\to 0} \frac{a^x-b^x}{x}$ for $a, b > 0$,

(g) $\lim_{x\to 0} \frac{x^2\sin(\frac{1}{x})}{\sin x}$.

(d) $\lim_{x\to 0} \frac{e^x-e^{-x}}{\sin x}$,

8.3   The difficult l'Hospital's theorem

In the second version of l'Hospital's theorem the assumptions that $\lim_{x\to a} f(x) = \lim_{x\to a} g(x) = 0$ is replaced by the assumption that $\lim_{x\to a} g(x) = \infty$.

Theorem 8.3.1. *Let f and g be differentiable on some interval $(a - \varepsilon, a + \varepsilon)$. We have*

$$\lim_{x\to a} \frac{f(x)}{g(x)} = \lim_{x\to a} \frac{f'(x)}{g'(x)}$$

whenever $\lim_{x\to a} g(x) = \infty$ *and the limit* $\lim_{x\to a} \frac{f'(x)}{g'(x)}$ *(proper or improper) exists.*

   This theorem is also true for the left-hand limit, the right-hand limit, as well as for the limit at infinity, that is, when a is replaced by $a-$, $a+$, or ∞, respectively. In the last case the interval is of the form (c, ∞).

Proof. First we prove the theorem under the assumption that $a = \infty$ and $\lim_{x\to\infty} f'(x)/g'(x) = L$, where L is a finite number. Let ε be an arbitrary positive number. There exists a number x_0 such that

$$g(x) > 0, \quad g'(x) \neq 0, \quad \text{and} \quad \frac{f'(x)}{g'(x)} < L + \frac{\varepsilon}{2} \quad \text{for } x \geq x_0.$$

By Theorem 4.11.5, the derivative $g'(x)$ cannot change sign for $x \geq x_0$ and thus $g'(x) > 0$, because $\lim_{x\to\infty} g(x) = \infty$. Hence

$$f'(x) - \left(L + \frac{\varepsilon}{2}\right)g'(x) < 0 \quad \text{for} \quad x \geq x_0, \tag{8.3}$$

and furthermore

$$f(x) - f(x_0) - \left(L + \frac{\varepsilon}{2}\right)(g(x) - g(x_0)) < 0 \quad \text{for} \quad x > x_0, \tag{8.4}$$

because the left-hand side of (8.4) equals 0 at $x = x_0$ and has (8.3) for its derivative. From (8.4) we obtain

$$\frac{f(x)}{g(x)} < L + \frac{\varepsilon}{2} + \frac{f(x_0) - (L + \frac{\varepsilon}{2})g(x_0)}{g(x)} \quad \text{for} \quad x > x_0.$$

Since $g(\infty) = \infty$, there exists a number $x_1 \geq x_0$ such that

$$\frac{f(x_0) - \left(L + \frac{\varepsilon}{2}\right)g(x_0)}{g(x)} < \frac{\varepsilon}{2}$$

for $x \geq x_1$. Consequently

$$\frac{f(x)}{g(x)} < L + \varepsilon \quad \text{for} \quad x \geq x_1. \tag{8.5}$$

Similarly we can prove existence of a number x_2 such that

$$\frac{f(x)}{g(x)} > L - \varepsilon \quad \text{for} \quad x \geq x_2. \tag{8.6}$$

To accomplish this it suffices to change, in the above argument, the sign $<$ to $>$ and replace $\frac{\varepsilon}{2}$ by $-\frac{\varepsilon}{2}$.

From (8.5) and (8.6) we have $\lim_{x\to\infty} \frac{f(x)}{g(x)} = L$, proving the theorem in the case when $a = \infty$.

Now we consider the case when $\lim_{x\to\infty} \frac{f'(x)}{g'(x)} = \infty$. Then there exists a number x_0 such that

$$g(x) > 0 \quad \text{and} \quad \frac{f'(x)}{g'(x)} > 1 \quad \text{for} \quad x \geq x_0.$$

Hence we have

$$f'(x) - g'(x) > 0 \quad \text{for} \quad x \geq x_0,$$

and consequently

$$f(x) - g(x) - (f(x_0) - g(x_0)) > 0 \quad \text{for} \quad x > x_0,$$

that is,

$$f(x) > g(x) - g(x_0) + f(x_0) \quad \text{for} \quad x > x_0.$$

This shows that $\lim_{x\to\infty} f(x) = \infty$. Since $\lim_{x\to\infty} \frac{g'(x)}{f'(x)} = 0$, we have $\lim_{x\to\infty} \frac{g(x)}{f(x)} = 0$, by the first part of this proof. But then $\lim_{x\to\infty} \frac{f(x)}{g(x)} = \infty$.

The case when $\lim_{x\to\infty} \frac{f(x)}{g(x)} = -\infty$ can be reduced to the previous one by considering the quotient $-f(x)/g(x)$. In this way the theorem is proved in the whole extent for the limit at infinity.

For the left-hand and right-hand limits, the assertion is obtained by considering the limits

$$\lim_{x\to a+} \frac{f(x)}{g(x)} = \lim_{x\to\infty} \frac{f(a+1/x)}{g(a+1/x)}$$

and

$$\lim_{x\to a-} \frac{f(x)}{g(x)} = \lim_{x\to\infty} \frac{f(a-1/x)}{g(a-1/x)}$$

and hence the assertion follows for two-sided limits. $\qquad \square$

For example, we have

$$\lim_{x \to \infty} \frac{\ln x}{x^a} = \lim_{x \to \infty} \frac{\frac{1}{x}}{ax^{a-1}} = \lim_{x \to \infty} \frac{1}{ax^a} = 0 \quad \text{for } a > 0$$

and

$$\lim_{x \to \infty} \frac{e^x}{x} = \lim_{x \to \infty} \frac{e^x}{1} = \infty.$$

More generally, for any real α we have

$$\lim_{x \to \infty} \frac{e^x}{x^\alpha} = \infty,$$

which is obvious for $\alpha \le 0$. For $\alpha > 0$ we first note that

$$\frac{e^x}{x^\alpha} = \left(\frac{e^{x/\alpha}}{x} \right)^\alpha.$$

Hence we get

$$\lim_{x \to \infty} \frac{e^x}{x^\alpha} = \left(\lim_{x \to \infty} \frac{e^{x/\alpha}}{x} \right)^\alpha = \left(\lim_{x \to \infty} \frac{\frac{1}{\alpha} e^{x/\alpha}}{1} \right)^\alpha = \infty.$$

In the above example we first use an auxiliary transformation and then l'Hospital's rule. A direct application of l'Hospital's rule leads to some difficulties.

Using induction we can also prove that

$$\lim_{x \to \infty} \frac{e^x}{x^n + a_{n-1}x^{n-1} + \cdots + a_1 x + a_0} = \infty,$$

for any $n \in \mathbb{N}$ and $a_{n-1}, \ldots, a_0 \in \mathbb{R}$, with at least one of the numbers $a_{n-1}, \ldots, a_0$ different form 0.

EXERCISES 8.3

(1) Find $\lim_{x \to \infty} \frac{(\ln x)^\beta}{x^\alpha}$ ($\alpha > 0$, β arbitrary).

(2) Find $\lim_{x \to 0+} \frac{\ln \tan x}{\ln \tan 2x}$.

(3) Find $\lim_{x \to \infty} \frac{x^2}{2^x}$.

(4) Prove that for any $n \in \mathbb{N}$ and any $a_{n-1}, \ldots, a_0 \in \mathbb{R}$, with at least one of the numbers $a_{n-1}, \ldots, a_0$ different form 0, we have

$$\lim_{x \to \infty} \frac{e^x}{x^n + a_{n-1}x^{n-1} + \cdots + a_1 x + a_0} = \infty.$$

(5) Justify the equality $\lim_{x \to \infty} \frac{e^x}{x^\alpha} = \left(\lim_{x \to \infty} \frac{e^{x/\alpha}}{x} \right)^\alpha$.

(6) In the proof of Theorem 8.3.1 we claim that $f'(x) - g'(x) > 0$ for $x \ge x_0$ implies $f(x) - g(x) - (f(x_0) - g(x_0)) > 0$ for $x > x_0$. Justify that claim.

8.4 Some applications of l'Hospital's rule

If $\lim f(x) = 0$ and $\lim g(x) = \infty$ (the sign lim may denote any of the symbols $\lim_{x\to a+}$, $\lim_{x\to a-}$, $\lim_{x\to a}$, or $\lim_{x\to\infty}$, but the same for both functions), then the limit $\lim f(x)g(x)$ can be computed by representing the product as

$$\frac{f(x)}{1/g(x)} \quad \text{or} \quad \frac{g(x)}{1/f(x)},$$

and then using the suitable form of l'Hospital's rule.

Example 8.4.1.

$$\lim_{x\to 0+} x\,(\ln x) = \lim_{x\to 0+} \frac{\ln x}{1/x} = \lim_{x\to 0+} \frac{1/x}{-1/x^2} = \lim_{x\to 0+} (-x) = 0.$$

Example 8.4.2.

$$\lim_{x\to\infty} x\left(\frac{\pi}{2} - \arctan x\right) = \lim_{x\to\infty} \frac{\frac{\pi}{2} - \arctan x}{1/x}$$

$$= \lim_{x\to\infty} \frac{-1/(1+x^2)}{-1/x^2} = \lim_{x\to\infty} \frac{x^2}{1+x^2} = 1.$$

Similarly, if $\lim f(x) = \infty$ and $\lim g(x) = \infty$, then the limit of the difference $f(x) - g(x)$ can be found by representing that difference in the form

$$\frac{\frac{1}{g(x)} - \frac{1}{f(x)}}{\frac{1}{f(x)}\frac{1}{g(x)}},$$

and then using l'Hospital's rule.

Example 8.4.3.

$$\lim_{x\to 0} \left(\frac{1}{x} - \frac{1}{\sin x}\right) = \lim_{x\to 0} \frac{\sin x - x}{x\sin x} = \lim_{x\to 0} \frac{\cos x - 1}{\sin x + x\cos x}$$

$$= \lim_{x\to 0} \frac{-\sin x}{2\cos x - x\sin x} = 0.$$

Example 8.4.4.

$$\lim_{x\to 1} \left(\frac{1}{\ln x} - \frac{1}{x - 1}\right) = \lim_{x\to 1} \frac{(x-1) - \ln x}{(x-1)\ln x} = \lim_{x\to 1} \frac{1 - 1/x}{\ln x + (x-1)/x}$$

$$= \lim_{x\to 1} \frac{1/x^2}{1/x + 1/x^2} = \lim_{x\to 1} \frac{1}{x+1} = \frac{1}{2}.$$

It often happens that the above method is not the best approach. For example,

$$\lim_{x\to\infty} (\ln(1+x) - \ln x) = \lim_{x\to\infty} \ln\left(\frac{1}{x} + 1\right) = \ln\left(\lim_{x\to\infty}\left(\frac{1}{x} + 1\right)\right) = \ln 1 = 0.$$

If $f(x) > 0$, $\lim f(x) = 0$, and $\lim g(x) = 0$, then

$$\lim f(x)^{g(x)} = \lim e^{g(x)\ln f(x)} = e^{\lim(g(x)\ln f(x))} \tag{8.7}$$

and we can apply l'Hospital's rule to the exponent. If we use exp to denote the exponential function, (8.7) can be written as

$$\lim f(x)^{g(x)} = \exp\left(\lim \frac{\ln f(x)}{1/g(x)}\right).$$

Example 8.4.5.

$$\lim_{x\to 0+} x^x = \exp\left(\lim_{x\to 0+} \frac{\ln x}{1/x}\right) = \exp\left(\lim_{x\to 0+} \frac{1/x}{-1/x^2}\right) = \exp(0) = 1.$$

If $\lim f(x) = \lim g(x) = 0$, then we have

$$\lim (1 + f(x))^{1/g(x)} = \exp\left(\lim \frac{f'(x)}{g'(x)}\right).$$

Indeed,

$$\lim (1 + f(x))^{1/g(x)} = \exp\left(\lim \frac{\ln(1 + f(x))}{g(x)}\right) = \exp\left(\lim \frac{\frac{f'(x)}{1+f(x)}}{g'(x)}\right)$$

$$= \exp\left(\lim \frac{1}{1 + f(x)} \lim \frac{f'(x)}{g'(x)}\right) = \exp\left(\lim \frac{f'(x)}{g'(x)}\right).$$

Example 8.4.6.

$$\lim_{x\to 0}(1 + x)^{1/x} = \exp\left(\lim_{x\to 0} \frac{1}{1}\right) = e.$$

Example 8.4.7.

$$\lim_{x\to 0}(\cos x)^{1/x^2} = \lim_{x\to 0}(1 + (\cos x - 1))^{1/x^2}$$

$$= \exp\left(\lim_{x\to 0} \frac{-\sin x}{2x}\right) = \exp(-\frac{1}{2}) = \frac{1}{\sqrt{e}}.$$

Example 8.4.8.

$$\lim_{x\to\infty}\left(1 - \frac{1}{x}\right)^x = \lim_{x\to 0+}(1 - x)^{1/x} = \exp\left(\lim_{x\to 0+} \frac{-1}{1}\right) = \exp(-1) = \frac{1}{e}.$$

EXERCISES 8.4

(1) Find the limits:

    (a) $\lim_{x\to 1}\left(\frac{x}{x-1} - \frac{1}{\ln x}\right)$,

    (b) $\lim_{x\to 0}\left(\frac{1}{2x} - \frac{1}{2x\tan x}\right)$,

    (c) $\lim_{x\to\frac{\pi}{2}}\left(x - \frac{\pi}{2}\right)\tan x$,

    (d) $\lim_{x\to 0+} x(-\ln x)^\alpha$,

    (e) $\lim_{x\to 0+} x^{\sin x}$,

    (f) $\lim_{x\to 0}(1 + mx)^{1/x}$,

    (g) $\lim_{x\to 1} x^{1/(1-x)}$,

    (h) $\lim_{x\to 0}\left(\frac{\tan x}{x}\right)^{1/x^2}$.

(2) Justify the equality: $\lim_{x\to\infty} \ln\left(\frac{1}{x} + 1\right) = \ln\left(\lim_{x\to\infty}\left(\frac{1}{x} + 1\right)\right)$.

8.5 A version of l'Hospital's rule for sequences

We know that if $\lim_{x \to \infty} f(x) = L$ and $a_n = f(n)$ for $n = 1, 2, 3, \ldots$, then $\lim_{n \to \infty} a_n = L$. Consequently, l'Hospital's rule can also serve in finding limits of sequences. For example, using this method it is easy to find the limit of the sequence $\left(1 + \frac{1}{n}\right)^n$. In fact, since

$$\lim_{x \to \infty} \left(1 + \frac{1}{x}\right)^x = \lim_{x \to 0+} (1 + x)^{1/x} = e,$$

we have

$$\lim_{n \to \infty} \left(1 + \frac{1}{n}\right)^n = e.$$

This formula is often used as a definition of the number e. However such a definition requires a proof of consistency: one has to prove that the limit exists. This can be accomplished by proving that the sequence is increasing and bounded, but the calculations are rather lengthy.

Changing to the limit of a function does not always work. For instance, it would be rather difficult to apply this method to the limit

$$\lim_{n \to \infty} \frac{1 + \frac{1}{2} + \cdots + \frac{1}{n}}{\ln n}.$$

In this and similar cases we can apply the following theorem which is somewhat similar to the difficult l'Hospital's rule.

Theorem 8.5.1. *The equality*

$$\lim_{n \to \infty} \frac{a_n}{b_n} = \lim_{n \to \infty} \frac{a_{n+1} - a_n}{b_{n+1} - b_n}$$

holds, whenever $b_{n+1} > b_n$ for all $n \in \mathbb{N}$, $\lim_{n \to \infty} b_n = \infty$ and the limit (proper or improper) on the right-hand side exists.

Proof. Let us first assume that the limit

$$\lim_{n \to \infty} \frac{a_{n+1} - a_n}{b_{n+1} - b_n} = \alpha$$

is finite. Let ε be an arbitrary positive number. There exists an index n_0 such that

$$\alpha - \frac{\varepsilon}{2} < \frac{a_{n+1} - a_n}{b_{n+1} - b_n} < \alpha + \frac{\varepsilon}{2} \quad \text{for all} \quad n \geq n_0.$$

Hence

$$\left(\alpha - \frac{\varepsilon}{2}\right)(b_{n+1} - b_n) < a_{n+1} - a_n < \left(\alpha + \frac{\varepsilon}{2}\right)(b_{n+1} - b_n) \quad \text{for all} \quad n \geq n_0. \quad (8.8)$$

Adding inequalities (8.8) with indices $n = n_0, n_0 + 1, \ldots, p - 1$ we obtain

$$\left(\alpha - \frac{\varepsilon}{2}\right)(b_p - b_{n_0}) < a_p - a_{n_0} < \left(\alpha + \frac{\varepsilon}{2}\right)(b_p - b_{n_0}).$$

Adding a_{n_0} and then dividing by b_p will give us

$$\alpha - \frac{\varepsilon}{2} + \frac{a_{n_0} - (\alpha - \frac{\varepsilon}{2})b_{n_0}}{b_p} < \frac{a_p}{b_p} < \alpha + \frac{\varepsilon}{2} + \frac{a_{n_0} - (\alpha + \frac{\varepsilon}{2})b_{n_0}}{b_p}$$

for all $p > n_0$. There exists an index $p_0 > n_0$ such that for all $p > p_0$ we have

$$\frac{a_{n_0} - (\alpha - \frac{\varepsilon}{2})b_{n_0}}{b_p} > -\frac{\varepsilon}{2} \quad \text{and} \quad \frac{a_{n_0} - (\alpha + \frac{\varepsilon}{2})b_{n_0}}{b_p} < \frac{\varepsilon}{2}.$$

Thus

$$\alpha - \varepsilon < \frac{a_p}{b_p} < \alpha + \varepsilon \quad \text{for all} \quad p > p_0,$$

which proves that $\lim_{n\to\infty} \frac{a_n}{b_n} = \alpha$.

If

$$\lim_{n\to\infty} \frac{a_{n+1} - a_n}{b_{n+1} - b_n} = \infty,$$

then $b_{n+1} > b_n$ and $\lim_{n\to\infty} b_n = \infty$ implies that there exists an index n_0 such that $a_{n+1} > a_n$ for all $n > n_0$ and $\lim_{n\to\infty} a_n = \infty$. Thus, in view of the case just considered, we have

$$\lim_{n\to\infty} \frac{b_n}{a_n} = \lim_{n\to\infty} \frac{b_{n+1} - b_n}{a_{n+1} - a_n} = 0$$

and consequently $\lim_{n\to\infty} \frac{a_n}{b_n} = \infty$. The case $\lim_{n\to\infty} \frac{a_{n+1} - a_n}{b_{n+1} - b_n} = -\infty$ reduces to the last considered case by an easy substitution. $\qquad\square$

Example 8.5.2.

$$\lim_{n\to\infty} \frac{1 + \frac{1}{2} + \cdots + \frac{1}{n}}{\ln n} = \lim_{n\to\infty} \frac{1/(n+1)}{\ln(n+1) - \ln n}$$

$$= \lim_{n\to\infty} \frac{n}{n+1} \frac{1}{n \ln(1 + 1/n)}$$

$$= \lim_{n\to\infty} \frac{1}{\ln(1 + 1/n)^n} = \frac{1}{\ln e} = 1.$$

Example 8.5.3.

$$\lim_{n\to\infty} \frac{\sqrt[n]{n!}}{n} = \lim_{n\to\infty} \exp\left(\frac{\ln n! - n \ln n}{n}\right)$$

$$= \exp\left(\lim_{n\to\infty} \frac{\ln(n+1)! - (n+1)\ln(n+1) - \ln n! + n \ln n}{(n+1) - n}\right)$$

$$= \exp\left(\lim_{n\to\infty} \left(-n \ln \frac{n+1}{n}\right)\right)$$

$$= \exp\left(\lim_{n\to\infty} \left(-\ln(1 + 1/n)^n\right)\right) = \exp(-\ln e) = \frac{1}{e}.$$

Example 8.5.4. If $\alpha > -1$, then

$$\lim_{n\to\infty} \frac{1^\alpha + 2^\alpha + \cdots + n^\alpha}{n^{\alpha+1}} = \lim_{n\to\infty} \frac{(n+1)^\alpha}{(n+1)^{\alpha+1} - n^{\alpha+1}}$$

$$= \lim_{n\to\infty} \left(\frac{n+1}{n}\right)^\alpha \lim_{n\to\infty} \frac{n^\alpha}{(n+1)^{\alpha+1} - n^{\alpha+1}}.$$

The first limit equals 1, whereas the second one equals

$$\lim_{x\to 0+} \frac{\left(\frac{1}{x}\right)^\alpha}{\left(\frac{1}{x} + 1\right)^{\alpha+1} - \left(\frac{1}{x}\right)^{\alpha+1}} = \lim_{x\to 0+} \frac{x}{(1+x)^{\alpha+1} - 1}$$

$$= \lim_{x\to +} \frac{1}{(\alpha+1)(1+x)^\alpha} = \frac{1}{\alpha+1}.$$

Consequently

$$\lim_{n\to\infty} \frac{1^\alpha + 2^\alpha + \cdots + n^\alpha}{n^{\alpha+1}} = \frac{1}{\alpha+1} \quad \text{for } \alpha > -1.$$

EXERCISES 8.5

(1) Find $\lim_{n\to\infty} \frac{\ln 1 + \ln 2 + \cdots + \ln n}{n \ln n}$.

(2) Find $\lim_{n\to\infty} \frac{\sqrt[n]{(n+1)(n+2)\ldots 2n}}{n}$.

(3) Complete the proof of Theorem 8.5.1 by considering the case when $\lim_{n\to\infty} \frac{a_{n+1}-a_n}{b_{n+1}-b_n} = -\infty$.

8.6 D'Alembert's test

D'Alembert's test (called also the *ratio test*) is used to determine convergence of series. It is attributed to the French mathematician Jean le Rond d'Alembert (1717–1783).

Theorem 8.6.1 (D'Alembert's Test). *The series*

$$\sum_{n=1}^{\infty} a_n \tag{8.9}$$

is absolutely convergent if $\lim\limits_{n\to\infty} \left|\frac{a_{n+1}}{a_n}\right| < 1$, *and divergent if* $\lim\limits_{n\to\infty} \left|\frac{a_{n+1}}{a_n}\right| > 1$.

Proof. Let $\lim_{n\to\infty} \left|\frac{a_{n+1}}{a_n}\right| = \alpha$. If $\alpha < 1$, then there exists a number q and an index n_0 such that

$$\alpha < q < 1 \quad \text{and} \quad \left|\frac{a_{n+1}}{a_n}\right| < q \quad \text{for } n \geq n_0.$$

Consequently,

$$|a_{n_0+1}| < |a_{n_0}|q,$$

$$|a_{n_0+2}| < |a_{n_0+1}|q < |a_{n_0}|q^2,$$

and generally, by induction,

$$|a_{n_0+n}| < |a_{n_0}|q^n. \tag{8.10}$$

Since the series $\sum_{n=0}^{\infty} |a_{n_0}|q^n$ is convergent, so is the series $\sum_{n=0}^{\infty} |a_{n_0+n}|$, by (8.10). This implies absolute convergence of the series (8.9).

If $\alpha > 1$, then there is an index n_0 such that

$$\left| \frac{a_{n+1}}{a_n} \right| > 1 \text{ for } n \geq n_0.$$

By induction we get

$$|a_{n_0+n}| > |a_{n_0}| > 0,$$

and hence the sequence does not converge to zero, which implies that the series (8.9) diverges. □

If $\lim_{n \to \infty} \left| \frac{a_{n+1}}{a_n} \right| = 1$, then the test is inconclusive.

Example 8.6.2. For $a_n = n!/n^n$ we have

$$\left| \frac{a_{n+1}}{a_n} \right| = \frac{(n+1)!}{(n+1)^{n+1}} \cdot \frac{n^n}{n!} = \frac{1}{(1+1/n)^n}.$$

Since

$$\lim_{n \to \infty} \left| \frac{a_{n+1}}{a_n} \right| = \frac{1}{e} < 1,$$

the series $\sum_{n=1}^{\infty} \frac{n!}{n^n}$ is convergent.

Example 8.6.3. For $a_n = 3^n n!/n^n$, we have

$$\left| \frac{a_{n+1}}{a_n} \right| = \frac{3}{(1+1/n)^n} \to \frac{3}{e} > 1.$$

This proves that the series $\sum_{n=1}^{\infty} \frac{3^n n!}{n^n}$ is divergent.

EXERCISES 8.6

(1) Give example of a convergent series $\sum_{n=1}^{\infty} a_n$ for which $\lim_{n \to \infty} \left| \frac{a_{n+1}}{a_n} \right| = 1$.

(2) Give example of a divergent series $\sum_{n=1}^{\infty} a_n$ for which $\lim_{n \to \infty} \left| \frac{a_{n+1}}{a_n} \right| = 1$.

(3) Prove *Cauchy's test* (called also the *root test*):

(a) If $\lim_{n\to\infty} |a_n|^{1/n} < 1$, then the series $\sum_{n=1}^{\infty} a_n$ is absolutely convergent.

(b) If $\lim_{n\to\infty} |a_n|^{1/n} > 1$, then the series $\sum_{n=1}^{\infty} a_n$ is divergent.

(4) Give example of a convergent series $\sum_{n=1}^{\infty} a_n$ for which $\lim_{n\to\infty} |a_n|^{1/n} = 1$.

(5) Give example of a divergent series $\sum_{n=1}^{\infty} a_n$ for which $\lim_{n\to\infty} |a_n|^{1/n} = 1$.

(6) Test the following series for convergence:

(a) $\sum_{n=1}^{\infty} \frac{n^3}{(-e)^n}$,

(b) $\sum_{n=1}^{\infty} \frac{(n!)^2}{(2n)!}$,

(c) $\sum_{n=1}^{\infty} \frac{n!}{2^n+1}$,

(d) $\sum_{n=1}^{\infty} \frac{7n+n^7+7^n}{n!+\ln(n+7)}$,

(e) $\sum_{n=1}^{\infty} \frac{\sin \pi^n}{\pi^n}$,

(f) $\sum_{n=1}^{\infty} \frac{2^n \, n!}{n^n}$.

8.7   Theorem of Abel

Conditional convergence can never be proved by the comparison test (Theorem 5.5.1) or D'Alembert's Test (Theorem 8.6.1). Leibniz's criterion (Theorem 5.5.4) can be often successfully used. For example the series

$$\sum_{n=1}^{\infty} \frac{(-1)^{n+1}}{n} \quad \text{and} \quad \sum_{n=1}^{\infty} \frac{(-1)^{n+1}}{2n - 1} \tag{8.11}$$

are convergent, by Leibniz's theorem. That theorem however does not help in finding the sums of these series. Evaluation of those sums is much more difficult than proving convergence. In some cases this can be done by using the following theorem due to the Norwegian mathematician Niels Henrik Abel (1802–1829).

Theorem 8.7.1. *If the series* $\sum_{n=1}^{\infty} a_n$ *is convergent, then the function*

$$f(x) = \sum_{n=0}^{\infty} a_n x^n$$

is left-hand continuous at $x = 1$, *that is,*

$$\lim_{x\to 1-} f(x) = \sum_{n=0}^{\infty} a_n.$$

Proof. Since the sequence of partial sums $s_n = a_0 + \cdots + a_n$ is convergent, it is bounded and thus the series

$$\sum_{n=0}^{\infty} s_n x^n$$

is convergent for $|x| < 1$. Moreover,

$$(1 - x) \sum_{n=0}^{\infty} s_n x^n = \sum_{n=0}^{\infty} s_n x^n - \sum_{n=0}^{\infty} s_n x^{n+1} = \sum_{n=0}^{\infty} a_n x^n,$$

and thus

$$f(x) = (1 - x) \sum_{n=0}^{\infty} s_n x^n. \tag{8.12}$$

Let $s = \sum_{n=0}^{\infty} a_n$. Since $\frac{1}{1-x} = \sum_{n=0}^{\infty} x^n$, we have

$$s = (1 - x) \sum_{n=0}^{\infty} s x^n. \tag{8.13}$$

From (8.12) and (8.13) we get

$$f(x) - s = (1 - x) \sum_{n=0}^{\infty} (s_n - s) x^n. \tag{8.14}$$

Let ε be an arbitrary positive number. Since $\lim_{n\to\infty} s_n = s$, there exists an index m such that $|s_n - s| < \varepsilon/2$ for all $n > m$. Equality (8.14) can be rewritten in the form

$$f(x) - s = (1 - x) \left(\sum_{n=0}^{m} (s_n - s) x^n + \sum_{n=m+1}^{\infty} (s_n - s) x^n \right).$$

Letting $p = \sum_{n=0}^{m} |s_n - s|$ we have

$$|f(x) - s| < (1 - x)p + (1 - x) \sum_{n=m+1}^{\infty} \frac{\varepsilon}{2} x^n < (1 - x)p + \frac{\varepsilon}{2},$$

for all $0 < x < 1$. There is a number $\delta > 0$ such that $(1 - x)p < \frac{\varepsilon}{2}$ for $1 - \delta < x < 1$. Thus

$$|f(x) - s| < \varepsilon \quad \text{for } 1 - \delta < x < 1,$$

which proves that $\lim_{x\to 1-} f(x) = s$. $\qquad\qquad\qquad\qquad\qquad\qquad\qquad\square$

In Section 7.4 we found that

$$\ln(1+x) = \frac{x}{1} - \frac{x^2}{2} + \frac{x^3}{3} - \frac{x^4}{4} + \cdots = \sum_{n=1}^{\infty} \frac{(-1)^{n+1}x^n}{n} \quad \text{for } |x| < 1. \quad (8.15)$$

In view of Abel's theorem the equality must also hold for $x = 1$, since the function $\ln(1+x)$ is continuous at $x = 1$. Thus

$$\ln 2 = \frac{1}{1} - \frac{1}{2} + \frac{1}{3} - \frac{1}{4} + \cdots.$$

Similarly, from the equation

$$\arctan x = \frac{x}{1} - \frac{x^3}{3} + \frac{x^5}{5} - \frac{x^7}{7} + \cdots = \sum_{n=1}^{\infty} \frac{(-1)^{n+1}x^{2n-1}}{2n-1} \quad \text{for } |x| < 1,$$

it follows by Abel's theorem that

$$\frac{\pi}{4} = \frac{1}{1} - \frac{1}{3} + \frac{1}{5} - \frac{1}{7} + \cdots.$$

These series, known already by Leibniz, have a very elegant form but they cannot be used to calculate approximate values of $\ln 2$ or π. For example, if we wanted to compute π with a 5 decimal place accuracy, we would have to use about 100,000 terms of the series and evaluate each of them with a 10 decimal place accuracy.

EXERCISES 8.7

(1) Prove that $\frac{1}{2\cdot1!} - \frac{1}{2^2\cdot2!} + \frac{1\cdot3}{2^3\cdot3!} - \frac{1\cdot3\cdot5}{2^4\cdot4!} + \cdots = \sqrt{2} - 1.$

(2) Prove the following version of Abel's theorem:

If the series $\displaystyle\sum_{n=0}^{\infty} a_n r^n$ *is convergent for some* $r > 0$, *then the function*

$$f(x) = \sum_{n=0}^{\infty} a_n x^n$$

is left-hand continuous at $x = r$, *that is,*

$$\lim_{x \to r-} f(x) = \sum_{n=0}^{\infty} a_n r^n.$$

(3) Prove the following version of Abel's theorem:

If $\displaystyle\sum_{n=0}^{\infty} a_n$ *converges, then* $\displaystyle\sum_{n=0}^{\infty} a_n x^n$ *converges uniformly on* $[0, 1]$.

Chapter 9

ANTIDERIVATIVES

9.1 Antiderivatives

We say that F is an *antiderivative* of f on an interval (finite or infinite), if for every x in that interval we have

$$F'(x) = f(x). \tag{9.1}$$

For instance, $\sin x$ is an antiderivative of $\cos x$ on $(-\infty, \infty)$ because $(\sin x)' = \cos x$ for every $x \in (-\infty, \infty)$.

If F is an antiderivative of f, then

   (a) $F + C$ is an antiderivative of f for any constant $C \in \mathbb{R}$;

   (b) Each antiderivative of f is of the form $F + C$, since the functions with the same derivative differ by a constant.

Consequently, it suffices to know a single antiderivative for f in order to obtain every one of them by adding a constant. We often say that antiderivatives are determined up to an additive constant.

Antiderivatives are also called *indefinite integrals* and denoted by

$$\int f(x)\, dx. \tag{9.2}$$

The word "indefinite" refers to the fact that this symbol does not denote any specific function, but one of an infinite collection of functions differing from one another by an additive constant. It is customary to indicate that by adding a constant to a specific antiderivative we find. For example, we may write

$$\int x\, dx = \frac{1}{2}x^2 + C.$$

The calculation of antiderivatives, called *antidifferentiation* or *integration*, is the inverse of differentiation. For this reason we can use formulas derived for differentiation to obtain formulas for antiderivatives.

We provide some antiderivatives in the following table. For purely practical purposes we supply these formulas in a slightly modified form, adjusted to make

179

calculations more convenient. The formulas are to be understood so that the function on the right represents one of the possible antiderivatives of the function on the left.

$f(x)$	$\int f(x)dx$
a (a constant function)	ax
x^a $(a \neq -1)$	$\dfrac{x^{a+1}}{a+1}$
$\dfrac{1}{x}$	$\ln|x|$
e^x	e^x
a^x $(a > 0$ and $a \neq 1)$	$\dfrac{a^x}{\ln a}$
$\dfrac{1}{a^2 + x^2}$ $(a > 0)$	$\dfrac{1}{a} \arctan \dfrac{x}{a}$
$\dfrac{1}{\sqrt{a^2 - x^2}}$ $(a > 0)$	$\arcsin \dfrac{x}{a}$
$\dfrac{1}{\sqrt{a + x^2}}$ $(a > 0)$	$\ln\left(x + \sqrt{a + x^2}\right)$
$\cos x$	$\sin x$
$\sin x$	$-\cos x$
$\dfrac{1}{\cos^2 x}$	$\tan x$
$\dfrac{1}{\sin^2 x}$	$-\dfrac{\cos x}{\sin x}$
$\dfrac{1}{\sin x}$	$\ln\left|\tan \dfrac{x}{2}\right|$

Verification of these formulas consists of differentiating the functions on the right-hand side to yield the functions on the left-hand side.

There are several formulas in the above list that look strange to those educated in certain parts of the world. It is because in this book we have not yet introduced the functions

$$\sec x = \frac{1}{\cos x}, \quad \csc x = \frac{1}{\sin x}, \quad \text{and} \quad \cot x = \frac{1}{\tan x}.$$

In terms of these functions the last three formulas in the table can be rewritten as follows:

$f(x)$	$\int f(x)dx$
$\sec^2 x$	$\tan x$
$\csc^2 x$	$-\cot x$
$\csc x$	$\ln\left|\tan \dfrac{x}{2}\right|$

If a function can be expressed in terms of elementary functions, then it can be effectively differentiated. It suffices to know the derivatives of basic functions and the differentiation rules. The process is mostly mechanical. The situation is very different when it comes to integration. First of all, not every function expressed in terms of elementary functions can be effectively integrated. For instance $\int \sin x^2\, dx$ cannot be calculated at all. This does not mean that the function $f(x) = \sin x^2$ does not have an antiderivative. In Section 11.1 it will be proved that every continuous function has an antiderivative, so there exists a function f such that $f'(x) = \sin x^2$. But that function cannot be expressed in terms of elementary functions, which means that, in practice, we cannot find $\int \sin x^2\, dx$. Quite often the indefinite integral of a simple function, although expressible in terms of elementary functions, is rather complicated. For example, we will show that

$$\int \frac{dx}{1+x^4} = \frac{1}{4\sqrt{2}} \ln \frac{x^2 + x\sqrt{2} + 1}{x^2 - x\sqrt{2} + 1} + \frac{1}{2\sqrt{2}} \arctan \frac{x\sqrt{2}}{1 - x^2} + C.$$

Unlike differentiation, integration is not a matter of mechanically applying some integration formulas and formulas for the indefinite integrals of basic functions listed above. Finding an indefinite integral often requires a certain level of ingenuity and what one might call tricks. There are some useful "tricks" that have been discovered by famous mathematicians, for example "Euler's substitutions" (see Section 9.5).

The purpose of this chapter is to provide some general methods of integration. With these methods we will be able to find the indefinite integrals of most functions that come up in applications.

In this chapter instead of "indefinite integral" we will say simply "integral." In the next chapter we will define a different notion of integral and then it will be necessary to keep the distinction clear.

We start with the following simple formulas:

$$\int (f(x) + g(x))\, dx = \int f(x)\, dx + \int g(x)\, dx,$$

$$\int (f(x) - g(x))\, dx = \int f(x)\, dx - \int g(x)\, dx,$$

$$\int a\, f(x)\, dx = a \int f(x)\, dx.$$

They can be expressed in words as follows. *The integral of a sum equals the sum of the integrals. The integral of a difference equals the difference of the integrals. The integral of a constant times a function equals that constant times the integral of that function.*

The equalities in the above (and all similar) formulas should be understood to hold to within an additive constant. They can be easily proved by differentiation of both sides. In fact, from the definition of indefinite integral it follows that

$$\left(\int f(x)\, dx \right)' = f(x).$$

Example 9.1.1.

$$\int \left(x^2 + \frac{1}{3x} \right) dx = \int x^2\, dx + \frac{1}{3} \int \frac{dx}{x} = \frac{x^3}{3} + \frac{1}{3} \ln |x| + C.$$

Example 9.1.2.

$$\int (e^x - 1)\, dx = \int e^x dx - \int dx = e^x - x + C.$$

EXERCISES 9.1

(1) Find the following antiderivatives:

(a) $\int (3 \cdot 2^x + 2 \cdot 3^x)\, dx,$

(b) $\int (a \sin x + b \cos x)\, dx,$

(c) $\int \left(\frac{1}{1+x^2} + \frac{1}{2+x^2} + \frac{1}{3+x^2} \right) dx,$

(d) $\int \left(\frac{1}{\sqrt{1+4x^2}} - \frac{1}{\sqrt{1-x^2}} \right) dx,$

(e) $\int \frac{1}{4x}\, dx.$

9.2   Integration by substitution

In evaluating integrals we often apply the formula

$$F(g(x)) = \int f(g(x))g'(x)\, dx \quad \text{where} \quad F' = f, \tag{9.3}$$

which can be easily verified by differentiating both sides. Note that this formula is a version of the chain rule for integrals.

Formula (9.3) is the basis for a method of calculating integrals called *integration by substitution*. In practice, when integrating by substitution we proceed as follows. First, we choose a function g such that the original integral can be written in the form $\int f(g(x))g'(x)\,dx$. Then we introduce a new variable $u = g(x)$ and find $du = g'(x)dx$. Now we can write

$$\int f(g(x))g'(x)\,dx = \int f(u)\,du.$$

We expect the new integral to be simpler than the original one. We find a function F such that $F' = f$ and complete the calculations by writing

$$\int f(g(x))g'(x)\,dx = \int f(u)\,du = F(u) + C = F(g(x)) + C.$$

Example 9.2.1. We illustrate the above method by calculating the integral

$$\int \frac{dx}{ax + b} \quad (a \neq 0).$$

We put $u = ax + b$ and find $du = a\,dx$. Then

$$\int \frac{dx}{ax + b} = \frac{1}{a} \int \frac{a\,dx}{ax + b} = \frac{1}{a} \int \frac{du}{u} = \frac{1}{a} \ln |u| + C = \frac{1}{a} \ln |ax + b| + C.$$

When deciding what we should use for the substitution, we need to plan ahead. While a good choice for u can make calculating the integral simple, a bad choice can complicate things and lead to a dead end. Let's examine another example.

Example 9.2.2. We want to find the integral

$$\int \frac{dx}{1 + x + x^2}.$$

We first note that it can be written in the form

$$\int \frac{dx}{\frac{3}{4} + \left(x + \frac{1}{2}\right)^2}$$

which looks similar to an integral from the table of integrals, namely,

$$\int \frac{dx}{a^2 + x^2} = \frac{1}{a} \arctan \frac{x}{a} + C.$$

In order to use this formula, we choose $u = x + \frac{1}{2}$ and find that $du = dx$. Now we are ready to complete the calculations:

$$\int \frac{dx}{1 + x + x^2} = \int \frac{dx}{\frac{3}{4} + \left(x + \frac{1}{2}\right)^2} = \int \frac{du}{\frac{3}{4} + u^2}$$

$$= \frac{1}{\sqrt{\frac{3}{4}}} \arctan \frac{u}{\sqrt{\frac{3}{4}}} + C$$

$$= \frac{1}{\sqrt{\frac{3}{4}}} \arctan \frac{x + \frac{1}{2}}{\sqrt{\frac{3}{4}}} + C$$

$$= \frac{2}{\sqrt{3}} \arctan \frac{2x + 1}{\sqrt{3}} + C.$$

EXERCISES 9.2

(1) Find the following integrals:

(a) $\int e^{ax+b}\, dx$,

(b) $\int \cos 2x\, dx$,

(c) $\int \sin (3x + 1)\, dx$,

(d) $\int \frac{dx}{1-x+2x^2}$,

(e) $\int \frac{dx}{\sqrt{1-x+2x^2}}$,

(f) $\int \frac{dx}{\sqrt{1-x-2x^2}}$.

9.3 Integration by parts

The product rule for derivatives gives us the following formula for integrals:

$$\int f(x)g'(x)\, dx = f(x)g(x) - \int f'(x)g(x)\, dx. \tag{9.4}$$

This formula is the basis for a method of calculating integrals called *integration by parts*. We illustrate this method by some examples.

Example 9.3.1. In order to find $\int x \cos x\, dx$ we write

$$f(x) = x, \qquad g'(x) = \cos x,$$
$$f'(x) = 1, \qquad g(x) = \sin x.$$

In view of (9.4) we have

$$\int x \cos x\, dx = x \sin x - \int \sin x\, dx = x \sin x + \cos x + C.$$

Example 9.3.2. To find $\int \ln x\, dx$ we write

$$f(x) = \ln x, \qquad g'(x) = 1,$$
$$f'(x) = \frac{1}{x}, \qquad g(x) = x,$$

and by (9.4),

$$\int \ln x\, dx = x \ln x - \int \frac{1}{x} x\, dx = x \ln x - x + C.$$

Systematic use of the scheme

$$
\begin{array}{cc}
f & g' \\
f' & g
\end{array}
$$

facilitates the work and helps to avoid errors. In the first line of the table the factors of the given integrand are found, on the main diagonal we have the factors already integrated, and in the bottom line we find the factors to be put under the integral sign on the right-hand side. Using this scheme we do not even have to write the letters f and g.

Example 9.3.3. To calculate the integral $\int x^2 e^x \, dx$ we write

$$\begin{matrix} x^2 & e^x \\ 2x & e^x \end{matrix} \qquad \int x^2 e^x \, dx = x^2 e^x - 2 \int x e^x \, dx.$$

To find the remaining integral we write

$$\begin{matrix} x & e^x \\ 1 & e^x \end{matrix} \qquad \int x e^x \, dx = x e^x - \int e^x \, dx = x e^x - e^x + C.$$

Thus we have

$$\int x^2 e^x \, dx = (x^2 - 2x + 2) e^x + C.$$

Example 9.3.4. For the integral $\int \frac{\ln x}{x} \, dx$ we write

$$\begin{matrix} \ln x & \dfrac{1}{x} \\[2mm] \dfrac{1}{x} & \ln x \end{matrix} \qquad \int \frac{\ln x}{x} \, dx = (\ln x)^2 - \int \frac{\ln x}{x} \, dx = (\ln x)^2 - \int \frac{\ln x}{x} \, dx.$$

Hence

$$\int \frac{\ln x}{x} \, dx = \frac{1}{2} (\ln x)^2 + C.$$

EXERCISES 9.3

(1) Find the following integrals:

(a) $\int \arctan x \, dx$,

(b) $\int \arcsin x \, dx$,

(c) $\int \arccos x \, dx$,

(d) $\int e^x \cos x \, dx$,

(e) $\int x^3 \sin x \, dx$,

(f) $\int \arctan(2x - 1) \, dx$,

(g) $\int \frac{\arctan x}{1+x^2} \, dx$,

(h) $\int \frac{\arcsin x}{\sqrt{1-x^2}} \, dx$,

(i) $\int x \, e^{2x} \sin \frac{x}{2} \, dx$.

9.4 Integration of rational functions

In this section we discuss techniques for integration of rational functions. Decomposition of rational functions into partial fractions plays an important role. The method is based on the fact that every rational function can be written as a sum of a polynomial and functions of the form

$$\frac{\alpha}{(ax + b)^n} \quad \text{or} \quad \frac{\alpha x + \beta}{(ax^2 + bx + c)^n},$$

where n is a natural number and the quadratic polynomial $ax^2 + bx + c$ is not reducible to a product of two linear factors. Such a representation of a rational function is called the *partial fraction decomposition*.

Example 9.4.1. To find the integral $\int \frac{dx}{1-x^2}$ we first decompose the integrand into partial fractions, that is,

$$\frac{1}{1-x^2} = \frac{\frac{1}{2}}{1+x} + \frac{\frac{1}{2}}{1-x}.$$

Hence

$$\int \frac{dx}{1-x^2} = \frac{1}{2} \int \frac{dx}{1+x} + \frac{1}{2} \int \frac{dx}{1-x}$$

$$= \frac{1}{2} \ln|1+x| - \frac{1}{2} \ln|1-x| + C = \frac{1}{2} \ln\left|\frac{1+x}{1-x}\right| + C.$$

Example 9.4.2. To evaluate the integral $\int \frac{x^2\,dx}{x^2-2x-3}$ we first decompose the integrand into partial fractions, that is,

$$\frac{x^2}{x^2-2x-3} = 1 + \frac{2x+3}{x^2-2x-3} = 1 + \frac{2x+3}{(x-3)(x+1)} = 1 + \frac{\alpha}{x-3} + \frac{\beta}{x+1}.$$

We find numbers α and β that satisfy the last equality, or equivalently, the equation

$$2x + 3 = \alpha(x+1) + \beta(x-3),$$

by comparing the coefficients. We find that $\alpha = \frac{9}{4}$ and $\beta = -\frac{1}{4}$. Thus

$$\int \frac{x^2\,dx}{x^2-2x-3} = \int dx + \frac{9}{4} \int \frac{dx}{x-3} - \frac{1}{4} \int \frac{dx}{x+1} = x + \frac{9}{4} \ln|x-3| - \frac{1}{4} \ln|x+1| + C.$$

Example 9.4.3. To find the integral $\int \frac{x^3\,dx}{x^2-2x+3}$ we write

$$\frac{x^3}{x^2-2x+3} = x + 2 + \frac{x-6}{x^2-2x+3}.$$

S: And I presume that the last fraction should be decomposed into simple fraction as in the preceding examples.

T: No, it cannot be done.

S: Why not?

T: Because $x^2 - 2x + 3$ cannot be decomposed into linear factors. It depends on the discriminant. By the discriminant of $ax^2 + bx + c$ we mean $\Delta = b^2 - 4ac$. If $\Delta > 0$, then the decomposition into two factors is possible:

$$ax^2 + bx + c = a\left(x - \frac{-b-\sqrt{\Delta}}{2a}\right)\left(x - \frac{-b+\sqrt{\Delta}}{2a}\right).$$

If $\Delta < 0$, then we make the following transformation:

$$\frac{\alpha x + \beta}{ax^2 + bx + c} = \frac{\alpha}{2a} \frac{(ax^2+bx+c)'}{(ax^2+bx+c)} + \frac{2a\beta - b\alpha}{2a^2} \frac{1}{\left(x+\frac{b}{2a}\right)^2 + \frac{-\Delta}{4a^2}}.$$

S: This formula is very complicated; almost impossible to remember.

T: It would be useless to remember it. We should rather remember the general method and not the details.

In the case of our example, we have

$$\frac{x-6}{x^2-2x+3} = \frac{1}{2}\frac{2x-2}{x^2-2x+3} - \frac{5}{(x-1)^2+2}.$$

Hence

$$\int \frac{x-6}{x^2-2x+3}\,dx = \frac{1}{2}\ln(x^2-2x+3) - \frac{5}{\sqrt{2}}\arctan\frac{x-1}{\sqrt{2}} + C,$$

and, in turn,

$$\int \frac{x^3\,dx}{x^2-2x+3} = \frac{1}{2}x^2 + 2x + \frac{1}{2}\ln(x^2-2x+3) - \frac{5}{\sqrt{2}}\arctan\frac{x-1}{\sqrt{2}} + C.$$

Example 9.4.4. Now we find the integral

$$\int \frac{dx}{x^3+2x^2+3x+6}.$$

Since

$$x^3 + 2x^2 + 3x + 6 = (x+2)(x^2+3),$$

we have

$$\frac{1}{x^3+2x^2+3x+6} = \frac{\alpha}{x+2} + \frac{\beta x+\gamma}{x^2+3},$$

and, from the equation

$$1 = \alpha(x^2+3) + (\beta x+\gamma)(x+2),$$

we find

$$\alpha = \frac{1}{7}, \quad \beta = -\frac{1}{7}, \quad \gamma = \frac{2}{7}.$$

Hence

$$\int \frac{dx}{x^3+2x^2+3x+6} = \frac{1}{7}\int \frac{dx}{x+2} - \frac{1}{14}\int \frac{2x\,dx}{x^2+3} + \frac{2}{7}\int \frac{dx}{x^2+3}$$

$$= \frac{1}{7}\ln|x+2| - \frac{1}{14}\ln|x^2+3| + \frac{2}{7\sqrt{3}}\arctan\frac{x}{\sqrt{3}} + C.$$

As the above examples illustrate, to integrate a rational function we have to be able to factor its denominator and then we need to know how to evaluate integrals of the following type:

$$\int \frac{dx}{(ax+b)^n} \quad \text{and} \quad \int \frac{ax+\beta}{(ax^2+bx+c)^n}$$

where n is a natural number and the quadratic polynomial $ax^2 + bx + c$ is not reducible to a product of two linear factors. If $n = 1$, we have

$$\int \frac{dx}{ax+b} = \frac{1}{a}\ln|ax+b| + C.$$

If $n \geq 2$, we have

$$\int \frac{dx}{(ax+b)^n} = \frac{1}{a(1-n)(ax+b)^{n-1}} + C.$$

It thus remains to consider the second integral. The assumption that the denominator does not decompose into linear factors implies that $\Delta = b^2 - 4ac < 0$. Thus using the substitution $x + \frac{b}{2a} = t$, we obtain after a few simple algebraic calculations

$$\frac{\alpha x + \beta}{(ax^2 + bx + c)^n} = A\frac{2t}{(t^2 + r^2)^n} + B\frac{1}{(t^2 + r^2)^n},$$

where $n \in \mathbb{N}$, $r^2 = -\Delta/4a^2$ and the coefficients A and B are properly chosen. Now, the first fraction is the derivative of

$$\frac{1}{(1-n)(t^2 + r^2)^{n-1}}$$

so that

$$\int \frac{2t\, dt}{(t^2 + r^2)^n} = \frac{1}{(1-n)(t^2 + r^2)^{n-1}} + C.$$

We still have to find the integral

$$J_n = \int \frac{dt}{(t^2 + r^2)^n}. \tag{9.5}$$

We will show that

$$J_{n+1} = \frac{1}{2nr^2}\frac{t}{(t^2 + r^2)^n} + \frac{2n-1}{2n}\frac{1}{r^2}J_n. \tag{9.6}$$

This formula allows us to reduce the degree of the power in the denominator by 1 and after n steps we only have to evaluate J_1:

$$J_1 = \int \frac{dt}{t^2 + r^2} = \frac{1}{r}\arctan\frac{t}{r} + C.$$

Applying integration by parts to (9.5) we write

$$\frac{1}{(t^2 + r^2)^n} \qquad\qquad 1$$

$$-\frac{2nt}{(t^2 + r^2)^{n+1}} \qquad\qquad t$$

and

$$J_n = \frac{t}{(t^2 + r^2)^n} + 2n\int \frac{t^2\, dt}{(t^2 + r^2)^{n+1}}. \tag{9.7}$$

The last integral may be transformed in the following way

$$\int \frac{t^2\, dt}{(t^2 + r^2)^{n+1}} = \int \frac{(t^2 + r^2) - r^2}{(t^2 + r^2)^{n+1}}\, dt$$

$$= \int \frac{dt}{(t^2 + r^2)^n} - r^2\int \frac{dt}{(t^2 + r^2)^{n+1}}$$

$$= J_n - r^2 J_{n+1}.$$

Substituting this into (9.7) we get

$$J_n = \frac{t}{(t^2 + r^2)^n} + 2nJ_n - 2nr^2 J_{n+1},$$

whence (9.6) follows.

It is interesting that, when applying a general method, a simple example can lead to complicated calculations. Consider, for example,

$$\int \frac{x\,dx}{x^4 + 1}.$$

Since

$$x^4 + 1 = (x^2 - \sqrt{2}\,x + 1)(x^2 + \sqrt{2}\,x + 1),$$

we look for a decomposition of the form

$$\frac{x}{x^4 + 1} = \frac{\alpha x + \beta}{x^2 - \sqrt{2}\,x + 1} + \frac{\gamma x + \delta}{x^2 + \sqrt{2}\,x + 1}.$$

Comparing coefficients in the equation

$$x = (\alpha x + \beta)(x^2 + \sqrt{2}\,x + 1) + (\gamma x + \delta)(x^2 - \sqrt{2}\,x + 1),$$

we find $\alpha = 0$, $\beta = \frac{1}{2\sqrt{2}}$, $\gamma = 0$, and $\delta = -\frac{1}{2\sqrt{2}}$. Thus

$$\frac{x}{x^4 + 1} = \frac{1}{2\sqrt{2}} \left(\frac{1}{x^2 - \sqrt{2}\,x + 1} - \frac{1}{x^2 + \sqrt{2}\,x + 1} \right)$$

and finally

$$\int \frac{x\,dx}{x^4 + 1} = \frac{1}{2} \left(\arctan(\sqrt{2}\,x - 1) - \arctan(\sqrt{2}\,x + 1) \right) + C.$$

S: I can find this integral in a much simpler way: I substitute $x^2 = t$ and obtain

$$\int \frac{x\,dx}{x^4 + 1} = \frac{1}{2} \int \frac{dt}{t^2 + 1} + C = \frac{1}{2}\arctan t + C = \frac{1}{2}\arctan x^2 + C.$$

T: Indeed. Your work is more like the work of an artist, whereas the use of standard methods is like the work of a craftsman. In integration we should always look for special "tricks" and only when they do not work, apply general methods.

S: And how do you explain that we got different results?

T: They only look different. In fact one can check, using the formula in Section 7.11, that they differ by an additive constant.

S: The above trick does not work if we erase the x in the numerator. What do we do in this case, $\int \frac{dx}{x^4 + 1}$?

T: It is said that Leibniz could not evaluate this integral, because he did not know how to factor the polynomial $x^4 + 1$. Knowing the decomposition and using the general method one can obtain

$$\int \frac{dx}{x^4+1} = \frac{1}{4\sqrt{2}} \ln \left| \frac{x^2+\sqrt{2}\,x+1}{x^2-\sqrt{2}\,x-1} \right| + \frac{1}{2\sqrt{2}} \arctan \frac{\sqrt{2}\,x}{1-x^2} + C.$$

EXERCISES 9.4

(1) Find the integrals:

(a) $\int \frac{x^4}{x^4-1}\, dx,$

(b) $\int \frac{dx}{x^3+1},$

(c) $\int \frac{dx}{x^4+x^2+1},$

(d) $\int \frac{dx}{(1+x^2)^2},$

(e) $\int \frac{3x+5}{(x^2+2x+2)^2}\, dx,$

(f) $\int \frac{dx}{x\,(x^7+1)},$

(g) $\int \frac{dx}{x\,(x^5+1)^2},$

(h) $\int \frac{x^2\, dx}{(x-1)^{10}}.$

(2) Prove that $\arctan(\sqrt{2}\,x - 1) - \arctan(\sqrt{2}\,x + 1) = \arctan x^2 - \frac{\pi}{2}.$

9.5   Integration of some irrational expressions

The integral of a rational function of x and $\sqrt{ax^2+bx+c}$ can always be expressed by elementary functions. This theorem is due to Euler (Leonhard Euler 1707–1783) who introduced three types of substitutions, called *Euler's substitutions*. However that general theory is seldom applied in practice, because with some ingenuity one can often find a better way to obtain the result.

Example 9.5.1. To find the integral $I = \int \frac{dx}{\sqrt{x^2-6x+15}}$ we substitute $u = x - 3$ and obtain

$$I = \int \frac{du}{\sqrt{u^2+6}} = \ln\left(u + \sqrt{u^2+6}\right) + C = \ln\left(x - 3 + \sqrt{x^2+6x+15}\right) + C.$$

Example 9.5.2. For the integral $I = \int \frac{dx}{\sqrt{4-2x-x^2}}$ we substitute $\sqrt{5}\,u = x+1$ and get

$$I = \int \frac{du}{\sqrt{1-u^2}} = \arcsin u + C = \arcsin \frac{x+1}{\sqrt{5}} + C.$$

Example 9.5.3.

$$\int \frac{x-5}{\sqrt{x^2-2x+5}}\, dx = \frac{1}{2} \int \frac{2x-2}{\sqrt{x^2-2x+5}}\, dx - 4 \int \frac{dx}{\sqrt{x^2-2x+5}}$$

$$= \sqrt{x^2-2x+5} - 4 \int \frac{dx}{\sqrt{(x-1)^2+4}}$$

$$= \sqrt{x^2-2x+5} - 4\ln\left| x-1 + \sqrt{x^2-2x+5} \right| + C.$$

Example 9.5.4. To evaluate the integral $I = \int \sqrt{x^2 - 2x + 5}\, dx$ we start by integrating by parts

$$I = \int \sqrt{x^2 - 2x + 5}\, dx = x\sqrt{x^2 - 2x + 5} - I_1,$$

where

$$I_1 = \int \frac{x^2 - x}{\sqrt{x^2 - 2x + 5}}\, dx = I + \int \frac{x - 5}{\sqrt{x^2 - 2x + 5}}\, dx.$$

Hence

$$I = \frac{1}{2}\left(x\sqrt{x^2 - 2x + 5} - \int \frac{x - 5}{\sqrt{x^2 - 2x + 5}}\, dx\right)$$

$$= \frac{1}{2}(x - 1)\sqrt{x^2 - 2x + 5} + 2\ln\left|x - 1 + \sqrt{x^2 - 2x + 5}\right| + C,$$

where the last equality follows from Example 9.5.3.

Now we consider the integral

$$\int x^p(a + x)^q dx, \tag{9.8}$$

which is sometimes called the integral of a *binomial differential*. Already Newton knew that this integral can be reduced, by a suitable substitution, to an integral of a rational function, whenever one of the three rational numbers p, q, and $p+q$ is an integer. However, we owe to the Russian mathematician Pafnuti Lvovich Chebyshev (1821–1894) a proof that these are the only cases in which such a reduction is possible. Although suitable substitutions suggest themselves, we will consider a special case and leave as exercises the proof of Newton's observation in the other two cases.

We will show how to reduce (9.8) to the integral of a rational function in the case that p and q are rational and $p + q$ is an integer. We can write $p = \frac{m}{n}$ and $q = \frac{k}{n}$ where k, m, n are all integers and $n > 0$. We substitute $u = \left(\frac{x}{a+x}\right)^{1/n}$ and find that $x = \frac{au^n}{1-u^n}$ and $dx = \frac{nau^{n-1}}{(1-u^n)^2}du$. It follows that

$$\int x^p(a + x)^q dx = \int \left(\frac{au^n}{1 - u^n}\right)^{\frac{m}{n}}\left(a + \frac{au^n}{1 - u^n}\right)^{\frac{k}{n}}\frac{nau^{n-1}}{(1 - u^n)^2}\, du$$

$$= \int \frac{na^{p+q+1}u^{m+n-1}}{(1 - u^n)^{p+q+2}}\, du,$$

which is the integral of a rational function.

Example 9.5.5. The integral $I = \int \frac{\sqrt[3]{1 + \sqrt[4]{x}}}{\sqrt{x}}\, dx$ has the form $\int x^m(a + x^n)^q dx$ which, as one can easily show, can always be reduced to the form (9.8). In our case we substitute $u = \sqrt[3]{1 + \sqrt[4]{x}}$ and find that $x = (u^3 - 1)^4$ and $dx = 12u^2(u^3 - 1)^3 du$.

Hence

$$I = 12 \int (u^6 - u^3)\, du = \frac{3}{7} u^4 (4u^3 - 7) + C = \frac{3}{7} (1 + \sqrt[4]{x})^{\frac{4}{3}} (4\sqrt[4]{x} - 3) + C.$$

Example 9.5.6. To find $\int \frac{dx}{\sqrt[4]{1+x^4}}$ we first substitute $u = x^4$. Then $x = u^{1/4}$, $dx = \frac{1}{4} u^{-3/4} du$, and

$$\int \frac{dx}{\sqrt[4]{1+x^4}} = \int \frac{1}{\sqrt[4]{1+u}} \frac{1}{4} u^{-\frac{3}{4}}\, du = \frac{1}{4} \int u^{-\frac{3}{4}} (1+u)^{-\frac{1}{4}}\, du.$$

Next we substitute $t = \left(\frac{u}{1+u} \right)^{1/4}$ and find that $u = \frac{t^4}{1-t^4}$ and $du = \frac{4t^3}{(1-t^4)^2} dt$. Consequently,

$$\int \frac{dx}{\sqrt[4]{1+x^4}} = \frac{1}{4} \int \left(\frac{t^4}{1-t^4} \right)^{-\frac{3}{4}} \left(1 + \frac{t^4}{1-t^4} \right)^{-\frac{1}{4}} \frac{4t^3}{(1-t^4)^2}\, dt$$

$$= \int \frac{dt}{1-t^4} = \frac{1}{2} \int \frac{dt}{1-t^2} + \frac{1}{2} \int \frac{dt}{1+t^2}$$

$$= \frac{1}{4} \int \frac{dt}{1-t} + \frac{1}{4} \int \frac{dt}{1+t} + \frac{1}{2} \arctan t$$

$$= -\frac{1}{4} \ln |1-t| + \frac{1}{4} \ln |1+t| + \frac{1}{2} \arctan t + C$$

$$= \frac{1}{4} \ln \left| \frac{1+t}{1-t} \right| + \frac{1}{2} \arctan t + C$$

$$= \frac{1}{4} \ln \left| \frac{x + \sqrt[4]{1+x^4}}{x - \sqrt[4]{1+x^4}} \right| + \frac{1}{2} \arctan \frac{x}{\sqrt[4]{1+x^4}} + C.$$

Example 9.5.7. To find $I = \int \frac{dx}{x\sqrt[3]{1+x^5}}$ we substitute $x = (u^3 - 1)^{\frac{1}{5}}$. Since $dx = \frac{3}{5} t^2 (t^3 - 1)^{-\frac{4}{5}}$, we have

$$I = \frac{3}{5} \int \frac{u\, du}{u^3 - 1} = \frac{1}{5} \int \left(\frac{1}{u-1} - \frac{u-1}{u^2+u+1} \right) du$$

$$= \frac{1}{10} \ln \frac{(u-1)^2}{u^2+u+1} + \frac{\sqrt{3}}{5} \arctan \frac{2u+1}{\sqrt{3}} + C.$$

Substituting back $u = \sqrt[3]{1+x^5}$ we obtain

$$I = \frac{1}{10} \ln \frac{(\sqrt[3]{1+x^5} - 1)^2}{(\sqrt[3]{1+x^5})^2 + \sqrt[3]{1+x^5} + 1} + \frac{\sqrt{3}}{5} \arctan \frac{2\sqrt[3]{1+x^5} + 1}{\sqrt{3}} + C.$$

EXERCISES 9.5

(1) Show how to reduce (9.8) to the integral of a rational function in the case when p is an integer and q is rational.

(2) Show how to reduce (9.8) to the integral of a rational function in the case when q is an integer and p is rational.

(3) Make a substitution in the integral $\int x^m (a + x^n)^q \, dx$ to reduce it to the form (9.8).

(4) Find the following integrals:

(a) $\int x^2 (4 + x)^{\frac{2}{3}} \, dx$,

(b) $\int \frac{x^6}{\sqrt{1-x^2}} \, dx$,

(c) $\int \frac{dx}{(x+1)^3 \sqrt{x^2+2x}}$,

(d) $\int \frac{dx}{x^4 \sqrt{1+x^2}}$,

(e) $\int \frac{dx}{x^2 (2+x^3)^{\frac{5}{3}}}$,

(f) $\int x^{-\frac{3}{2}} (1 + x^{\frac{3}{4}})^{-\frac{1}{3}} \, dx$.

9.6 Integration of trigonometric functions

In this section we consider some integrals with trigonometric functions.

Consider the integral $\int \sin x \cos x \, dx$. Applying the general formula

$$\int f(x) f'(x) \, dx = \frac{1}{2} (f(x))^2 + C,$$

we obtain

$$\int \sin x \cos x \, dx = \frac{1}{2} \sin^2 x + C.$$

The same integral can be solved with the use of the trigonometric identity $\sin x \cos x = \frac{1}{2} \sin 2x$:

$$\int \sin x \cos x \, dx = \frac{1}{2} \int \sin 2x \, dx = -\frac{1}{4} \cos 2x + C.$$

Note that the obtained answers look different. This simple example shows that a variety of methods can be used to solve integrals of trigonometric functions and that different methods often lead to answers that appear different.

Example 9.6.1.

$$\int \frac{\sin^5 x}{\cos^4 x} \, dx = -\int \frac{(1 - \cos^2 x)^2 (\cos x)'}{\cos^4 x} \, dx$$

$$= -\int \frac{(\cos x)'}{\cos^4 x} \, dx + 2 \int \frac{(\cos x)'}{\cos^2 x} \, dx + \int \sin x \, dx$$

$$= \frac{1}{3 \cos^3 x} - \frac{2}{\cos x} - \cos x + C.$$

Example 9.6.2. To find the integral $\int \sin^2 x \cos^4 x \, dx$ we can substitute $u = \tan x$ and get the integral of a rational function

$$\int \frac{u^2 \, du}{(1+u^2)^4}.$$

The same integral can be evaluated in another way. Since

$$\sin^2 x \cos^4 x = \frac{1}{8} \sin^2 2x \, (\cos 2x + 1) = \frac{1}{8} \sin^2 2x \cos 2x + \frac{1}{16}(1 - \cos 4x),$$

we have

$$\int \sin^2 x \cos^4 x \, dx = \frac{1}{48} \sin^3 2x + \frac{1}{16} x - \frac{1}{64} \sin 4x + C.$$

Example 9.6.3.

$$\int \sin 7x \cos 4x \, dx = \int \frac{1}{2}(\sin 11x + \sin 3x) \, dx = -\frac{1}{22} \cos 11x - \frac{1}{6} \cos 3x + C.$$

In the above example we used the following trigonometric identity:

$$\sin mx \cos nx = \frac{1}{2}\left(\sin(m+n)x + \sin(m-n)x\right).$$

Here are other identities that may be useful in similar integrals:

$$\cos mx \cos nx = \frac{1}{2}\left(\cos(m+n)x + \cos(m-n)x\right),$$

$$\sin mx \sin nx = \frac{1}{2}\left(\cos(m-n)x - \cos(m+n)x\right).$$

Example 9.6.4. To find the integral $\int \frac{\sin^2 x \cos x}{\sin x + \cos x} \, dx$ we substitute $u = \tan x$:

$$\int \frac{\sin^2 x \cos x}{\sin x + \cos x} \, dx = \int \frac{u^2 \, du}{(1+u)(1+u^2)^2}$$

$$= \int \left(\frac{1}{4}\frac{1}{u+1} - \frac{1}{4}\frac{u-1}{u^2+1} + \frac{1}{2}\frac{u-1}{(u^2+1)^2}\right) du$$

$$= \frac{1}{4} \ln \frac{1+u}{\sqrt{1+u^2}} - \frac{1}{4}\frac{1+u}{1+u^2} + C$$

$$= \frac{1}{4} \ln |\cos x + \sin x| - \frac{1}{4} \cos x \, (\sin x + \cos x) + C.$$

There is a general method of solving integrals of the form $R(\sin x, \cos x)$, where R denotes a rational function of two variables. The general substitution which works for every R is $u = \tan \frac{x}{2}$ $(-\pi < x < \pi)$, because then

$$\sin x = \frac{2 \tan \frac{x}{2}}{1 + \left(\tan \frac{x}{2}\right)^2} = \frac{2u}{1+u^2}$$

and

$$\cos x = \frac{1 - \left(\tan \frac{x}{2}\right)^2}{1 + \left(\tan \frac{x}{2}\right)^2} = \frac{1 - u^2}{1 + u^2}.$$

This leads to an integral of the form

$$\int R(\sin x, \cos x)\, dx = \int R\left(\frac{2u}{1+u^2}, \frac{1-u^2}{1+u^2}\right) \frac{2\, du}{1+u^2},$$

which is the integral of a rational function, hence always solvable. Although this method always applies, it often leads to unduly complicated calculations. Before applying this general method we should always look for another, more suitable approach. This leaves an open field for creativity and individuality. It seems that in the problem of integrating trigonometric functions knowing trigonometry is at least as important as knowing calculus.

EXERCISES 9.6

(1) Explain why when calculating $\int \sin x \cos x\, dx$ using two different methods at the beginning of this section we obtained two different answers.

(2) Find the following integrals:

(a) $\int \frac{\cos^5 x}{\sin^3 x}\, dx$,

(b) $\int \frac{dx}{\sin^2 x \cos^4 x}$,

(c) $\int \frac{dx}{\sqrt{\tan x}}$,

(d) $\int \sin x \sin 2x \sin 3x\, dx$,

(e) $\int \frac{dx}{\sqrt{\sin x \cos^3 x}}$,

(f) $\int \cos \frac{x}{2} \cos \frac{x}{3}\, dx$,

(g) $\int \frac{dx}{\sin x + \cos x}$,

(h) $\int \frac{dx}{\cos x + 2\sin x + 3}$.

Chapter 10

THE LEBESGUE INTEGRAL

10.1 Introduction

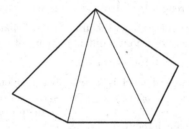

Fig. 10.1   Area calculated by triangulation.

Measuring the area of a geometrical figure in the plane in elementary geometry often consists of dividing it into a finite number of smaller parts like rectangles or triangles, whose areas are easy to calculate, and then summing up the areas of all the parts, see Fig. 10.1. This method fails, however, when the figure is bounded by a curve like a circle. In such a case the figure might be decomposed into parts such that each of them is contained between a horizontal line and an arc of a curve. For instance, in Fig. 10.2 there are six such parts. This procedure reduces the evaluation of the area of an arbitrary figure to the evaluation of the area between an arc representing a function and a horizontal line segment.

The last task may be undertaken using various methods leading to the concept of integral. For continuous functions the integral was already considered by Isaac Barrow (1630–1677), the teacher of Newton. The precise definition is due to Cauchy. Nowadays Riemann (Georg Friedrich Bernhard Riemann 1826–1866) and Lebesgue (Henri Léon Lebesgue 1875–1941) are the best known names in the theory of integration. Riemann extended Cauchy's definition to include some discontinuous functions. However the Riemann integral is unsatisfactory in certain respects. Lebesgue made an essential change in the approach to integration which allows us to deal conveniently with infinite sequences and series of functions.

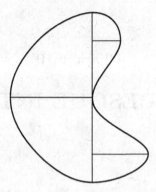

Fig. 10.2 Area calculated by integrals.

S: Then what is the reason for introducing the Riemann integral?

T: Because its definition is easy and requires little preparation. On the other hand, Lebesgue's is usually based on a difficult theory of measure.

S: And what is the approach in this book?

T: We introduce an integral which is completely equivalent to the Lebesgue integral. But instead of using measure theory as the basis for this integral, we use the theory of absolutely convergent series. The definition is actually simpler than that of the Riemann integral but as general as Lebesgue's. It produces the Lebesgue integral, but is introduced in an elementary way.

10.2 Integrable functions

By a *brick* we mean a bounded half-closed interval of the form $[a, b)$, closed on the left and open on the right. Since bricks are all of this form, they form a single brick when placed end-to-end, one after another. A function that is equal to 1 at every point of a brick J and 0 elsewhere is called a *brick function* and J its *support*. By the integral $\int f$ of a brick function f we mean the length of its support, that is, if the support is $[a, b)$, then $\int f = b - a$.

S: The symbol $\int f$ was used before to denote an antiderivative of f, a function. Now $\int f$ is a number. Is this correct?

T: No, it is certainly an inconsistency to use the same symbol for different objects, but we will follow this long tradition which originates from a connection between the integral and antiderivative. It does not lead to any misunderstanding here since it will always be clear from the context which one we have in mind. In this chapter $\int f$ will always denote the integral of f as a number.

Definition 10.2.1. Given a function f on $\mathbb{R}$ we will write

$$f \simeq \lambda_1 f_1 + \lambda_2 f_2 + \dots \quad \text{or} \quad f \simeq \sum_{n=1}^{\infty} \lambda_n f_n, \tag{10.1}$$

where $f_1, f_2, \dots$ are brick functions and $\lambda_1, \lambda_2, \dots$ are real numbers, if

$$\mathbb{I} \quad \sum_{n=1}^{\infty} |\lambda_n| \int f_n < \infty$$

and

$$\mathbb{III} \quad f(x) = \sum_{n=1}^{\infty} \lambda_n f_n(x) \quad \text{at every point } x \text{ at which } \sum_{n=1}^{\infty} |\lambda_n| f_n(x) < \infty.$$

If $f \simeq \lambda_1 f_1 + \lambda_2 f_2 + \dots$, then we say that f *expands into a series of brick functions*. A function expandable into a series of brick functions is called *Lebesgue integrable* or simply *integrable*.

We cannot use the symbol $=$ instead of $\simeq$ because the series need not converge at every point. Note that $\mathbb{I}$ means simply that the series $\sum_{n=1}^{\infty} |\lambda_n| \int f_n$ converges. The integral of an integrable function $f \simeq \lambda_1 f_1 + \lambda_2 f_2 + \dots$ will be defined as $\int f = \lambda_1 \int f_1 + \lambda_2 \int f_2 + \dots$.

S: This definition is remarkably short and easy to remember. But a function can expand in various series of brick functions, so how do we know whether the value of the integral will be always the same?

T: At first we don't know. The proof that it is so is not short, not easy to remember, and will not be concluded prior to Section 10.5. To make the matter easier we will start with a few heuristic remarks. The core of the reasoning is the fact that the integral of a nonnegative function is a nonnegative number. Before we prove it we need to establish some properties of the so-called step functions. They appear naturally as partial sums of series of brick functions.

EXERCISES 10.2

(1) Prove: If $f \simeq \lambda_1 f_1 + \lambda_2 f_2 + \dots$ and $\alpha \in \mathbb{R}$, then $\alpha f \simeq \alpha \lambda_1 f_1 + \alpha \lambda_2 f_2 + \dots$.

(2) Prove: If $f \simeq \lambda_1 f_1 + \lambda_2 f_2 + \dots$ and $g \simeq \gamma_1 g_1 + \gamma_2 g_2 + \dots$, then $f + g \simeq \lambda_1 f_1 + \gamma_1 g_1 + \lambda_2 f_2 + \gamma_2 g_2 + \dots$.

(3) Let f be an integrable function and let $g(x) = f(ax + b)$ for some $a, b \in \mathbb{R}$, $a \neq 0$. Show that g is an integrable function.

10.3   Examples of expansions of integrable functions

We want to show what an expansion into a series of brick functions looks like in a few particular cases. To make the exposition clearer we use the following notation. A brick function whose support is $[\alpha, \beta)$ is denoted by $\chi_{[\alpha,\beta)}$. In general, if S is a subset of $\mathbb{R}$, then by χ_S we denote the *characteristic function* of S, that is,

$$\chi_S(x) = \begin{cases} 1, & \text{for } x \in S, \\ 0, & \text{for } x \notin S. \end{cases}$$

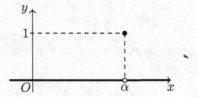

Fig. 10.3   A discontinuous integrable function.

Example 10.3.1. The function

$$\chi_{\{\alpha\}}(x) = \begin{cases} 1, & \text{for } x = \alpha, \\ 0, & \text{for } x \neq \alpha, \end{cases}$$

(Fig. 10.3) is integrable. It is easy to construct an expansion which converges to $\chi_{\{\alpha\}}$ at every point. Indeed, let (β_n) be a decreasing sequence of numbers such that $\lim_{n\to\infty} \beta_n = \alpha$. Then

$$\chi_{\{\alpha\}} \simeq \chi_{[\alpha,\beta_1)} - \chi_{[\beta_2, \beta_1)} - \chi_{[\beta_3,\beta_2)} - \cdots.$$

Note that $(\beta_1 - \alpha) - (\beta_1 - \beta_2) - (\beta_2 - \beta_3) - \cdots = 0$, as expected.

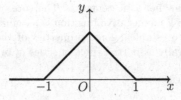

Fig. 10.4   A continuous integrable function.

Example 10.3.2. The continuous function

$$f(x) = \begin{cases} 1 + x & \text{for } -1 \leq x < 0, \\ 1 - x & \text{for } 0 \leq x < 1, \\ 0 & \text{elsewhere,} \end{cases}$$

(see Fig. 10.4) can be expanded as follows:

$$f \simeq \frac{1}{2}\chi_{[-\frac{1}{2},\frac{1}{2})} + \frac{1}{4}\chi_{[-\frac{3}{4},-\frac{2}{4})} + \frac{1}{4}\chi_{[-\frac{1}{4},\frac{1}{4})} + \frac{1}{4}\chi_{[\frac{2}{4},\frac{3}{4})} + \frac{1}{8}\chi_{[-\frac{7}{8},-\frac{6}{8})} + \frac{1}{8}\chi_{[-\frac{5}{8},-\frac{4}{8})}$$

$$+ \frac{1}{8}\chi_{[-\frac{3}{8},-\frac{2}{8})} + \frac{1}{8}\chi_{[-\frac{1}{8},\frac{1}{8})} + \frac{1}{8}\chi_{[\frac{2}{8},\frac{3}{8})} + \frac{1}{8}\chi_{[\frac{4}{8},\frac{5}{8})} + \frac{1}{8}\chi_{[\frac{6}{8},\frac{7}{8})} + \cdots$$

The partial sums consisting of one, four, and eleven terms are shown in Fig. 10.5.

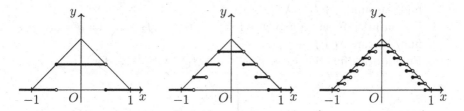

Fig. 10.5 Examples of partial sums of an expansion of the function in Example 10.3.2.

Note that

$$\frac{1}{2} + \frac{1}{16} + \frac{1}{8} + \frac{1}{16} + \frac{1}{64} + \frac{1}{64} + \frac{1}{64} + \frac{1}{32} + \frac{1}{64} + \frac{1}{64} + \frac{1}{64} + \cdots$$

$$= \frac{1}{2} + \frac{1}{4} + \frac{1}{8} + \cdots = 1,$$

as expected.

A similar construction can be used for any integrable continuous function. Of course, the construction is, in general, much more complicated.

Example 10.3.3. If $\sum_{n=1}^{\infty} |\lambda_n| < \infty$, then for the function

$$f(x) = \begin{cases} \lambda_n, & \text{for } n \le x \le n+1, \\ 0, & \text{elsewhere,} \end{cases}$$

we have

$$f \simeq \lambda_1 \chi_{[1,2)} + \lambda_2 \chi_{[2,3)} + \cdots.$$

EXERCISES 10.3

(1) Verify whether the following are valid expansions:

(a) $\chi_{\{0\}} \simeq \chi_{[0,\frac{1}{2})} - \chi_{[0,\frac{1}{2})} + \chi_{[0,\frac{1}{4})} - \chi_{[0,\frac{1}{4})} + \cdots + \chi_{[0,2^{-n})} - \chi_{[0,2^{-n})} + \cdots$,

(b) $\chi_{\{0\}} \simeq \chi_{[0,\frac{1}{2})} - \chi_{[0,\frac{1}{2})} + 2\chi_{[0,\frac{1}{4})} - 2\chi_{[0,\frac{1}{4})} + \cdots + n\chi_{[0,2^{-n})} - n\chi_{[0,2^{-n})} + \cdots$,

(c) $\chi_{\{0\}} \simeq \chi_{[0,1)} - \chi_{[0,1)} + \chi_{[0,\frac{1}{2})} - \chi_{[0,\frac{1}{2})} + \cdots + \chi_{[0,\frac{1}{n})} - \chi_{[0,\frac{1}{n})} + \cdots$,

(d) $\chi_{\{0\}} \simeq \chi_{[0,1)} - \chi_{[0,1)} + \frac{1}{2}\chi_{[0,\frac{1}{2})} - \frac{1}{2}\chi_{[0,\frac{1}{2})} + \cdots + \frac{1}{n}\chi_{[0,\frac{1}{n})} - \frac{1}{n}\chi_{[0,\frac{1}{n})} + \cdots$,

(e) $\chi_{\{0\}} \simeq \frac{1}{2}\chi_{[0,1)} - \frac{1}{2}\chi_{[0,1)} + \frac{1}{4}\chi_{[0,\frac{1}{2})} - \frac{1}{4}\chi_{[0,\frac{1}{2})} + \cdots + \frac{1}{2^n}\chi_{[0,\frac{1}{n})} - \frac{1}{2^n}\chi_{[0,\frac{1}{n})} + \cdots$.

(2) Can this be a valid expansion for some function f?

$$f \simeq \chi_{[0,1)} - \chi_{[0,\frac{1}{\sqrt{2}})} + \chi_{[0,\frac{1}{\sqrt{3}})} - \cdots + (-1)^{n+1}\chi_{[0,\frac{1}{\sqrt{n}})} + \cdots.$$

(3) Find a brick function expansion $f \simeq \lambda_1 f_1 + \lambda_2 f_2 + \ldots$ for $f = \chi_{[a,b]}$ and find the sum $\lambda_1 \int f_1 + \lambda_2 \int f_2 + \ldots$.

(4) Find a brick function expansion $f \simeq \lambda_1 f_1 + \lambda_2 f_2 + \ldots$ for $f = \chi_{(a,b)}$ and find the sum $\lambda_1 \int f_1 + \lambda_2 \int f_2 + \ldots$.

(5) Find a brick function expansion $f \simeq \lambda_1 f_1 + \lambda_2 f_2 + \ldots$ for $f = \chi_{(a,b]}$ and find the sum $\lambda_1 \int f_1 + \lambda_2 \int f_2 + \ldots$.

(6) Find a brick function expansion $f \simeq \lambda_1 f_1 + \lambda_2 f_2 + \ldots$ for

$$f(x) = \begin{cases} 0, & \text{if } x < 0, \\ x, & \text{if } 0 \leq x \leq 1, \\ 0, & \text{if } x > 1, \end{cases}$$

and find the sum $\lambda_1 \int f_1 + \lambda_2 \int f_2 + \ldots$.

10.4 Step functions

Definition 10.4.1. A linear combination of a finite number of brick functions is called a *step function*.

According to the above definition, a function f is a step function if there exist numbers $\lambda_1, \ldots, \lambda_n$ and brick functions $f_1, \ldots, f_n$ such that

$$f(x) = \lambda_1 f_1(x) + \cdots + \lambda_n f_n(x) \quad \text{for every } x \in \mathbb{R}. \tag{10.2}$$

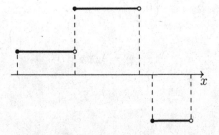

Fig. 10.6   A step function.

An example of a step function is shown in Fig. 10.6. Clearly, every step function can be represented in the form (10.2) in infinitely many ways. On the other hand, it should be evident that there is exactly one representation such that the supports of the brick functions $f_1, \ldots, f_n$ are disjoint and a minimal number of brick functions is used. To obtain such a representation, denote by $a_0, \ldots, a_m$ all points of discontinuity of f arranged into an increasing sequence, that is, $a_0 < a_1 < \cdots < a_m$. Then, for $k = 1, \ldots, m$, define

$$g_k = \chi_{[a_{k-1}, a_k)} \quad \text{and} \quad \gamma_k = f(a_{k-1}).$$

If each $\gamma_k \neq 0$, then $f = \gamma_1 g_1 + \cdots + \gamma_m g_m$ is the desired representation. If some $\gamma_k = 0$, we merely delete each term $\gamma_k g_k$ for which $\gamma_k = 0$ to obtain the desired representation of f. Such a representation will be called the *basic representation* of a step function f.

From the definition it follows that $f \equiv 0$ is a step function. However in this case the question of a basic representation does not make much sense. For convenience, $f = 0$ will be called the basic representation for this function.

The set of all points $x \in \mathbb{R}$ for which $f(x) \neq 0$ is called the *support* of f and denoted by supp f. If f is a step function, a point $x_0 \in$ supp f is an interior point of supp f if it is not an endpoint of the support of any of the bricks used in the basic representation of f.

It is easily seen that a linear combination of a finite number of step functions is a step function. The absolute value of a step function is a step function. Since

$$\min\{f, g\} = \frac{1}{2}(f + g - |f - g|) \quad \text{and} \quad \max\{f, g\} = \frac{1}{2}(f + g + |f - g|),$$

the minimum and maximum of two step functions are step functions too.

The next lemma looks like an obvious statement that should not require a proof. However, it is necessary to establish this fact before we can define the integral of a step function.

Lemma 10.4.2. *Let $f_1, \ldots, f_n$ be brick functions. If $\lambda_1 f_1 + \cdots + \lambda_n f_n = 0$, then*

$$\lambda_1 \int f_1 + \cdots + \lambda_n \int f_n = 0.$$

Proof. Let $f_k = \chi_{[a_k, b_k)}$ for $k = 1, \ldots, n$ and let $c_1, \ldots, c_m$ be a complete listing of all different points in the set $\{a_1, \ldots, a_n, b_1, \ldots, b_1\}$ arranged so that $c_1 < c_2 < \cdots < c_m$. For each $i = 1, \ldots, m - 1$, let $g_i = \chi_{[c_i, c_{i+1})}$. Then every brick function f_k can be represented in the form

$$f_k = g_{i_k} + g_{i_k+1} + \cdots + g_{i_k+l_k},$$

where $c_{i_k} = a_k$ and $c_{i_k+l_k} = b_k$. Observe that

$$
\begin{aligned}
0 &= \lambda_1 f_1 + \cdots + \lambda_n f_n \\
&= \lambda_1 \left(g_{i_1} + g_{i_1+1} + \cdots + g_{i_1+l_1} \right) + \cdots + \lambda_n \left(g_{i_n} + g_{i_n+1} + \cdots + g_{i_n+l_n} \right) \quad (10.3) \\
&= \eta_1 g_1 + \cdots + \eta_{m-1} g_{m-1},
\end{aligned}
$$

where in the last line we merely collected the coefficients of each g_i and called their sum η_i. Since the supports of the g_i's are disjoint, we conclude from (10.3) that each $\eta_i = 0$. Consequently,

$$\lambda_1 \int f_1 + \cdots + \lambda_n \int f_n$$

$$= \lambda_1 \int (g_{i_1} + g_{i_1+1} + \cdots + g_{i_1+l_1}) + \cdots + \lambda_n \int (g_{i_n} + g_{i_n+1} + \cdots + g_{i_n+l_n})$$

$$= \eta_1 \int g_1 + \cdots + \eta_{m-1} \int g_{m-1} = 0 \cdot \int g_1 + \cdots + 0 \cdot \int g_{m-1} = 0,$$

proving the lemma. $\square$

Lemma 10.4.3. *Let*

$$\sum_{k=1}^{m} \lambda_k f_k = \sum_{k=1}^{n} \gamma_k g_k, \tag{10.4}$$

where f_k and g_k are brick functions. Then

$$\sum_{k=1}^{m} \lambda_k \int f_k = \sum_{k=1}^{n} \gamma_k \int g_k. \tag{10.5}$$

Proof. Since

$$\sum_{k=1}^{m} \lambda_k f_k - \sum_{k=1}^{n} \gamma_k g_k = 0,$$

we have, by the Lemma 10.4.2,

$$\sum_{k=1}^{m} \lambda_k \int f_k - \sum_{k=1}^{n} \gamma_k \int g_k = 0.$$

$\square$

The above lemma ensures that the following definition is correct, that is, that the defined integral does not depend on a particular representation of f as a linear combination of brick functions.

Definition 10.4.4. The *integral of a step function $f = \lambda_1 f_1 + \cdots + \lambda_n f_n$* is defined as

$$\int f = \lambda_1 \int f_1 + \cdots + \lambda_n \int f_n.$$

S: The symbol $\int f$ is used in this definition even though f is not a brick function. Is that correct?

T: Yes, it is correct. A brick function f is a step function and $f = f$ is its basic representation. Lemma 10.4.3 guarantees that the integral of the brick function f and the step function f are the same. Thus Definition 10.4.4 extends the use of the symbol $\int$ in a consistent way to a larger class of functions.

Theorem 10.4.5. *Let f and g be step functions.*

(a) $\displaystyle\int (f+g) = \int f + \int g,$

(b) $\displaystyle\int \lambda f = \lambda \int f$ *for any* $\lambda \in \mathbb{R}$,

(c) $f \leq g$ *implies* $\displaystyle\int f \leq \int g,$

(d) $\displaystyle\left| \int f \right| \leq \int |f|,$

(e) *If* $|f| \leq M$ *and the support of* f *is contained in the interval* $[a,b]$, *then*
$$\left| \int f \right| \leq (b-a)M.$$

Proof. The above properties follow easily from the definition of the integral of step functions and Lemma 10.4.3. □

We close this section with three auxiliary lemmas. The first technical lemma is only used to prove the second one. The second one is used to prove the third one, but it is also interesting in itself. The result of the third lemma is crucial in the proof of consistency of our general definition of the integral.

Lemma 10.4.6. *Let f be a nonnegative step function and let* $a_0 < a_1 < \cdots < a_n$ *be all the points of discontinuity of f. For* $\varepsilon > 0$ *we define a number* η *and a function g as follows:*

$$\eta = \min \left\{ \frac{a_1 - a_0}{4}, \frac{a_2 - a_1}{4}, \ldots, \frac{a_n - a_{n-1}}{4}, \frac{\varepsilon}{2 \cdot \sum_{k=1}^{n} f(a_{k-1})} \right\}$$

and

$$g = \sum_{k=1}^{n} f(a_{k-1}) \chi_{[a_{k-1}+\eta, a_k - \eta)}.$$

Then

$$g(x) = 0 \quad \text{whenever } f(x) \neq g(x), \tag{10.6}$$

and

$$\int (f - g) \leq \varepsilon. \tag{10.7}$$

Proof. The method of construction of the function g is shown in Fig. 10.7. The gray lines represent the graph of f and the black lines on top of the gray lines show

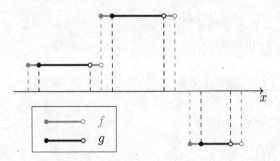

Fig. 10.7   Step functions f and g in Lemma 10.4.6.

how the support of f is shrunk to produce the graph of g. On the shrunken support f and g agree so that (10.6) is satisfied. To see that (10.7) also holds notice that

$$\int g = \sum_{k=1}^{n} f(a_{k-1})(a_k - a_{k-1} - 2\eta)$$

$$= \sum_{k=1}^{n} f(a_{k-1})(a_k - a_{k-1}) - 2\eta \sum_{k=1}^{n} f(a_{k-1})$$

$$= \int f - 2\eta \sum_{k=1}^{n} f(a_{k-1}).$$

Thus,

$$\int (f - g) = 2\eta \cdot \sum_{k=1}^{n} f(a_{k-1}) \le \varepsilon.$$

$\square$

The statement in the following lemma may seem obvious. However, careful analysis of the described property reveals that it is not obvious at all. The proof of this lemma is arguably the most difficult proof in this book.

Lemma 10.4.7. *If (f_n) is a non-increasing sequence of step functions which converges to 0 at every point, then $\int f_n \to 0$ as $n \to \infty$.*

Proof. Suppose (f_n) is a non-increasing sequence of step functions which converges to 0 at every point, but the sequence $(\int f_n)$ does not converge to 0. First notice that, since each step function f_n is nonnegative and the sequence $(\int f_n)$ is non-increasing and bounded below by 0, there is some $M > 0$ such that $\int f_n \to M$ as $n \to \infty$. For each n define a nonnegative step function g_n for f_n and $\varepsilon_n = M/2^{n+1}$ as in Lemma 10.4.6 so that we have

$$g_n(x) = 0 \quad \text{whenever } f_n(x) \neq g_n(x), \tag{10.8}$$

and

$$\int (f_n - g_n) \le \frac{M}{2^{n+1}}.$$

Next define

$$h_1 = g_1$$

and,

$$h_n = g_n - (f_1 - g_1) - (f_2 - g_2) - \cdots - (f_{n-1} - g_{n-1}) \qquad (10.9)$$

for $n = 2, 3, \ldots$. Then

$$\int (f_n - h_n) = \int (f_n - g_n) + \int (f_1 - g_1) + \cdots + \int (f_{n-1} - g_{n-1})$$

$$< \frac{M}{2^{n+1}} + \frac{M}{2^2} + \frac{M}{2^3} + \cdots + \frac{M}{2^n}$$

$$= \frac{M}{2} \left[\frac{1}{2} + \frac{1}{2^2} + \cdots + \frac{1}{2^{n-1}} + \frac{1}{2^n} \right] < \frac{M}{2}.$$

It now follows that

$$M \leq \int f_n < \int h_n + \frac{M}{2}. \qquad (10.10)$$

There is some closed bounded interval $[a, b]$ that contains the support of each f_n. Let $x_n \in [a, b]$ be chosen such that h_n assumes its largest value at x_n. Then from (10.10) and Theorem 10.4.5(e) it follows that

$$\frac{M}{2} \leq \int h_n \leq h_n(x_n)(b - a).$$

Consequently, for each n,

$$h_n(x_n) \geq \frac{M}{2(b - a)}. \qquad (10.11)$$

By the Bolzano–Weierstrass theorem there exists a subsequence (x_{p_n}) of (x_n) that converges to some $x_0 \in [a, b]$. We claim that x_0 is an interior point of supp f_k for each $k \in \mathbb{N}$. To see this, fix a $k \in \mathbb{N}$. For $n > k$, it follows from the fact that the sequence (f_n) is non-increasing and from (10.8), (10.9), and (10.11) that

$$f_k(x_{p_n}) \geq g_{p_n}(x_{p_n}) \geq h_{p_n}(x_{p_n}) \geq \frac{M}{2(b - a)} > 0. \qquad (10.12)$$

If x_0 were not an interior point of suppf_k, then for each $n > k$ we would have $|x_{p_n} - x_0| > \eta_k > 0$, where η_k is defined as in Lemma 10.4.6. This, however, is a contradiction because (x_{p_n}) converges to x_0.

Now, since f_k is continuous at each interior point of its support, from (10.12) we get

$$f_k(x_0) = \lim_{n \to \infty} f_k(x_{p_n}) \geq \lim_{n \to \infty} g_{p_n}(x_{p_n}) \geq \lim_{n \to \infty} h_{p_n}(x_{p_n}) \geq \frac{M}{2(b - a)}.$$

Finally,

$$\lim_{k \to \infty} f_k(x_0) \geq \frac{M}{2(b - a)} > 0,$$

a contradiction. This completes the proof. $\qquad \square$

Lemma 10.4.8. *If (f_n) and (g_n) are non-decreasing sequences of step functions and*

$$\lim_{n\to\infty} f_n(x) \le \lim_{n\to\infty} g_n(x)$$

for every $x \in \mathbb{R}$, then

$$\lim_{n\to\infty} \int f_n \le \lim_{n\to\infty} \int g_n.$$

(In both inequalities we allow infinity as the limit.)

Proof. Let $m \in \mathbb{N}$ be fixed for now. Set

$$h_n = g_n - f_m. \tag{10.13}$$

We decompose h_n into its positive and negative parts

$$h_n = h_n^+ - h_n^-, \tag{10.14}$$

where $h_n^+ = \max\{h_n, 0\}$ and $h_n^- = \max\{-h_n, 0\}$. Then the sequence (h_n^-) is non-increasing and $\lim_{n\to\infty} h_n^- = 0$. By Lemma 10.4.7, we thus have

$$\lim_{n\to\infty} \int h_n^- = 0. \tag{10.15}$$

From (10.14) it follows that $\int h_n = \int h_n^+ - \int h_n^-$, and, by (10.15),

$$\lim_{n\to\infty} \int h_n = \lim_{n\to\infty} \int h_n^+ \ge 0.$$

But (10.13) implies $\int h_n = \int (g_n - f_m)$, and thus

$$\lim_{n\to\infty} \int g_n - \int f_m \ge 0,$$

that is,

$$\int f_m \le \lim_{n\to\infty} \int g_n.$$

Now, by letting $m \to \infty$, we obtain

$$\lim_{n\to\infty} \int f_n \le \lim_{n\to\infty} \int g_n,$$

which is the desired inequality. □

Corollary 10.4.9. *If (f_n) and (g_n) are non-decreasing sequences of step functions and*

$$\lim_{n\to\infty} f_n(x) = \lim_{n\to\infty} g_n(x)$$

for every $x \in \mathbb{R}$, then

$$\lim_{n\to\infty} \int f_n = \lim_{n\to\infty} \int g_n.$$

(In both equalities we allow infinity as the limit.)

EXERCISES 10.4

(1) Find basic representations for the following step functions:

    (a) $f = \chi_{[-2,2)} - 2\chi_{[-1,1)}$,

    (b) $g = \chi_{[0,2)} - 3\chi_{[1,3)}$,

    (c) $h = \chi_{[0,5)} + 2\chi_{[1,4)} - \chi_{[2,3)}$.

(2) True or false: If $f = \lambda_1 f_1 + \cdots + \lambda_n f_n$, then $|f| = |\lambda_1| f_1 + \cdots + |\lambda_n| f_n$?

(3) True or false: If $f = \lambda_1 f_1 + \cdots + \lambda_n f_n$ is the basic representation of f, then $|f| = |\lambda_1| f_1 + \cdots + |\lambda_n| f_n$?

(4) True or false: If $f = \lambda_1 f_1 + \cdots + \lambda_n f_n$ is the basic representation of f, then $|f| = |\lambda_1| f_1 + \cdots + |\lambda_n| f_n$ is the basic representation of $|f|$?

(5) Prove the following properties of step functions:

    (a) $\int (f + g) = \int f + \int g$.

    (b) $\int \lambda f = \lambda \int f$, for any $\lambda \in \mathbb{R}$.

    (c) $f \le g$ implies $\int f \le \int g$.

    (d) $|\int f| \le \int |f|$.

    (e) If $|f| \le M$ and supp $f \subset [a, b)$, then $|\int f| \le (b - a)M$.

(6) Let f be a step function and let $g(x) = f(ax + b)$ for some $a, b \in \mathbb{R}$, $a \ne 0$. Show that g is a step function and find $\int g$.

(7) Let f be a step function and let $g(x) = f(x + h)$ for some $h \in \mathbb{R}$. Show that $\int g = \int f$.

(8) Prove: If (f_n) and (g_n) are non-increasing sequences of step functions and $\lim_{n \to \infty} f_n(x) \le \lim_{n \to \infty} g_n(x)$ for every $x \in \mathbb{R}$, then $\lim_{n \to \infty} \int f_n \le \lim_{n \to \infty} \int g_n$.

(9) Prove: If (f_n) and (g_n) are non-decreasing sequences of step functions and $\lim_{n \to \infty} f_n(x) = \lim_{n \to \infty} g_n(x)$ for every $x \in \mathbb{R}$, then $\lim_{n \to \infty} \int f_n = \lim_{n \to \infty} \int g_n$.

(10) True or false: If (f_n) is a sequence of step functions which converges to 0 at every point, then $\lim_{n \to \infty} \int f_n = 0$?

(11) Let f and g be arbitrary functions. Show the following properties:

    (a) supp $(f + g) \subseteq$ supp $f \cup$ supp g.

    (b) supp $fg =$ supp $f \cap$ supp g.

    (c) supp $|f| =$ supp f.

    (d) supp $\lambda f =$ supp f whenever $\lambda \ne 0$.

    (e) supp $f \subseteq$ supp g whenever $|f| \le |g|$.

10.5   Basic properties of the integral

First we extend the use of the symbol $\simeq$. Let f be an arbitrary function and let (f_n) be a sequence of step functions. We write

$$f \simeq f_1 + f_2 + \ldots \quad \text{or} \quad f \simeq \sum_{n=1}^{\infty} f_n \tag{10.16}$$

if

$$\text{I} \quad \sum_{n=1}^{\infty} \int |f_n| < \infty$$

and

$$\text{III} \quad f(x) = \sum_{n=1}^{\infty} f_n(x) \quad \text{at every point } x \text{ at which } \sum_{n=1}^{\infty} |f_n(x)| < \infty.$$

If $f \simeq f_1 + f_2 + \ldots$, then we say that f *expands into a series of step functions.*

Lemma 10.5.1. *Let $f \simeq f_1 + f_2 + \ldots$ be an expansion of f into a series of step functions. If $f \geq 0$, then $\int f_1 + \int f_2 + \cdots \geq 0$.*

Proof. Let $\varepsilon > 0$. By I there exists an $n_0 \in \mathbb{N}$ such that

$$\sum_{n=n_0+1}^{\infty} \int |f_n| < \varepsilon. \tag{10.17}$$

For $n = 1, 2, \ldots$ define

$$g_n = f_1 + \cdots + f_{n_0} + |f_{n_0+1}| + \cdots + |f_{n_0+n}|$$

and

$$h_n = \max\{g_n, 0\}.$$

Clearly, the g_n's and h_n's are step functions and the sequences (g_n) and (h_n) are non-decreasing. Moreover, since $f \geq 0$, III implies that the limit $\lim_{n \to \infty} g_n(x)$ is either a nonnegative number or ∞. Therefore $\lim_{n \to \infty} g_n(x) = \lim_{n \to \infty} h_n(x)$ for every $x \in \mathbb{R}$. Thus, by Corollary 10.4.9, we have

$$\lim_{n \to \infty} \int g_n = \lim_{n \to \infty} \int h_n \geq 0.$$

Consequently

$$\int f_1 + \cdots + \int f_{n_0} + \int |f_{n_0+1}| + \int |f_{n_0+2}| + \cdots \geq 0,$$

which, in light of (10.17), implies that

$$\int f_1 + \cdots + \int f_{n_0} \geq - \left(\int |f_{n_0+1}| + \int |f_{n_0+2}| + \ldots \right) > -\varepsilon.$$

Thus,

$$\sum_{n=1}^{\infty} \int f_n \geq \int f_1 + \cdots + \int f_{n_0} - \int |f_{n_0+1}| - \int |f_{n_0+2}| - \cdots$$

$$\geq -2 \left(\int |f_{n_0+1}| + \int |f_{n_0+2}| + \ldots \right) > -2\varepsilon.$$

Since ε is an arbitrary positive number, we conclude that $\sum_{n=1}^{\infty} \int f_n \geq 0$. $\quad\square$

Lemma 10.5.2. *If* $f \simeq f_1 + f_2 + \dots$ *and* $f \simeq g_1 + g_2 + \dots$ *are expansions of* f *into series of step functions, then*

$$\int f_1 + \int f_2 + \dots = \int g_1 + \int g_2 + \dots .$$

Proof. If $f \simeq f_1 + f_2 + \dots$ and $f \simeq g_1 + g_2 + \dots$, then

$$0 \simeq f_1 - g_1 + f_2 - g_2 + \dots .$$

In view of Lemma 10.5.1, we have

$$\int f_1 - \int g_1 + \int f_2 - \int g_2 + \dots \geq 0.$$

Thus

$$\int f_1 + \int f_2 + \dots \geq \int g_1 + \int g_2 + \dots .$$

By considering $0 \simeq g_1 - f_1 + g_2 - f_2 + \dots$, we obtain similarly

$$\int f_1 + \int f_2 + \dots \leq \int g_1 + \int g_2 + \dots ,$$

which proves the desired equality. $\qquad\square$

   Integrable functions are defined in terms of brick functions (Definition 10.2.1). Reasons for choosing brick functions in the original definition are that brick functions are simpler yet sufficient and the expansion into a series of brick functions corresponds better to the geometrical intuition behind the definition of the integral. On the other hand, in many arguments it is much more convenient to use expansions into series of step functions. It is possible to use step functions instead of brick functions in Definition 10.2.1. It turns out that both definitions are equivalent, as we show in the next theorem.

Theorem 10.5.3. *A function expands into a series of brick functions if and only if it expands into a series of step functions. Moreover, if*

$$f \simeq \lambda_1 f_1 + \lambda_2 f_2 + \dots$$

is an expansion of f *into a series of brick functions and*

$$f \simeq g_1 + g_2 + \dots$$

is an expansion of f *into a series of step functions, then*

$$\lambda_1 \int f_1 + \lambda_2 \int f_2 + \dots = \int g_1 + \int g_2 + \dots .$$

Proof. If $f \simeq \lambda_1 f_1 + \lambda_2 f_2 + \dots$, where f_n's are brick functions, then f expands into the series of step functions $\lambda_n f_n$, that is, we can take $g_n = \lambda_n f_n$.

   Suppose now, $f \simeq g_1 + g_2 + \dots$ is an expansion of f into a series of step functions. For $n = 1, 2, \dots$, let

$$g_n = \lambda_{n1} f_{n1} + \dots + \lambda_{np_n} f_{np_n}$$

be the basic representation of g_n. We will show that

$$f \simeq \lambda_{11}f_{11} + \cdots + \lambda_{1p_1}f_{1p_1} + \cdots + \lambda_{n1}f_{n1} + \cdots + \lambda_{np_n}f_{np_n} + \cdots.$$

In fact, since

$$\int |g_n| = |\lambda_{n1}| \int f_{n1} + \cdots + |\lambda_{np_n}| \int f_{np_n}$$

we see that condition I is satisfied. Moreover, if for some $x \in \mathbb{R}$,

$$|\lambda_{11}|f_{11}(x) + \cdots + |\lambda_{1p_1}|f_{1p_1}(x) +$$
$$\cdots + |\lambda_{n1}|f_{n1}(x) + \cdots + |\lambda_{np_n}|f_{np_n}(x) + \cdots < \infty,$$

then

$$|g_1(x)| + |g_2(x)| + \cdots < \infty,$$

and thus

$$f(x) = \lambda_{11}f_{11}(x) + \cdots + \lambda_{1p_1}f_{1p_1}(x) +$$
$$\cdots + \lambda_{n1}f_{n1}(x) + \cdots + \lambda_{np_n}f_{np_n}(x) + \cdots$$

which proves that condition III is also satisfied. Therefore f expands into a series of brick functions.

Now, if $f \simeq \lambda_1 f_1 + \lambda_2 f_2 + \ldots$ is an arbitrary expansion of f into a series of brick functions and $f \simeq g_1 + g_2 + \ldots$ is an arbitrary expansion of f into a series of step functions, then

$$\lambda_1 \int f_1 + \lambda_2 \int f_2 + \cdots = \int g_1 + \int g_2 + \ldots$$

by Lemma 10.5.2. $\qquad\qquad\qquad\qquad\qquad\qquad\qquad\qquad\qquad\qquad\qquad\qquad\qquad$ $\square$

We are finally in a position to give a consistent definition the integral of an integrable function. The work done up to this point was necessary to justify that the value of integral is independent of a particular expansion into a series of brick functions.

Definition 10.5.4. The *integral* $\int f$ of an integrable function f is defined as the sum

$$\int f = \lambda_1 \int f_1 + \lambda_2 \int f_2 + \ldots$$

where $f \simeq \lambda_1 f_1 + \lambda_2 f_2 + \ldots$ is an arbitrary expansion of f into a series of brick functions.

The uniqueness of the integral having been proved, it follows directly from the definition that

$$\int (\lambda f) = \lambda \int f \quad \text{and} \quad \int (f + g) = \int f + \int g.$$

From Lemma 10.5.1 we easily derive that

$$f \le g \text{ implies } \int f \le \int g.$$

S: In the definition of integrable functions the only connection between the function f and the series of brick functions is at those points at which the series is absolutely convergent. What if there are no such points?

T: This is impossible. It can be proved that in every brick there are points at which the series converges absolutely. In fact, suppose that $f \simeq \lambda_1 f_1 + \lambda_2 f_2 + \ldots$ and that there is a brick $[a, b)$ such that the series $\sum_{n=1}^{\infty} \lambda_n f_n(x)$ does not converge absolutely at any point of $[a, b)$. Let f_0 be the characteristic function of $[a, b)$. Then we have $f \simeq f_0 + \lambda_1 f_1 + \lambda_2 f_2 + \ldots$ and hence $\int f = \int f_0 + \lambda_1 \int f_1 + \lambda_2 \int f_2 + \ldots$. Since we know that $\int f = \lambda_1 \int f_1 + \lambda_2 \int f_2 + \ldots$, the uniqueness of the integral implies $\int f_0 = 0$, which contradicts the definition of f_0.

S: Thus the set of those "bad" points cannot be too large.

T: We can prove that if f_0 is the characteristic function of the set Z of all points x for which the series $\sum_{n=1}^{\infty} \lambda_n f_n(x)$ does not converge absolutely, then $\int f_0 = 0$. In fact, since $f - f_0 \simeq \lambda_1 f_1 + \lambda_2 f_2 + \ldots$, we have $\int (f - f_0) = \int f$. Hence $\int f_0 = \int f - \int (f - f_0) = 0$. In such a case, we say that Z is a null set. Generally, we say that a set is a null set if its characteristic function is integrable and the integral equals 0. Null sets will be discussed in Section 10.8.

EXERCISES 10.5

(1) True or false: If f is integrable and supp $f \subseteq J$ for some bounded interval J, then there are step functions $f_1, f_2, \ldots$ such that $f \simeq f_1 + f_2 + \ldots$ and supp $f_n \subseteq J$ for every $n \in \mathbb{N}$?

(2) If $f \simeq f_1 + f_2 + \ldots$, show that $f + g \simeq g + f_1 + f_2 + \ldots$ for any step function g.

(3) If $f \simeq f_1 + f_2 + \ldots$ and $g \simeq g_1 + g_2 + \ldots$, show that

$$f \simeq f_1 + g_1 - g_1 + f_2 + g_2 - g_2 + f_3 + g_3 - g_3 + \ldots.$$

(4) Let $f(x) = \begin{cases} 2^{-\lfloor x \rfloor}, & \text{for } x \geq 0, \\ 0, & \text{otherwise}, \end{cases}$ where $\lfloor x \rfloor$ is the greatest integer less than or equal to x. Prove that f is integrable and find $\int f$.

(5) What is wrong with the following "proof" of Lemma 10.4.7?

Let (f_n) be a non-increasing sequence of step functions that converges to 0 at every point. Then, for every $n \geq 1$, $f_n - f_{n-1}$ is a step function satisfying $f_n - f_{n-1} \leq 0$. Notice that

$$0 \simeq f_1 + (f_2 - f_1) + (f_3 - f_2) + \ldots$$

is a valid expansion of f into step functions. Consequently

$$0 = \int 0 = \int f_1 + \int (f_2 - f_1) + \int (f_3 - f_2) + \cdots = \lim_{n \to \infty} \int f_n.$$

This completes the proof.

10.6   The absolute value of an integrable function

The following theorem plays an important role in the theory of the Lebesgue integral.

Theorem 10.6.1. *If f is an integrable function, then the function $|f|$ is integrable and we have*

$$\left| \int f \right| \leq \int |f|. \tag{10.18}$$

Moreover, if $f \simeq f_1 + f_2 + \dots$ is an arbitrary expansion of f into a series of step functions, then

$$\int |f| \leq \int |f_1| + \int |f_2| + \dots. \tag{10.19}$$

Proof. The equality

$$f(x) = f_1(x) + f_2(x) + \dots$$

holds at every point at which the series converges absolutely. Let Z denote the set of all such points. Then

$$f(x) = \lim_{n \to \infty} s_n(x) \quad \text{for all } x \in Z,$$

where

$$s_n = f_1 + \dots + f_n.$$

Consequently

$$|f(x)| = \lim_{n \to \infty} |s_n(x)| \quad \text{for all } x \in Z.$$

Let $g_1 = |s_1| = |f_1|$, $g_2 = |s_2| - |s_1|$, and generally

$$g_n = |s_n| - |s_{n-1}| \quad \text{for } n \geq 2.$$

Note that all the g_n's (as well as the s_n's) are step functions. Since

$$|g_n| \leq \big||s_n| - |s_{n-1}|\big| \leq |s_n - s_{n-1}| = |f_n|, \tag{10.20}$$

we have

$$\sum_{n=1}^{\infty} \int |g_n| \leq \sum_{n=1}^{\infty} \int |f_n| < \infty. \tag{10.21}$$

Although

$$|f(x)| = \sum_{n=1}^{\infty} g_n(x) \quad \text{for all } x \in Z,$$

we cannot claim that $|f| \simeq g_1 + g_2 + \dots$. because the series may converge to some extraneous values at some points outside Z. In order to "spoil" convergence of the series at those points we modify the series by adding and subtracting terms of the series $f_1 + f_2 + \ldots$ to get

$$|f| \simeq g_1 + f_1 - f_1 + g_2 + f_2 - f_2 + \ldots.$$

Clearly, condition $\mathbb{I}$ is still satisfied. Moreover, the above series and the series $\sum_{n=1}^{\infty} f_n$ converge absolutely at exactly the same points. Therefore, condition $\mathbb{III}$ is also satisfied, which proves that $|f|$ is integrable, in view of Theorem 10.5.3.

To prove (10.18) note that $f \leq |f|$ and $-f \leq |f|$, and hence $\int f \leq \int |f|$ and $-\int f \leq \int |f|$, which gives us $|\int f| \leq \int |f|$. Finally

$$\int |f| = \int g_1 + \int g_2 + \cdots \leq \int |f_1| + \int |f_2| + \ldots,$$

by (10.20). $\qquad\qquad\qquad\qquad\qquad\qquad\qquad\qquad\qquad\qquad\qquad\square$

If $f \simeq f_1 + f_2 + \ldots$, there is no reason to expect that $\int |f| = \int |f_1| + |f_2| + \ldots$. However, as we show in the next theorem, by choosing appropriately the expansion $f \simeq f_1 + f_2 + \ldots$ we can make the difference between $\int |f|$ and $\int |f_1| + |f_2| + \ldots$ as small as we like.

Theorem 10.6.2. *For any integrable function f and any positive number ε there exists an expansion into a series of step functions $f \simeq f_1 + f_2 + \ldots$ such that*

$$\int |f_1| + \int |f_2| + \cdots < \int |f| + \varepsilon.$$

Proof. Let us first take an arbitrary expansion

$$f \simeq g_1 + g_2 + \ldots. \qquad\qquad\qquad (10.22)$$

By $\mathbb{I}$, there exists a number $n_0 \in \mathbb{N}$ such that

$$\sum_{k=n_0}^{\infty} \int |g_k| < \frac{1}{2}\varepsilon. \qquad\qquad\qquad (10.23)$$

Define

$$f_1 = g_1 + \cdots + g_{n_0}, \quad f_2 = g_{n_0+1}, \quad f_3 = g_{n_0+2},$$

and in general

$$f_n = g_{n_0+n-1} \text{ for } n \geq 2. \qquad\qquad\qquad (10.24)$$

Then clearly

$$f \simeq f_1 + f_2 + \ldots$$

and hence also

$$f - f_1 \simeq f_2 + f_3 + \ldots. \qquad\qquad\qquad (10.25)$$

Moreover, by (10.23) and (10.24), we have

$$\int |f_2| + \int |f_3| + \cdots < \frac{1}{2}\varepsilon. \qquad\qquad\qquad (10.26)$$

Since

$$|f_1| \leq |f| + |f - f_1|,$$

we have

$$\int |f_1| \leq \int |f| + \int |f - f_1|.$$

Now, applying Theorem 10.6.1 to the function $f - f_1$ and using (10.25), we obtain

$$\int |f_1| \leq \int |f| + \int |f_2| + \int |f_3| + \cdots.$$

Hence, by (10.26),

$$\int |f_1| \leq \int |f| + \frac{1}{2}\varepsilon.$$

By adding the corresponding parts of this inequality and (10.26) we get the desired result

$$\int |f_1| + \int |f_2| + \cdots < \int |f| + \varepsilon.$$

$\square$

EXERCISES 10.6

(1) Find an example of an integrable function f and an expansion of $f \simeq f_1 + f_2 + \ldots$ such that both inequalities (10.18) and (10.19) are strict.

(2) Find an example of an integrable function f and an expansion $f \simeq f_1 + f_2 + \ldots$ such that the inequality (10.18) is strict, but the inequality (10.19) is actually an equality.

(3) If f is integrable, show that the functions $\max\{0, f\}$ and $\min\{0, f\}$ are integrable.

(4) Prove that $\int |f + g| \leq \int |f| + \int |g|$ for any integrable functions f and g.

(5) Show that the relation

$$f \sim g \text{ if and only if } \int |f - g| = 0$$

is an equivalence relation in the space of integrable functions.

(6) If f is a nonnegative integrable function, is it always possible to find nonnegative step functions $f_1, f_2 \ldots$ such that $f \simeq f_1 + f_2 + \ldots$?

(7) Show that if f is a continuous integrable function, then there are step functions $f_1, f_2, \ldots$ such that $f \simeq f_1 + f_2 + \ldots$ and $|f| \simeq |f_1| + |f_2| + \cdots$.

10.7 Series of integrable functions

In this section we extend the use of the symbol $\simeq$ once again. This will be the final extension.

Let f be an arbitrary function and let (f_n) be a sequence of integrable functions. We write

$$f \simeq f_1 + f_2 + \dots \quad \text{or} \quad f \simeq \sum_{n=1}^{\infty} f_n \qquad (10.27)$$

if

$$\text{II} \quad \sum_{n=1}^{\infty} \int |f_n| < \infty$$

and

$$\text{III} \quad f(x) = \sum_{n=1}^{\infty} f_n(x) \quad \text{at every point } x \text{ at which } \sum_{n=1}^{\infty} |f_n(x)| < \infty.$$

Note that the only difference between the definitions of $\simeq$ given in Sections 10.2 and 10.5 and the above definition is the class of functions which are allowed to be used in the series. At the beginning only brick functions were allowed, then step functions, and finally any integrable functions.

The following theorem is of fundamental importance in the theory of the integral.

Theorem 10.7.1. *If a function f expands into a series of integrable functions*

$$f \simeq f_1 + f_2 + \dots, \qquad (10.28)$$

then f is integrable and

$$\int f = \int f_1 + \int f_2 + \dots. \qquad (10.29)$$

Proof. Let $\sum_{n=1}^{\infty} \varepsilon_n$ be a convergent series of positive numbers. By Theorem 10.6.2, for every $n = 1, 2, \dots$ there are step functions $f_{n1}, f_{n2}, \dots$ such that

$$f_n \simeq f_{n1} + f_{n2} + \dots, \qquad (10.30)$$

and

$$\int |f_{n1}| + \int |f_{n2}| + \dots < \int |f_n| + \varepsilon_n. \qquad (10.31)$$

Let $\sum_{n=1}^{\infty} g_n$ be a series of step functions arranged from all f_{nk}'s $(n, k \in \mathbb{N})$. Then, by (10.31) and Theorem 5.6.5, we have

$$\int |g_1| + \int |g_2| + \dots < \int |f_1| + \int |f_2| + \dots + \varepsilon_1 + \varepsilon_2 + \dots < \infty. \qquad (10.32)$$

Moreover, by Theorem 5.6.4, if $\sum\limits_{n=1}^{\infty} g_n(x)$ converges absolutely at a point x, then all the series in (10.30) as well as the series in (10.28) converge absolutely at that x and we have

$$g_1(x) + g_2(x) + \cdots = f_1(x) + f_2(x) + \cdots = f(x).$$

This together with (10.32) shows that

$$f \simeq g_1 + g_2 + \cdots$$

which proves that f is integrable. Moreover, since for every $n \in \mathbb{N}$ we have

$$\int f_n = \int f_{n1} + \int f_{n2} + \cdots,$$

we conclude that

$$\int f = \int g_1 + \int g_2 + \cdots$$
$$= \int f_{11} + \int f_{12} + \cdots + \int f_{21} + \int f_{22} + \cdots$$
$$= \int f_1 + \int f_2 + \cdots$$

proving (10.29).   □

The following corollary is often useful.

Corollary 10.7.2. *If $f_1, f_2, \ldots$ are integrable functions and*

$$\int |f_1| + \int |f_2| + \cdots < \infty,$$

then there exists an integrable function f such that

$$f \simeq f_1 + f_2 + \cdots.$$

EXERCISES 10.7

(1) Prove Corollary 10.7.2.
(2) Let f be the characteristic function of the set of all rational numbers. Show that f is integrable and find $\int f$.
(3) Prove that any function which vanishes outside a countable number of points is integrable and its integral is 0.
(4) Let g be the characteristic function of the set of all irrational numbers. Is g integrable? If it is, find $\int g$. If it is not, prove it is not.
(5) Let g be the characteristic function of the set of all irrational numbers in the interval $[0, 1]$. Is g integrable? If it is, find $\int g$. If it is not, prove it is not.

(6) In this exercise we construct a bounded function with bounded support that is not integrable. We use equivalence classes in the construction (see Section 12.8).

For any $a, b \in [0, 1]$, we say that a is equivalent to b, and write $a \sim b$, if and only if $a - b$ is a rational number. The relation is an equivalence relation. For any $x \in [0, 1]$, let $[x]$ denote the equivalence class containing x. Since any two equivalence classes are either disjoint or identical, the interval $[0, 1]$ can be regarded as a disjoint union of these equivalence classes. Also, each equivalence class is nonempty. Choose exactly one point from each equivalence class and let A be the set of all selected points. Let $r_1, r_2, r_3, \ldots$ be an enumeration of the set of all rational numbers in $[0, 1]$ such that $r_i \neq r_j$ unless $i = j$. Let $A_n = \{r_n + x : x \in A\}$.

(a) Prove: If χ_A is integrable, then χ_{A_n} is integrable and $\int \chi_A = \int \chi_{A_n}$.

(b) Prove that $[0, 1] \subseteq \bigcup_{k=1}^{\infty} A_k \subseteq [0, 2]$.

(c) Prove that A_i and A_j are disjoint for $i \neq j$.

(d) Prove that $\int \chi_{[0,1]} \leq \int \chi_{A_1} + \int \chi_{A_2} + \cdots \leq \int \chi_{[0,2]}$ if we assume that χ_A is integrable.

(e) Prove that χ_A is not integrable, but it is a bounded function with bounded support.

10.8   Null functions and null sets

Definition 10.8.1. A function f is called a *null function* if it is integrable and $\int |f| = 0$. Similarly, a set is called a *null set* if its characteristic function is a null function.

The role of null functions in the theory of the integral is similar to that of the number 0 in the theory of real numbers. The main difference is that the number 0 is unique whereas the class of null functions is infinite. The function which is identically equal to 0 is a null function. Also any function which vanishes outside a single point, or a finite number of points, or a countable number of points is a null function.

We start with the following simple observation.

10.8.2. *A function f is a null function if and only if it vanishes everywhere except for a null set. In other words, f is a null function if and only if $\operatorname{supp} f$ is a null set.*

Proof. Let f be an arbitrary function and let Z be the set of all points x such that $f(x) \neq 0$, that is, $Z = \operatorname{supp} f$. Denote by g the characteristic function of Z.

If f is a null function, then $g \simeq f + f + \ldots$. Since $\int f = 0$, we have $\int g = 0$. Thus Z is a null set.

If Z is a null set, then $f \simeq g + g + \ldots$. Since $\int g = 0$, we have $\int f = 0$. Thus f is a null function.   □

The properties of null functions described in the following proposition can be easily obtained from the definition. The proofs are left as exercises.

10.8.3. *The sum of two null functions is a null function, the product of a number and a null function is a null function. Moreover, the absolute value of a null function is a null function.*

Definition 10.8.4. Two integrable functions f and g are called *equivalent* if there exists a null function h such that $f = g + h$. If f and g are equivalent we write

$$f = g \text{ a.e.}$$

which is read f *equals* g *almost everywhere.*

The notation $f = g$ a.e. is very popular and convenient because it suggests that equality almost everywhere has properties similar to those of ordinary equality, namely:

   $f = f$ a.e..
   If $f = g$ a.e., then $g = f$ a.e..
   If $f = g$ a.e. and $g = h$ a.e., then $f = h$ a.e..

Note that f is a null function if and only if $f = 0$ a.e..

EXERCISES 10.8

(1) Show that the sum of two null functions is a null function.
(2) Show that the product of a number and a null function is a null function.
(3) Show that the absolute value of a null function is a null function.
(4) Prove that $f = g$ a.e. if and only if $\int |f - g| = 0$.
(5) Prove that equality almost everywhere is an equivalence relation, that is, it has the following properties.

   (a) $f = f$ a.e..
   (b) If $f = g$ a.e., then $g = f$ a.e..
   (c) If $f = g$ a.e. and $g = h$ a.e., then $f = h$ a.e..

(6) Prove: If f is a null function and $|g| \leq f$, then g is a null function.
(7) Prove that every subset of a null set is a null set.
(8) Let f be an arbitrary function and let g be a null function. Prove that the product fg is a null function.

10.9 Norm convergence

We have already considered two types of convergence of sequences of functions: pointwise convergence and uniform convergence (see Section 6.2). In this and the following section we introduce two new types of convergence that are more natural and more important in the theory of the integral.

Definition 10.9.1. We say that a sequence of integrable functions $f_1, f_2, \ldots$ converges to an integrable function f *in norm*, and we write $f_n \to f$ i.n., if $\int |f_n - f| \to 0$ as $n \to \infty$.

This convergence is called norm convergence, because it is defined in terms of the integral $\int |f|$ which, in the theory of integrable functions, is called the norm of f.

Norm convergence has similar algebraic properties to those of pointwise or uniform convergence:

10.9.2. *If $f_n \to f$ i.n. and $g_n \to g$ i.n., then*

- (a) $f_n + g_n \to f + g$ *i.n.,*
- (b) $f_n - g_n \to f - g$ *i.n.,*
- (c) $\lambda f_n \to \lambda f$ *i.n., $(\lambda \in \mathbb{R})$,*
- (d) $|f_n| \to |f|$ *i.n..*

Here is a simple but important property of the convergence in norm:

10.9.3. *If $f_n \to f$ i.n., then $\int f_n \to \int f$.*

The above property follows directly from the inequality

$$\left| \int f_n - \int f \right| \leq \int |f_n - f|.$$

A series of integrable functions $\sum_{n=1}^{\infty} f_n$ is said to converge to f in norm if the sequence of partial sums $f_1 + \cdots + f_n$ converges to f in norm. In such a case we write

$$f_1 + f_2 + \cdots = f \text{ i.n.} \quad \text{or} \quad \sum_{n=1}^{\infty} f_n = f \text{ i.n..}$$

Theorem 10.9.4. *If $f \simeq f_1 + f_2 + \ldots$, then the series converges to f in norm, that is, $f_1 + f_2 + \cdots = f$ i.n..*

Proof. Given any $\varepsilon > 0$ we choose an index n_0 such that

$$\int |f_{n+1}| + \int |f_{n+2}| + \cdots < \varepsilon \quad \text{for all } n > n_0.$$

Since

$$f - f_1 - \ldots - f_n \simeq f_{n+1} + f_{n+2} + \ldots$$

we have

$$\int |f - f_1 - \cdots - f_n| \le \int |f_{n+1}| + \int |f_{n+2}| + \cdots < \varepsilon$$

for all $n > n_0$. $\square$

Unlike in the case of pointwise convergence or uniform convergence, the limit of a sequence convergent in norm is not unique. The following theorem describes all possible limits of a sequence convergent in norm.

Theorem 10.9.5. *If $f_n \to f$ i.n. and $f = g$ a.e., then $f_n \to g$ i.n.. Conversely, if $f_n \to f$ i.n. and $f_n \to g$ i.n., then $f = g$ a.e.. In other words: the limit of a sequence convergent in norm is defined up to a null function.*

Proof. Let $f_n \to f$ i.n. and let $g = f + h$ for some null function h. Then

$$\int |f_n - g| = \int |f_n - (f + h)| \le \int |f_n - f| + \int |h|.$$

Since $\int |f_n - f| \to 0$ and $\int |h| = 0$, we have $f_n \to g$ i.n..

Now, assume that $f_n \to f$ i.n. and $f_n \to g$ i.n.. Then

$$\int |f - g| = \int |f - f_n + f_n - g| \le \int |f_n - f| + \int |f_n - g|.$$

Since the left-hand side does not depend on n and the right-hand side tends to 0 as $n \to \infty$, we obtain $\int |f - g| = 0$, which proves that $f = g$ a.e.. $\square$

In the next theorem it is necessary to indicate what function is being integrated. To accomplish this, we write $\int |f(t + h) - f(t)| \, dt$ in order to communicate to the reader that we are integrating the function whose value at t is $|f(t + h) - f(t)|$.

Theorem 10.9.6. *If f is an integrable function, then*

$$\lim_{h \to 0} \int |f(t + h) - f(t)| \, dt = 0.$$

Proof. The assertion is obviously true if f is a brick function and it can be easily extended for an arbitrary step function. Let $f \simeq f_1 + f_2 + \ldots$ be an arbitrary integrable function expanded into a series of step functions f_n. Given any $\varepsilon > 0$, there exists an index $n_0 \in \mathbb{N}$ such that

$$\sum_{n=n_0+1}^{\infty} \int |f_n| < \frac{\varepsilon}{2}.$$

Since the theorem is true for step functions and $f_1 + \cdots + f_{n_0}$ is a step function, we have

$$\lim_{h \to 0} \int \left| \sum_{n=1}^{n_0} (f_n(t + h) - f_n(t)) \right| dt = 0. \tag{10.33}$$

For any $h \in \mathbb{R}$, we have

$$\int \left| \sum_{n=n_0+1}^{\infty} (f_n(t+h) - f_n(t)) \right| dt \leq \sum_{n=n_0+1}^{\infty} \int |f_n(t+h) - f_n(t)| \, dt$$

$$\leq \sum_{n=n_0+1}^{\infty} \int |f_n(t+h)| \, dt + \sum_{n=n_0+1}^{\infty} \int |f_n(t)| \, dt$$

$$= 2 \sum_{n=n_0+1}^{\infty} \int |f_n| < \varepsilon$$

and

$$\int |f(t+h) - f(t)| dt = \int \left| \sum_{n=1}^{n_0} (f_n(t+h) - f_n(t)) + \sum_{n=n_0+1}^{\infty} (f_n(t+h) - f_n(t)) \right| dt$$

$$\leq \int \left| \sum_{n=1}^{n_0} (f_n(t+h) - f_n(t)) \right| dt + \int \left| \sum_{n=n_0+1}^{\infty} (f_n(t+h) - f_n(t)) \right| dt$$

$$< \int \left| \sum_{n=1}^{n_0} (f_n(t+h) - f_n(t)) \right| dt + \varepsilon.$$

Thus, in view of (10.33), we obtain

$$\lim_{h \to 0} \int |f(t+h) - f(t)| \, dt \leq \lim_{h \to 0} \int \left| \sum_{n=1}^{n_0} (f_n(t+h) - f_n(t)) \right| dt + \varepsilon = \varepsilon.$$

Hence the assertion follows. $\qquad\qquad\qquad\qquad\qquad\qquad\qquad\qquad\qquad$ $\square$

EXERCISES 10.9

(1) If $f_n \to f$ i.n. and $g_n \to g$ i.n., show that

    (a) $f_n + g_n \to f + g$ i.n.,
    (b) $f_n - g_n \to f - g$ i.n.,
    (c) $\lambda f_n \to \lambda f$ i.n., $(\lambda \in \mathbb{R})$,
    (d) $|f_n| \to |f|$ i.n..

(2) Prove Theorem 10.9.6 for brick functions.

(3) Prove Theorem 10.9.6 for step functions.

(4) Let f be an integrable function and let $f_h(x) = f(x+h)$ for $h \in \mathbb{R}$. Prove that $\int f_h = \int f$.

(5) True or false: If (f_n) is a sequence of integrable functions convergent to 0 uniformly on $\mathbb{R}$, then $f_n \to 0$ i.n.?

(6) Let (f_n) be a sequence of integrable functions vanishing outside a bounded interval $[a, b]$. Prove that, if (f_n) converges to 0 uniformly on $[a, b]$, then $f_n \to 0$ i.n..

10.10 Convergence almost everywhere

In Section 7.5 we proved that, if $f \simeq \lambda_1 f_1 + \lambda_2 f_2 + \ldots$, then the series converges at every point outside a null set. This result can be easily generalized:

Theorem 10.10.1. *If the series $\sum_{n=1}^{\infty} \int |f_n|$ converges, then the set of points x where the series $\sum_{n=1}^{\infty} f_n(x)$ fails to converge absolutely is a null set.*

Proof. Let Z be the set of all points x at which the series $\sum_{n=1}^{\infty} f_n(x)$ fails to converge absolutely and let g be the characteristic function of Z. Then

$$g \simeq f_1 - f_1 + f_2 - f_2 + \ldots.$$

Hence $\int g = 0$ and Z is a null set.   □

The type of convergence in the above theorem is called convergence almost everywhere.

Definition 10.10.2. We say that a sequence of functions f_n converges to a function f *almost everywhere* and write $f_n \to f$ a.e. if $\lim_{n\to\infty} f_n(x) = f(x)$ for all x except those in a null set. In other words, $f_n \to f$ a.e. if there exists a null set Z such that $f_n(x) \to f(x)$ for every $x \notin Z$.

The next theorem is similar to Theorem 10.9.5 about convergence in norm.

Theorem 10.10.3. *A limit of a sequence convergent almost everywhere is determined up to a null function, that is, if $f_n \to f$ a.e., then $f_n \to g$ a.e. if and only if $f = g$ a.e..*

Proof. To prove this theorem it is useful to note that the union of two null sets is a null set. In order to see this, let χ_X and χ_Y be the characteristic functions of null sets X and Y, and $\chi_{X\cup Y}$ the characteristic function of their union. The function $\chi_{X\cup Y} = \frac{1}{2}(\chi_X + \chi_Y + |\chi_X - \chi_Y|)$ is clearly integrable. Since $\chi_{X\cup Y} \leq \chi_X + \chi_Y$, we have $\int \chi_{X\cup Y} \leq \int \chi_X + \int \chi_Y = 0$.

Let $f_n \to f$ a.e., that is, $f_n \to f$ outside a null set X. If $f = g$ outside a null set Y, then $f_n \to g$ outside $X \cup Y$, which is a null set. Thus $f_n \to g$ a.e..

Now assume that $f_n \to f$ outside a null set X and $f_n \to g$ outside a null set Y. Then $f = g$ outside $X \cup Y$. Since $X \cup Y$ is a null set, $f = g$ a.e..   □

The following auxiliary lemma will be useful in proofs of further properties of convergence almost everywhere.

Lemma 10.10.4. *A sequence (f_n) converges to f almost everywhere if and only if there exist null functions h_n such that the sequence $(f_n + h_n)$ converges to f at every point.*

Proof. Assume first that $f_n \to f$ a.e.. Denote by Z the set of all points where the sequence $(f_n(x))$ does not converge to $f(x)$. Let $h_n(x) = f(x) - f_n(x)$ for every

$x \in Z$ and $h_n(x) = 0$ outside Z. Then h_n's are null functions and the sequence of functions $f_n + h_n$ converges to f everywhere.

Assume now, that $g_n = f_n + h_n$ and that the sequence (g_n) converges to f at every point. If h_n's are null functions, then

$$\int |g_1 - f_1| + \int |g_2 - f_2| + \cdots = 0.$$

Hence, by Theorem 10.10.1, the series $\sum_{n=1}^{\infty}(g_n - f_n)$ converges everywhere except on a null set Z. Therefore $g_n - f_n \to 0$ outside Z and, consequently, $f_n = g_n - (g_n - f_n) \to f$ outside Z. □

In view of Lemma 10.10.4, the following properties of convergence almost everywhere can be easily derived from the analogous properties of pointwise convergence.

10.10.5. If $f_n \to f$ a.e. and $g_n \to g$ a.e., then

   (a) $f_n + g_n \to f + g$ a.e.,
   (b) $f_n - g_n \to f - g$ a.e.,
   (c) $\lambda f_n \to \lambda f$ a.e., $(\lambda \in \mathbb{R})$,
   (d) $|f_n| \to |f|$ a.e..

The next theorem is analogous to Theorem 10.9.4 about convergence in norm.

Theorem 10.10.6. If $f \simeq f_1 + f_2 + \ldots$, then the series converges to f almost everywhere, that is, $f_1 + f_2 + \cdots = f$ a.e..

Proof. Let f_0 be the characteristic function of the set Z of all points x for which the series $\sum_{n=1}^{\infty} f_n(x)$ does not converge absolutely. Then $f - f_0 \simeq f_1 + f_2 + \ldots$ and consequently $\int (f - f_0) = \int f$. Hence $\int f_0 = \int f - \int (f - f_0) = 0$, which means that Z is a null set. Clearly, $f(x) = f_1(x) + f_2(x) + \ldots$ outside of Z, which proves the theorem. □

The next theorem describes a special case when convergence in norm implies convergence almost everywhere and conversely. As we can see in the two examples following the theorem, in general neither implication holds.

Theorem 10.10.7. Let $f_1, f_2, \ldots$ be integrable functions such that

$$\int |f_1| + \int |f_2| + \cdots < \infty.$$

Then, $f_1 + f_2 + \cdots = f$ i.n. if and only if $f_1 + f_2 + \cdots = f$ a.e..

Proof. Let $f_1, f_2, \ldots$ be integrable functions such that $\int |f_1| + \int |f_2| + \cdots < \infty$. By Corollary 10.7.2, there exists a function g such that $g \simeq f_1 + f_2 + \ldots$. Then, by Theorems 10.9.4 and 10.10.6 we have $f_1 + f_2 + \cdots = g$ i.n. and $f_1 + f_2 + \cdots = g$ a.e..

If $f_1 + f_2 + \cdots = f$ i.n., then f and g differ by a null function, by Theorem 10.9.5. Hence, by Theorem 10.10.3, we have $f_1 + f_2 + \cdots = f$ a.e..

The proof of the other implication is similar. □

Example 10.10.8. Consider the following sequence functions:

$$f_n(x) = \begin{cases} n, & \text{if } x \in [n, n+1), \\ 0, & \text{otherwise.} \end{cases}$$

Since $\lim_{n\to\infty} f_n(x) = 0$ for every $x \in \mathbb{R}$, we have $f_n \to 0$ a.e.. On the other hand, since $\int f_n = n$, the sequence is not convergent in norm.

Example 10.10.9. Consider the following sequence functions:

$$f_n(x) = \begin{cases} 1, & \text{if } x \in \left[\frac{n}{2^{k-1}} - 1, \frac{n+1}{2^{k-1}} - 1 \right) \text{ and } 2^{k-1} \le n < 2^k \text{ for some } k \in \mathbb{N}, \\ 0, & \text{otherwise.} \end{cases}$$

Although functions $f_1, f_2, f_3, \ldots$ are defined in a complicated manner, the idea is simple as one can see on Fig. 10.8.

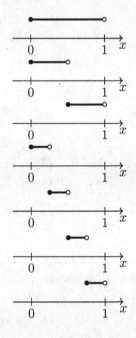

Fig. 10.8   Graphs of the first seven functions f_n in Example 10.10.9.

Since $\int f_n = \frac{1}{2^k}$ for every n such that $2^{k-1} \le n < 2^k$, we have $\int |f_n| = \int f_n \to 0$ as $n \to \infty$. Thus $f_n \to 0$ i.n.. On the other hand, for every $x \in [0, 1)$, $f_n(x) = 0$ for infinitely many $n \in \mathbb{N}$, and $f_n(x) = 1$ for infinitely many $n \in \mathbb{N}$. Consequently, the sequence $(f_n(x))$ fails to converge at every $x \in [0, 1)$. Since $[0, 1)$ is not a null set, $f_n \not\to 0$ a.e..

In Theorem 10.10.7, the functions $f_1, f_2, \ldots$ were assumed to be integrable. This example shows that this assumption cannot be dropped because it is possible to construct functions $f, f_1, f_2, \ldots$ satisfying $\int |f_1| + \int |f_2| + \cdots < \infty$ and $f_1 + f_2 + \cdots = f$ a.e. without satisfying $f_1 + f_2 + \cdots = f$ i.n..

Example 10.10.10. Let $A_1, A_2, A_3, \ldots$ be the disjoint sets constructed in the directions for Exercise 10.7.6. For each $n \in \mathbb{N}$, let B_n be the set of all points in $[0, 2]$ that are not in A_n. For each $n \in \mathbb{N}$, define f, f_n, and g_n via

$$f = \sum_{k=1}^{\infty} \frac{1}{2^{k-1}} \chi_{A_k} - \chi_{[0,2]},$$

$$f_n = \frac{1}{2^n} \left(\chi_{A_n} - \chi_{B_n} \right),$$

and

$$g_n = \frac{1}{2^n} \chi_{[0,2]} - \sum_{k=n+1}^{\infty} \frac{1}{2^{k-1}} \chi_{A_k}.$$

It is left as an exercise to show that $\int |f_1| + \int |f_2| + \cdots < \infty$, $f_1 + f_2 + \cdots = f$ a.e., and $|f_1 + f_2 + \cdots + f_n - f| = g_n$, but it is not true that $f_1 + f_2 + \cdots = f$ i.n..

EXERCISES 10.10

(1) Prove that a countable union of null sets is a null set.

(2) If $f_n \to f$ a.e. and $g_n \to g$ a.e., show that

    (a) $f_n + g_n \to f + g$ a.e.,

    (b) $f_n - g_n \to f - g$ a.e.,

    (c) $\lambda f_n \to \lambda f$ a.e., $(\lambda \in \mathbb{R})$,

    (d) $|f_n| \to |f|$ a.e..

(3) Complete the proof of Theorem 10.10.7.

(4) True or false: If $f_n \to f$ a.e., then $\int f_n \to \int f$?

(5) Modify Example 10.10.8 so that all functions in the sequence (f_n) vanish outside of $[0, 1]$ and we still have $f_n \to 0$ a.e. and $\int f_n \to \infty$.

(6) Modify Example 10.10.9 so that the sequence $(f_n(x))$ does not converge at any $x \in \mathbb{R}$ and we still have $f_n \to 0$ i.n..

(7) Prove the claims made in Example 10.10.10, that $\int |f_1| + \int |f_2| + \cdots < \infty$, $f_1 + f_2 + \cdots = f$ a.e., and $|f_1 + f_2 + \cdots + f_n - f| = g_n$, but it is not true that $f_1 + f_2 + \cdots = f$ i.n..

10.11 Theorems of Riesz

The following two theorems are attributed to the Hungarian mathematician Frigyes Riesz (1880–1956). They are existence theorems. The first of them asserts existence of integrable functions which are limits in norm of sequences satisfying a certain condition. The second one asserts existence of subsequences convergent almost everywhere for sequences convergent in norm.

Theorem 10.11.1. *Let (f_n) be a sequence of integrable functions. If for every increasing sequence of indices $p_1, p_2, \ldots$ we have*

$$f_{p_{n+1}} - f_{p_n} \to 0 \ i.n. \quad as \ n \to \infty,$$

then there is an integrable function f such that $f_n \to f$ i.n..

Proof. Let (ε_n) be a sequence of positive numbers such that $\sum_{n=1}^{\infty} \varepsilon_n < \infty$, and let (p_n) be an increasing sequence of natural numbers such that

$$\int |f_k - f_m| < \varepsilon_n \quad \text{for all } k, m > p_n. \tag{10.34}$$

(Why does such a sequence (p_n) always exist?) Then

$$\int |f_{p_1}| + \int |f_{p_2} - f_{p_1}| + \int |f_{p_3} - f_{p_2}| + \cdots < \infty.$$

Hence, by Corollary 10.7.2 there exists an integrable function f such that

$$f \simeq f_{p_1} + (f_{p_2} - f_{p_1}) + (f_{p_3} - f_{p_2}) + \cdots.$$

Thus, by Theorem 10.9.4, we have $\int |f_{p_n} - f| \to 0$. Since, by (10.34),

$$\int |f_k - f_{p_m}| < \varepsilon_n \quad \text{for all } k, p_m > p_n,$$

we have

$$\int |f_k - f| \leq \int |f_k - f_{p_m}| + \int |f_{p_m} - f| < \varepsilon_n + \int |f_{p_m} - f|.$$

By letting $m \to \infty$ we obtain

$$\int |f_k - f| \leq \varepsilon_n \quad \text{for all } k > p_n.$$

Since $\varepsilon_n \to 0$, the above implies $f_n \to f$ i.n.. $\qquad \square$

Example 10.10.9 shows that convergence in norm does not imply convergence almost everywhere. On the other hand we have the following important theorem.

Theorem 10.11.2. *If $f_n \to f$ i.n., then there is a subsequence (f_{p_n}) of (f_n) such that $f_{p_n} \to f$ a.e..*

Proof. Let (ε_n) be a sequence of positive numbers such that $\sum_{n=1}^{\infty} \varepsilon_n < \infty$. Since $\int |f_n - f| \to 0$, there exists an increasing sequence of natural numbers $p_1, p_2, \ldots$ such that

$$\int |f_{p_n} - f| < \varepsilon_n \quad \text{for all } n \in \mathbb{N}.$$

Then

$$\int |f_{p_{n+1}} - f_{p_n}| \leq \int |f_{p_{n+1}} - f| + \int |f - f_{p_n}| < \varepsilon_{n+1} + \varepsilon_n,$$

and consequently

$$\int |f_{p_1}| + \int |f_{p_2} - f_{p_1}| + \int |f_{p_3} - f_{p_2}| + \cdots < \infty.$$

Hence, by Corollary 10.7.2 there is a function g such that

$$g \simeq f_{p_1} + (f_{p_2} - f_{p_1}) + (f_{p_3} - f_{p_2}) + \cdots.$$

This implies $f_{p_n} \to g$ a.e., by Theorem 10.10.6, and also $f_{p_n} \to g$ i.n., by Theorem 10.10.7. In view of Theorem 10.9.5 and the fact that $f_n \to f$ i.n., we get $f = g$ a.e.. Therefore, by Theorem 10.10.3, $f_{p_n} \to f$ a.e.. $\qquad\square$

EXERCISES 10.11

(1) Complete the proof of Theorem 10.11.1 by proving the existence of a sequence (p_n) satisfying (10.34).

(2) Find a subsequence (f_{p_n}) of the sequence (f_n) defined in Example 10.10.9 such that $f_{p_n} \to 0$ a.e..

(3) True or false: If $f_n \to f$ a.e., then there is a subsequence (f_{p_n}) of (f_n) such that $f_{p_n} \to f$ i.n.?

(4) A sequence (f_n) of integrable functions is called a Cauchy sequence if for every $\varepsilon > 0$ there exists an n_0 such that $\int |f_m - f_n| < \varepsilon$ for all $m, n > n_0$. Show that every Cauchy sequence of integrable functions is convergent in norm.

10.12 The Monotone Convergence Theorem and the Dominated Convergence Theorem

One of the most important properties of convergence in norm is that it implies convergence of integrals, that is, $f_n \to f$ i.n. implies $\int f_n \to \int f$. Convergence almost everywhere does not have that property. On the other hand, for a particular sequence (f_n) it is usually much easier to verify convergence almost everywhere than convergence in norm. Therefore, theorems giving conditions under which convergence almost everywhere implies convergence in norm are of great importance. In this section we prove two such theorems. They are among the most important theorems in the theory of the Lebesgue integral. The first one is due to the Italian mathematician Beppo Levi (1875–1961).

Theorem 10.12.1 (Monotone Convergence Theorem). *If (f_n) is a monotone sequence of integrable functions and $\int |f_n| \leq M$, for some M and all $n \in \mathbb{N}$, then there exists an integrable function f such that $f_n \to f$ i.n. and $f_n \to f$ a.e..*

Proof. Without loss of generality, we may assume that the sequence is non-negative and non-decreasing, otherwise we would consider $(f_n - f_1)$ or $(f_1 - f_n)$. Then

$$\int |f_1| + \int |f_2 - f_1| + \cdots + \int |f_n - f_{n-1}| = \int |f_n| \leq M.$$

Hence, by letting $n \to \infty$, we obtain

$$\int |f_1| + \int |f_2 - f_1| + \cdots \leq M < \infty.$$

By Corollary 10.7.2, there is an integrable function f such that

$$f \simeq f_1 + (f_2 - f_1) + \cdots .$$

Then, by Theorem 10.9.4, $f = f_1 + (f_2 - f_1) + \cdots$ i.n. and, by Theorem 10.10.6, $f = f_1 + (f_2 - f_1) + \cdots$ a.e.. $\qquad\square$

We are finally ready to prove the famous Dominated Convergence Theorem, proved by Lebesgue in his doctoral dissertation in 1902. Without much exaggeration, one can say that this theorem was the main goal of the development in the first 9 sections of this chapter.

We first prove a special case of the Dominated Convergence Theorem.

Lemma 10.12.2. *Let (f_n) be a sequence of integrable functions such that $|f_n| \leq g$ for some integrable function g and all $n \in \mathbb{N}$. If $f_n \to 0$ a.e., then $f_n \to 0$ i.n..*

Proof. For $m, n = 1, 2, \ldots$, define

$$h_{n,m} = \max\{|f_n|, \ldots, |f_{n+m}|\}.$$

Then, for every fixed $n \in \mathbb{N}$, the sequence $h_{n,1}, h_{n,2}, \ldots$ is non-decreasing because $h_{n,m+1} = \max\{h_{n,m}, |f_{n+m+1}|\} \geq h_{n,m}$. Since the function $h_{n,m}$ is integrable and for each x there is some $k \in \{n, n+1, \ldots, n+m\}$ such that

$$0 \leq |h_{n,m}(x)| = h_{n,m}(x) = |f_k(x)| \leq g(x),$$

we have

$$\left| \int h_{n,m} \right| = \int h_{n,m} \leq \int g < \infty.$$

By the Monotone Convergence Theorem, there exists an integrable function h_n such that $h_{n,m} \to h_n$ a.e. as $m \to \infty$. Note that each h_n can be chosen so that $h_n \geq h_{n+1} \geq 0$ for every $n \in \mathbb{N}$. Moreover, $h_n \to 0$ a.e., because $f_n \to 0$ a.e.. By the Monotone Convergence Theorem, $h_n \to 0$ i.n.. Since $\int |f_n| \leq \int h_n$, we obtain $f_n \to 0$ i.n.. $\qquad\square$

Theorem 10.12.3 (Dominated Convergence Theorem). *If a sequence of integrable functions (f_n) converges almost everywhere to a function f and there is an integrable function g such that $|f_n| \leq g$ for every $n \in \mathbb{N}$, then f is integrable and $f_n \to f$ i.n..*

Proof. For every increasing sequence of positive integers p_n we have

$$g_n = f_{p_{n+1}} - f_{p_n} \to 0 \text{ a.e.}$$

and

$$|g_n| = |f_{p_{n+1}} - f_{p_n}| \leq |f_{p_{n+1}}| + |f_{p_n}| \leq 2g$$

for every $n \in \mathbb{N}$. Thus, the sequence (g_n) satisfies the assumptions of Lemma 10.12.2, and consequently $g_n \to 0$ i.n.. Hence, by Theorem 10.11.1, the sequence (f_n) converges in norm to some integrable function f^*. By Theorem 10.11.2, there exists an increasing sequence of positive integers (q_n) such that $f_{q_n} \to f^*$ a.e.. But $f_{q_n} \to f$ a.e., so we must have $f^* = f$ a.e.. In view of Theorem 10.9.5, we conclude that $f_n \to f$ i.n.. $\square$

EXERCISES 10.12

(1) Find an example of a sequence of functions which shows that Theorem 10.12.1 does not hold if the monotonicity assumption is dropped.

(2) Find an example of a sequence of functions which shows that Theorem 10.12.1 does not hold if the assumption $\int |f_n| \leq M$ is dropped.

(3) Let (f_n) be a sequence of integrable functions such that $|f_n| \leq g$ for some integrable function g and every $n \in \mathbb{N}$. Is it true that, if $f_n \to 0$ i.n., then $f_n \to 0$ a.e..

(4) Prove that, if $f_1, \ldots, f_n$ are integrable, then so is $\max\{|f_1|, \ldots, |f_n|\}$.

(5) Find an example of a sequence of functions which shows that Theorem 10.12.3 does not hold if the assumption $|f_n| \leq g$ is dropped.

(6) Let f be an integrable function. Define

$$f_n(x) = \begin{cases} f(x), & \text{if } |x| \leq n, \\ 0, & \text{otherwise.} \end{cases}$$

Show that each f_n is integrable and that $f_n \to f$ i.n..

(7) In the proof of Lemma 10.12.2 we claim that each h_n can be chosen so that, for every $n \in \mathbb{N}$, $h_n \geq h_{n+1} \geq 0$. Justify that claim.

(8) In the proof of Lemma 10.12.2 we claim that $h_n \to 0$ a.e.. Justify that claim.

10.13   Integral over an interval

By $\int_a^b f$ we mean the integral $\int f\chi_{(a,b)}$ where $\chi_{(a,b)}$ is the characteristic function of the interval (a,b) (a and b may be finite or infinite). This definition makes sense whenever the product $f\chi_{(a,b)}$ is integrable. We assume that $f\chi_{(a,b)}$ vanishes outside the interval (a,b) no matter what is the value of f or whether f is defined there.

Instead of the interval (a, b) we may take intervals $[a, b)$, $[a, b]$, or $(a, b]$, because the change of the integrand at points a and b does not affect the existence or the value of the integral. The only reason for using (a, b) in the above definition is that it makes sense for both bounded and unbounded intervals. If f is an integrable function, then $\int_a^b f$ exists for every $a < b$. Indeed, if $f \simeq f_1 + f_2 + \dots$, then $f\chi_{(a,b)} \simeq f_1\chi_{(a,b)} + f_2\chi_{(a,b)} + \dots$. Similarly, if f is integrable over (A, B) and $A \le a < b \le B$, then f is integrable over (a, b).

10.13.1. *If $a < b < c$, then the integrability of f over (a, b) and (b, c) implies integrability of f over (a, c) and we have*

$$\int_a^c f = \int_a^b f + \int_b^c f.$$

Proof. Let χ_1, χ_2, and χ_3 be the characteristic functions of (a, b), (b, c), and (a, c), respectively. Then $\chi_1 + \chi_2 = \chi_3$ a.e. (everywhere except b) and hence

$$\int f\chi_3 = \int f(\chi_1 + \chi_2) = \int f\chi_1 + \int f\chi_2.$$

$\square$

It is convenient to define

$$\int_a^a f = 0 \quad \text{and} \quad \int_b^a f = -\int_a^b f.$$

Then 10.13.1 holds for all a, b, and c.

From the definition of the integral over an interval it follows immediately that

$$\int_a^b \lambda f = \lambda \int_a^b f \quad \text{for any } \lambda \in \mathbb{R}.$$

The Dominated Convergence Theorem applies to functions on an interval.

10.13.2. *If a sequence (f_n) of functions integrable over an interval (a, b) converges to f almost everywhere in (a, b) and is bounded by a function integrable over (a, b), then f is integrable over (a, b) and $\int_a^b |f_n - f| \to 0$, and hence also $\int_a^b f_n \to \int_a^b f$.*

Proof. We assume that f, all the f_n's, and the bounding function vanish outside (a, b) and then apply Theorem 10.12.3. $\square$

The following theorem is often useful.

Theorem 10.13.3. *If f is a bounded integrable function on an interval (a, b), bounded or unbounded, and g is integrable on (a, b), then the function fg is integrable on (a, b).*

Proof. If g is integrable on (a, b), then

$$g \simeq \lambda_1 g_1 + \lambda_2 g_2 + \ldots$$

where $g_n = \chi_{[a_n, b_n)}$ for some intervals $[a_n, b_n) \subset (a, b)$ and some numbers λ_n. Let f be an integrable function on (a, b) such that $|f(x)| \leq M$ for some constant M and all $x \in (a, b)$. Note that the functions $\lambda_n f g_n$ are integrable. We will show that

$$fg \simeq \lambda_1 f g_1 + \lambda_2 f g_2 + \ldots.$$

Since $\int |\lambda_n f g_n| \leq \int M |\lambda_n g_n|$ for every $n \in \mathbb{N}$, we have

$$\int |\lambda_1 f g_1| + \int |\lambda_2 f g_2| + \cdots < \infty.$$

Clearly, $f(x)g(x) = \lambda_1 f(x)g_1(x) + \lambda_2 f(x)g_2(x) + \ldots$ at every x such that $f(x) = 0$. For points where $f(x) \neq 0$, the series $\lambda_1 f(x)g_1(x) + \lambda_2 f(x)g_2(x) + \ldots$ converges absolutely at x if and only if the series $\lambda_1 g_1(x) + \lambda_2 g_2(x) + \ldots$ converges absolutely at x and we have $f(x)g(x) = \lambda_1 f(x)g_1(x) + \lambda_2 f(x)g_2(x) + \ldots$ at every point of absolute convergence. $\qquad\square$

The next proposition gives us another tool for proving integrability of function.

10.13.4. *If a function f defined on an interval (a, b), bounded or unbounded, is integrable over every bounded interval (α, β) inside (a, b), that is $a < \alpha < \beta < b$, and is bounded by an integrable function over the entire interval (a, b), then f is integrable over (a, b).*

Proof. Assume f is integrable over every bounded interval $(\alpha, \beta) \subset (a, b)$ and $|f| \leq g$ on (a, b) for some integrable function g. Let

$$a < \cdots < \alpha_n < \alpha_{n-1} < \cdots < \alpha_1 < \beta_1 < \cdots < \beta_{n-1} < \beta_n < \cdots < b$$

with $\alpha_n \to a$ and $\beta_n \to b$. Then the functions

$$f_n(x) = \begin{cases} f(x), & \text{for } x \in (\alpha_n, \beta_n), \\ 0, & \text{elsewhere}, \end{cases}$$

are integrable and $f_n(x) \to f(x)$ for every $x \in (a, b)$. Moreover, since $|f| \leq g$ on (a, b), then also $|f_n| \leq g$ for all $n \in \mathbb{N}$. Therefore, by the Dominated Convergence Theorem, $f_n \to f$ i.n., which implies integrability of f over (a, b). $\qquad\square$

Example 10.13.5. We will show that the function $f(x) = x^\alpha e^{-x}$ is integrable over $(0, \infty)$ for every $\alpha > -1$. First note that f is integrable over $(\varepsilon, 1)$ for any $0 < \varepsilon < 1$ and is bounded by the function $g(x) = x^\alpha$. Since g is integrable over the interval $(0, 1)$, so is f, by 10.13.4. It is thus integrable over the interval $(0, 1)$. To prove its integrability over $(1, \infty)$, observe that f is integrable over each of the intervals $(1, n)$, $n = 2, 3, \ldots$, because it is a continuous function. Let β be a number less than -1. Then there exists a number λ such that

$$0 < f(x) < \lambda x^\beta \quad \text{for all } x \in (1, \infty).$$

Since $\beta < -1$, the function $h(x) = \lambda x^\beta$ is integrable over $(1, \infty)$, and so is f. Since f is integrable over $(0, 1)$ and $(1, \infty)$, it is integrable over $(0, \infty)$, by proposition 10.13.1.

Theorem 10.13.6. *Every function continuous on a bounded closed interval is integrable over that interval.*

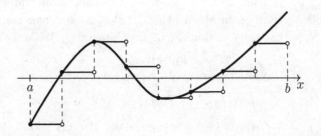

Fig. 10.9   Approximation of a continuous function by a step function.

Proof. Let f be a continuous function on the inetrval $[a, b]$. For every $n \in \mathbb{N}$ and every $i = 0, 1, 2, 3, \ldots, 2^n$ let

$$x_{n,i} = a + \frac{i(b - a)}{2^n}.$$

Then the union of the 2^n disjoint intervals $[x_{n,i-1}, x_{n,i})$ is $[a, b)$. For every $n \in \mathbb{N}$ let

$$f_n = \sum_{i=1}^{2^n} f(x_{n,i-1}) \chi_{[x_{n,i-1}, x_{n,i})},$$

see Fig 10.9.

Suppose x_0 is an arbitrary point of $[a, b)$. We want to show that $f_n(x_0) \to f(x_0)$. To this end, let $\varepsilon > 0$ be given. Since f is continuous at x_0, there is a $\delta > 0$ such that

$$|f(x_0) - f(x)| < \varepsilon \quad \text{whenever } |x_0 - x| < \delta \text{ and } x \in [a, b].$$

There exists an $N \in \mathbb{N}$ such that

$$\frac{b - a}{2^N} < \delta.$$

Now suppose $n > N$. There is exactly one integer i such that $1 \le i \le 2^n$ and $x_0 \in [x_{n,i-1}, x_{n,i})$. Clearly $|x_0 - x_{n,i-1}| < \delta$ and consequently $|f(x_0) - f(x_{n,i-1})| < \varepsilon$. Since $f_n(x_0) = f(x_{n,i-1})$, we have

$$|f(x_0) - f_n(x_0)| < \varepsilon.$$

This proves that $f_n \to f$ a.e. in the interval $[a, b]$ (namely, everywhere except at the point b). Each f_n is integrable and each is bounded by the integrable function $M\chi_{[a,b]}$, where M is the maximum of f on $[a, b]$. Thus, by the Dominated Convergence Theorem, f is integrable on $[a, b]$.   □

In the above theorem the closed interval $[a, b]$ cannot be replaced by the open interval (a, b) (see, for example, 11.1.8).

EXERCISES 10.13

(1) Prove the following generalization of proposition 10.13.1:

Let $x_0 < x_1 < \cdots < x_n$. If a function f is integrable over each of the intervals $(x_0, x_1), (x_1, x_2), \ldots, (x_{n-1}, x_n)$, then f is integrable over the interval (x_0, x_n) and

$$\int_{x_0}^{x_n} f = \int_{x_0}^{x_1} f + \cdots + \int_{x_{n-1}}^{x_n} f.$$

(2) True or false: If f is integrable over $[-n, n]$ for all $n \in \mathbb{N}$, then f is integrable over $\mathbb{R}$?

(3) Prove: If f is continuous on the closed bounded interval $[a, b]$ and g is integrable on $[a, b]$, then the function fg is integrable on $[a, b]$.

(4) A function $f: \mathbb{R} \to \mathbb{R}$ is called *locally integrable* if the product $f\chi_{(-n,n)}$ is an integrable function for every $n \in \mathbb{N}$.

  (a) Prove that integrable functions are locally integrable. Is the converse true?

  (b) Prove: If f and g are locally integrable, then $f + g$ and λf ($\lambda \in \mathbb{R}$) are locally integrable.

  (c) Prove: If f is a bounded locally integrable function and g is integrable, then the function fg is integrable.

  (d) Prove: If f is a bounded locally integrable function and g is locally integrable, then the function fg is locally integrable.

  (e) Prove: If a sequence of locally integrable functions (f_n) converges almost everywhere to a function f and $|f_n| \leq h$ for every $n \in \mathbb{N}$, where h is a locally integrable function, then f is locally integrable.

(5) A set $S \subset \mathbb{R}$ is called *measurable* if the function χ_S is locally integrable. The family of all measurable subsets of $\mathbb{R}$ will be denoted by $\mathcal{M}$. The *measure* of a measurable set S, denoted by $\mu(S)$, is defined as

$$\mu(S) = \begin{cases} \int \chi_S & \text{if } \chi_S \text{ is integrable,} \\ \infty & \text{otherwise.} \end{cases}$$

Prove the following properties of measurable sets.

  (a) The empty set, all single element sets, the set $\mathbb{Q}$ of all rational numbers, and $\mathbb{R}$ are measurable.

  (b) $[a, b], (a, b), [a, b), (a, b] \in \mathcal{M}$ for any $a < b$.

  (c) If $A, B \in \mathcal{M}$, then $A \cup B \in \mathcal{M}$ and $A \cap B \in \mathcal{M}$.

  (d) If S is measurable, then the complement of S is measurable.

  (e) If $A_1, A_2, \cdots \in \mathcal{M}$, then $\bigcup_{n=1}^{\infty} A_n \in \mathcal{M}$ and $\bigcap_{n=1}^{\infty} A_n \in \mathcal{M}$.

  (f) If $A, B \in \mathcal{M}$ and $A \subset B$, then $\mu(A) \leq \mu(B)$.

  (g) If $A \in \mathcal{M}$, $\mu(A) = 0$, and $B \subset A$, then $B \in \mathcal{M}$ and $\mu(B) = 0$.

  (h) If $A, B \in \mathcal{M}$ and A and B are disjoint, then $\mu(A \cup B) = \mu(A) + \mu(B)$.
  (We define: $a + \infty = \infty$, for any $a \in \mathbb{R}$, and $\infty + \infty = \infty$.)

(i) If $A_n \in \mathcal{M}$ for each $n \in \mathbb{N}$ and $A_k \cap A_m = \emptyset$ whenever $k \neq m$, then $\mu(\bigcup_{n=1}^{\infty} A_n) = \sum_{n=1}^{\infty} \mu(A_n)$.

Chapter 11

MISCELLANEOUS TOPICS

11.1 Integration and differentiation

If f is integrable on (a, b), then for any two points α and β in (a, b) the integral $\int_\alpha^\beta f$ exists. Consequently, the formula $F(x) = \int_c^x f$, where c is an arbitrary point in (a, b), defines a function on (a, b).

Theorem 11.1.1. *Let* $-\infty \le a < c < b \le \infty$ *and let* f *be an integrable function over* (a, b). *Then the function*

$$F(x) = \int_c^x f$$

is continuous in (a, b) *and the limits* $F(a+)$ *and* $F(b-)$ *exist.*

Proof. Let x be any fixed point such that $a \le x < b$ and let (x_n) be a decreasing sequence convergent to x, that is $x < x_{n+1} < x_n$ and $x_n \to x$. Denote by g_n the characteristic function of the interval (x, x_n). Then the sequence of products fg_n is convergent to zero at every point of (a, b). Since $|fg_n| \le |f|$, we have

$$F(x_n) - F(x) = \int_c^{x_n} f - \int_c^x f = \int_x^{x_n} f = \int_a^b fg_n \to 0,$$

by the Dominated Convergence Theorem (in the form 10.13.2). Thus for $a \le x < b$, the right-hand limit of F at x exists and is given by $F(x+) = \int_c^x f$. When $a < x < b$, F is right-hand continuous at x because $F(x+) = F(x)$. Similarly, we can prove left-hand continuity and the existence of $F(b-)$ by considering the characteristic functions of intervals (x_n, x) where $x_n < x_{n+1} < x$ and $x_n \to x$. Since F is right-hand and left-hand continuous, it is continuous in (a, b) and the limits $F(a+)$ and $F(b-)$ exist. $\qquad \square$

By the right-hand or left-hand derivative of a function f at a point x_0 we mean the limit

$$\lim_{h \to 0+} \frac{f(x_0 + h) - f(x_0)}{h} \quad \text{or} \quad \lim_{h \to 0-} \frac{f(x_0 + h) - f(x_0)}{h},$$

respectively, if it exists.

Theorem 11.1.2. *Let f be an integrable function over (a, b) and let $c \in (a, b)$.*

(a) *If f is right-hand continuous at a point $x_0 \in [a, b)$, then the function*
$F(x) = \int_c^x f$ *has a right-hand derivative at x_0 which is equal to $f(x_0+)$.*

(b) *If f is left-hand continuous at a point $x_0 \in (a, b]$, then the function*
$F(x) = \int_c^x f$, *has a left-hand derivative at x_0 which is equal to $f(x_0-)$.*

Proof. In view of full symmetry it suffices to prove the assertion for the right-hand limit. We have

$$\frac{F(x_0 + h) - F(x_0)}{h} = \frac{\int_c^{x_0+h} f - \int_c^{x_0} f}{h} = \frac{1}{h} \int_{x_0}^{x_0+h} f. \qquad (11.1)$$

Since f is right-hand continuous at x_0, for an arbitrary $\varepsilon > 0$ there is a $\delta > 0$ such that

$$f(x_0) - \varepsilon < f(t) < f(x_0) + \varepsilon \quad \text{for all } t \in (x_0, x_0 + \delta).$$

Integrating these inequalities from x_0 to $x_0 + h$, we get

$$h\,(f(x_0) - \varepsilon) < \int_{x_0}^{x_0+h} f \; < h\,(f(x_0) + \varepsilon) \quad \text{for all } t \in (x_0, x_0 + \delta).$$

Hence, by (11.1), we have

$$f(x_0) - \varepsilon < \frac{F(x_0 + h) - F(x_0)}{h} < f(x_0) + \varepsilon \quad \text{for } 0 < h < \delta,$$

which proves that

$$\lim_{h \to 0+} \frac{F(x_0 + h) - F(x_0)}{h} = f(x_0).$$

$\square$

From the above theorem we can easily obtain the following two important corollaries.

Corollary 11.1.3. *Every continuous function has an antiderivative.*

Corollary 11.1.4. *If the derivative of f is integrable and continuous in (a, b), then for every $c \in (a, b)$ the function*

$$g(x) = f(x) - \int_c^x f'$$

is constant in (a, b).

The next theorem reduces the evaluation of many integrals to the search for antiderivatives. This has been known for a long time. It is said that it was already used by Barrow, the teacher of Newton. It is the most important and most magical theorem of calculus. For that reason it is often called the fundamental theorem of calculus.

Theorem 11.1.5 (Fundamental Theorem of Calculus). *If f is continuous and integrable over a bounded or unbounded interval (a, b) and has an antiderivative F which has limits $\lim_{x \to a+} F(x) = F(a)$ and $\lim_{x \to b-} F(x) = F(b)$, then*

$$\int_a^b f = F(b) - F(a).$$

Proof. By Theorem 11.1.2 the function $G(x) = \int_c^x f$, where c is an arbitrary point in (a, b), is differentiable at every point of (a, b) and its derivative is f. It is thus an antiderivative of f. Since all antiderivatives differ from one another by a constant, we have $F(x) + C = G(x)$ for some constant C and all $x \in (a, b)$. Hence

$$\int_a^b f = \int_a^c f + \int_c^b f = \int_c^b f - \int_c^a f = G(b) - G(a) = F(b) - F(a).$$

$\square$

It must be emphasized that the integrability condition on f in the above theorem is very important. Just because a function f is continuous on some interval and has an antiderivative on that interval for which the limits $\lim_{x \to a+} F(x) = F(a)$ and $\lim_{x \to b-} F(x) = F(b)$ exist is not necessarily sufficient to guarantee that f is integrable on that interval. Exercises 12 of this section deals with this issue.

In calculations we often write $\int_a^b f(x) \, dx$ instead of $\int_a^b f$. It is very convenient if the integrand f is an algebraic expression of x. Another advantage is that it shows clearly which function is being integrated, which is especially important in the case of integrals with parameters. For example, the following two integrals are essentially different: $\int_0^1 x^\alpha \, dx$ and $\int_0^1 x^\alpha \, d\alpha$. In the first the function being integrated is the one whose value at x is x^α; while in the second, the one being integrated is the function whose value at α is x^α.

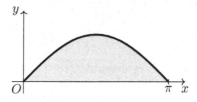

Fig. 11.1 $\int_0^\pi \sin x \, dx = 2$.

Example 11.1.6. To calculate the integral $\int_0^\pi \sin x \, dx$ (Fig. 11.1) we first find an anti-derivative

$$\int \sin x \, dx = -\cos x + C$$

and then use the Fundamental Theorem of Calculus:

$$\int_0^\pi \sin x \, dx = -\cos \pi + \cos 0 = 1 + 1 = 2.$$

Such a procedure is often written as

$$\int_0^\pi \sin x \, dx = [-\cos x]_0^\pi = 1 + 1 = 2.$$

Example 11.1.7.

$$\int_{-a}^a \sqrt{a^2 - x^2} \, dx = \left[\frac{x}{2}\sqrt{a^2 - x^2} + \frac{a^2}{2}\arcsin\frac{x}{a}\right]_{-a}^a = \frac{a^2}{2}\frac{\pi}{2} + \frac{a^2}{2}\frac{\pi}{2} = \frac{a^2\pi}{2}.$$

Note that in this example we have evaluated the area of a semicircle of radius a (see Fig. 11.2).

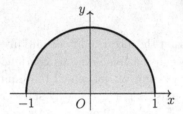

Fig. 11.2 $\int_{-a}^a \sqrt{a^2 - x^2} \, dx = \frac{a^2\pi}{2}.$

Using Theorem 11.1.5 we will investigate integrability of $f(x) = x^\alpha$ on the intervals $(0, 1)$ and $(1, \infty)$.

11.1.8. (a) *The function $f(x) = x^\alpha$ is integrable over $(0, 1)$ if and only if $\alpha > -1$, and in that case the integral is given by*

$$\int_0^1 x^\alpha \, dx = \frac{1}{\alpha + 1}.$$

(b) *The function $f(x) = x^\alpha$ is integrable over $(1, \infty)$ if and only if $\alpha < -1$, and in that case the integral is given by*

$$\int_1^\infty x^\alpha \, dx = -\frac{1}{\alpha + 1}.$$

This theorem has a good intuitive interpretation. If $\alpha \geq 0$ (Fig. 11.3), then the function $f(x) = x^\alpha$ is continuous in the whole closed interval $[0, 1]$ and thus it is integrable over that interval. If $\alpha < 0$ (Fig. 11.4), then the function is unbounded in the right neighborhood of the origin. The integral represents the area of the shaded region under the curve. If the region is "thin enough", which depends on α, then the area may be finite. Otherwise it is infinite and the function is not integrable. Similarly, if we consider the function on the interval $(1, \infty)$, then the area of the shaded region (Fig. 11.4) may be finite or infinite, depending on α.

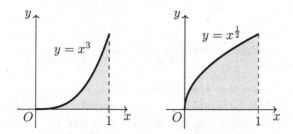

Fig. 11.3 Examples of graphs of $y = x^\alpha$ for $\alpha > 0$ and $0 < x \leq 1$.

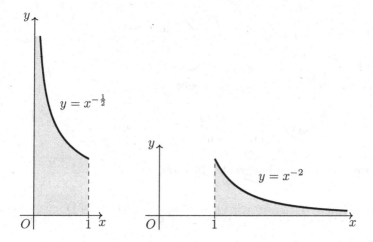

Fig. 11.4 Examples of graphs of $y = x^\alpha$ for $\alpha < 0$.

Proof of 11.1.8. To prove the first part of the theorem assume $\alpha > -1$. The sequence

$$g_n(x) = \begin{cases} x^\alpha, & \text{for } x \in \left(\frac{1}{n}, 1\right), \\ 0, & \text{otherwise,} \end{cases} \quad n = 1, 2, \ldots, \tag{11.2}$$

is a monotone sequence of integrable functions and

$$0 < \int_{\frac{1}{n}}^1 x^\alpha \, dx = \left[\frac{x^{\alpha+1}}{\alpha+1}\right]_{\frac{1}{n}}^1 = \frac{1}{\alpha+1}\left(1 - \frac{1}{n^{\alpha+1}}\right) \leq \frac{1}{\alpha+1}.$$

Hence the limit

$$g(x) = \begin{cases} x^\alpha, & \text{for } x \in (0, 1), \\ 0, & \text{otherwise,} \end{cases} \tag{11.3}$$

is an integrable function, by the Monotone Convergence Theorem. Consequently, the function $f(x) = x^\alpha$ is integrable over $(0, 1)$ for any $\alpha > -1$ and the value of the integral is

$$\int_0^1 x^\alpha \, dx = \lim_{n\to\infty} \frac{1}{\alpha+1}\left(1 - \frac{1}{n^{\alpha+1}}\right) = \frac{1}{\alpha+1}.$$

We still have to prove that the function $f(x) = x^\alpha$ is not integrable over $(0,1)$ if $\alpha \leq -1$. Suppose it is integrable. Then its integral would be a positive number $\int_0^1 x^\alpha \, dx = \int g = M < \infty$, where g is defined as before by (11.3). Since $g_n \leq g$, (where the g_n's are defined by (11.2)), we would have $\int g_n \leq \int g = M$. But for $\alpha < -1$, we have $\lim_{n\to\infty} \int g_n = \infty$. This contradiction shows that f is not integrable for $\alpha < -1$. Finally, if $\alpha = -1$, we have

$$\int g_n = \int_{\frac{1}{n}}^1 \frac{dx}{x} = \ln n \to \infty$$

and as before we get a contradiction.

Now assume that $\alpha < -1$. The sequence

$$h_n(x) = \begin{cases} x^\alpha, & \text{for } x \in (1,n), \\ 0, & \text{otherwise,} \end{cases} \qquad n = 1, 2, \ldots$$

is a monotone sequence of integrable functions and

$$0 < \int h_n = \int_1^n x^\alpha \, dx = \left[\frac{x^{\alpha+1}}{\alpha+1}\right]_1^n = \frac{1}{\alpha+1}\left(n^{\alpha+1} - 1\right) < -\frac{1}{\alpha+1}.$$

By the Monotone Convergence Theorem, the limit

$$h(x) = \begin{cases} x^\alpha, & \text{for } x \in (1,\infty), \\ 0, & \text{otherwise,} \end{cases}$$

is an integrable function and we have

$$\int_1^\infty x^\alpha \, dx = \int h = \lim_{n\to\infty} \frac{1}{\alpha+1}\left(n^{\alpha+1} - 1\right) = -\frac{1}{\alpha+1}.$$

If $\alpha \geq -1$, then $\int h_n \to \infty$, which shows that h is not integrable, because $\int h_n \leq \int h$ for all $n \in \mathbb{N}$. Finally, if $\alpha = -1$, then

$$\int h_n = \int_1^n \frac{dx}{x} = \ln n \to \infty,$$

and again the limit function is not integrable. $\qquad\qquad\qquad\square$

EXERCISES 11.1

(1) Explain how Corollary 11.1.3 is obtained from Theorem 11.1.2.

(2) Explain how Corollary 11.1.4 is obtained from Theorem 11.1.2.

(3) Prove the mean value theorem for integrals: If f is a continuous function on $[a, b]$, then there exists a $c \in (a, b)$ such that

$$\int_a^b f = f(c)\,(b - a).$$

(4) Find an integrable function f such that f^2 is not integrable.

(5) Find a function f such that f^2 is integrable but f is not.

(6) Prove: If f and g are continuous functions on $[a, b]$ and $\int_a^b f = \int_a^b g$, then there exists a $c \in [a, b]$ such that $f(c) = g(c)$.

(7) Let f be a continuous function on $[a, b]$. Define $g(x) = \int_x^b f$ for $x \in [a, b]$. Prove that g is differentiable. Express the derivative of g in terms of f. Caution: The points a and b require special treatment.

(8) Let f be a continuous function on $[a, b]$. Prove that, if $\int_a^x f = \int_x^b f$ for all $x \in [a, b]$, then $f(x) = 0$ for all $x \in [a, b]$.

(9) Find an example of an unbounded continuous function on $\mathbb{R}$ that is integrable.

(10) Evaluate $\displaystyle\int_{-\infty}^{\infty} \frac{dx}{x^2 + 1}$.

(11) Evaluate $\int_0^1 \ln x \, dx$.

(12) (a) Verify that $K(x) = \frac{-\cos x}{x}$ is an antiderivative for $k(x) = \frac{x \sin x + \cos x}{x^2}$ on $(1, \infty)$, and then find $\lim_{x \to 1+} K(x)$ and $\lim_{x \to \infty} K(x)$.

    (b) Let $f(x) = \frac{\cos x}{x^2}$. Show that f is integrable on $[1, \infty)$.

    (c) Let $g(x) = \frac{\sin x}{x}$. Show that g is not integrable on $[1, \infty)$.

    (d) Show that the function k is not integrable on $[1, \infty)$ even though

$$\lim_{t \to \infty} \int_1^t k(x)\,dx = \cos 1.$$

11.2 The change of variables theorem

The following theorem is one of the most important tools in integral calculus.

Theorem 11.2.1 (Change of variables). *Let $-\infty \le a < b \le \infty$ and let g be a non-decreasing differentiable function defined on the interval (a, b) such that g' is continuous on (a, b). By $g(a+)$ and $g(b-)$ we denote the right-hand and left-hand limits of g at a and b, respectively, (finite or infinite). If f is an integrable function over $(g(a+), g(b-))$, then the product $f(g(x))g'(x)$ is integrable over (a, b) and*

$$\int_{g(a+)}^{g(b-)} f(x)\,dx = \int_a^b f(g(x))g'(x)\,dx.$$

Proof. First we prove that the product $f(g(x))g'(x)$ is integrable over (a,b). To this end, suppose

$$f \simeq \lambda_1 f_1 + \lambda_2 f_2 + \dots \tag{11.4}$$

where $f_n = \chi_{[a_n, b_n)}$ for some intervals $[a_n, b_n) \subset (g(a+), g(b-))$. Let

$$\Phi(x) = f\left(g(x)\right)g'(x) \quad \text{and} \quad \Phi_n(x) = f_n(g(x))g'(x).$$

We will show that

$$\Phi \simeq \lambda_1 \Phi_1 + \lambda_2 \Phi_2 + \dots \tag{11.5}$$

Since $g(a+) < a_n < b_n < g(b-)$ and g is continuous, there are numbers $\alpha_n < \beta_n$ such that $g(\alpha_n) = a_n$ and $g(\beta_n) = b_n$, for all $n \in \mathbb{N}$. Hence

$$\int \Phi_n = \int_{\alpha_n}^{\beta_n} g'(x)\, dx = g(\beta_n) - g(\alpha_n) = b_n - a_n = \int f_n. \tag{11.6}$$

Since $|\lambda_1| \int f_1 + |\lambda_2| \int f_2 + \dots < \infty$, we also have $\int |\lambda_1 \Phi_1| + \int |\lambda_2 \Phi_2| + \dots < \infty$. Moreover, whenever the series $\lambda_1 \Phi_1(x) + \lambda_2 \Phi_2(x) + \dots$ converges absolutely for some x, then it converges to $\Phi(x)$. This is trivially true if $g'(x) = 0$. If $g'(x) = m \neq 0$, then letting $g(x) = t$, we have

$$\lambda_1 \Phi_1(x) + \lambda_2 \Phi_2(x) + \dots = (\lambda_1 f_1(t) + \lambda_2 f_2(t) + \dots)\, m = f(t)m = \Phi(x),$$

with the series being absolutely convergent. Thus (11.5) holds, which proves that $f(g(x))g'(x)$ is integrable. Now, by (11.5), (11.6), and (11.4), we have

$$\int_a^b f(g(x))g'(x)\, dx = \lambda_1 \int \Phi_1 + \lambda_2 \int \Phi_2 + \dots$$
$$= \lambda_1 \int f_1 + \lambda_2 \int f_2 + \dots$$
$$= \int_{g(a+)}^{g(b-)} f(x)\, dx.$$

$\square$

Theorem 11.2.1 was formulated for non-decreasing functions, but it can easily be reformulated for non-increasing functions. To obtain that result we first prove a simple lemma.

Lemma 11.2.2. *Let* $-\infty \leq a < b \leq \infty$. *If* f *is integrable over the interval* (a,b), *then*

$$\int_a^b f(x)\, dx = - \int_{-a}^{-b} f(-x)\, dx.$$

Proof. Let $g(x) = f(-x)$ for $-b < x < -a$. Then, for all $x \in \mathbb{R}$, we have

$$\left(g\chi_{(-b,-a)}\right)(x) = \left(f\chi_{(a,b)}\right)(-x)$$

and consequently

$$\int_a^b f(x)\,dx = \int f\chi_{(a,b)} = \int g\chi_{(-b,-a)}$$

$$= \int_{-b}^{-a} g(x)\,dx = \int_{-b}^{-a} f(-x)\,dx = -\int_{-a}^{-b} f(-x)\,dx.$$

$\square$

Corollary 11.2.3. *Let $-\infty \leq a < b \leq \infty$ and let h be a non-increasing differentiable function defined on the interval (a,b) such that h' is continuous on (a,b). By $h(a+)$ and $h(b-)$ we denote the right-hand and left-hand limits of h at a and b, respectively, (finite or infinite). If f is an integrable function over $(h(b-), h(a+))$, then the product $f(h(x))h'(x)$ is integrable over (a,b) and*

$$\int_{h(b-)}^{h(a+)} f(x)\,dx = -\int_a^b f(h(x))h'(x)\,dx.$$

Proof. Let $k(x) = f(-x)$ and let $g(x) = -h(x)$. Then, using Lemma 11.2.2 and Theorem 11.2.1, we obtain

$$\int_{h(b-)}^{h(a+)} f(x)\,dx = \int_{-h(a+)}^{-h(b-)} f(-x)\,dx = \int_{g(a+)}^{g(b-)} k(x)\,dx$$

$$= \int_a^b k(g(x))g'(x)\,dx = \int_a^b k(-h(x))(-h'(x))\,dx$$

$$= -\int_a^b f(h(x))h'(x)\,dx.$$

$\square$

Corollary 11.2.4. *If f is integrable over (a,b) and $\lambda \neq 0$, then*

$$\int_a^b f(x)\,dx = \lambda \int_{a/\lambda}^{b/\lambda} f(\lambda x)\,dx.$$

Proof. If $\lambda > 0$, we take $g(x) = \lambda x$ and use Theorem 11.2.1. For $\lambda < 0$ we use Lemma 11.2.2 and then Theorem 11.2.1. $\square$

Example 11.2.5. *Euler's gamma function* is defined by the integral

$$\Gamma(p) = \int_0^\infty x^{p-1} e^{-x}\,dx, \quad p > 0.$$

It is well-defined by Example 10.13.5. Substituting $x = ty$ for $t > 0$ we obtain

$$\Gamma(p) = t^p \int_0^\infty y^{p-1} e^{-ty}\,dy.$$

EXERCISES 11.2

(1) Evaluate the following integrals

(a) $\int_0^3 \frac{dx}{9+x}$,

(b) $\int_1^2 \frac{dx}{x^2+5x+4}$,

(c) $\int_{-1}^1 \arccos x \, dx$,

(d) $\int_0^1 \sqrt{x^2+1} \, dx$,

(e) $\int_{-\infty}^\infty e^{-|x|} dx$,

(f) $\int_0^3 \frac{dx}{\sqrt{3-x}}$.

(2) Prove the following properties of Euler's gamma function.

(a) Γ is a continuous function on $(0,\infty)$.

(b) $\Gamma(1) = 1$.

(c) $\Gamma(p+1) = p\Gamma(p)$ for all $p \in (0,\infty)$.

(d) $\Gamma(n+1) = n!$ for all $n \in \mathbb{N}$.

11.3   Cauchy's formula

Let $[a,b]$ be an interval containing 0. For any function f, integrable over $[a,b]$, the function $F_1(x) = \int_0^x f$ is continuous in $[a,b]$ and thus integrable. Consequently it can be integrated again

$$F_2(x) = \int_0^x F_1 = \int_0^x \int_0^x f$$

and again

$$F_3(x) = \int_0^x F_2 = \int_0^x \int_0^x \int_0^x f$$

and so on. In this way obtain a sequence of functions defined on $[a,b]$:

$$F_1(x) = \int_0^x f,$$

$$F_n(x) = \int_0^x F_{n-1} = \underbrace{\int_0^x \cdots \int_0^x}_{n \text{ times}} f, \quad \text{for } n = 2, 3, 4, \ldots.$$

We are going to show that each of these iterated integrals can be expressed as a single integral. The obtained formula is usually called *Cauchy's formula*.

Theorem 11.3.1 (Cauchy's formula). *If $a < 0 < b$ and f is integrable over $[a,b]$, then*

$$\underbrace{\int_0^x \cdots \int_0^x}_{n \text{ times}} f = \frac{1}{(n-1)!} \int_0^x (x-t)^{n-1} f(t) \, dt \qquad (11.7)$$

for $x \in [a,b]$ and any $n \in \mathbb{N}$.

Proof. We will use mathematical induction. Formula (11.7) is trivially true for $n = 1$. Assume that it holds for some $m \in \mathbb{N}$.

First we show that the function

$$G(x) = \frac{1}{m!} \int_0^x (x - t)^m f(t)\, dt, \tag{11.8}$$

is an antiderivative of the function defined by the integral

$$\frac{1}{(m-1)!} \int_0^x (x - t)^{m-1} f(t)\, dt, \tag{11.9}$$

for $x \in [a, b]$. By Theorem 10.13.3, the integrand in (11.8) is integrable on the interval of integration. Note that

$$\frac{G(x+h) - G(x)}{h} = \frac{1}{hm!} \int_0^{x+h} (x + h - t)^m f(t)\, dt - \frac{1}{hm!} \int_0^x (x - t)^m f(t)\, dt$$

$$= \frac{1}{m!} \int_0^x \frac{(x + h - t)^m - (x - t)^m}{h} f(t)\, dt + \frac{1}{hm!} \int_x^{x+h} (x + h - t)^m f(t)\, dt.$$

Let $x \in [a, b]$ be fixed. For every $t \in [a, b]$ we have

$$\lim_{h \to 0} \frac{(x + h - t)^m - (x - t)^m}{h} = m(x - t)^{m-1}.$$

Since

$$\frac{(x + h - t)^m - (x - t)^m}{h} = m\xi^{m-1},$$

for some ξ between $x - t$ and $x + h - t$, there exists a constant M such that

$$\left| \frac{(x + h - t)^m - (x - t)^m}{h} \right| \leq mM^{m-1}$$

for all $t \in [a, b]$ and all sufficiently small h. Therefore, by the Dominated Convergence Theorem, we have

$$\lim_{h \to 0} \frac{1}{m!} \int_0^x \frac{(x + h - t)^m - (x - t)^m}{h} f(t)\, dt = \frac{1}{(m-1)!} \int_0^x (x - t)^{m-1} f(t)\, dt.$$

Moreover, since $|x + h - t|^m \leq |h|^m$ for t between x and $x + h$, we have

$$\left| \frac{1}{h\, m!} \int_x^{x+h} (x + h - t)^m f(t)\, dt \right| \leq \frac{|h|^{m-1}}{m!} \int_x^{x+h} |f(t)|\, dt \to 0$$

as $h \to 0$. Consequently,

$$\lim_{h \to 0} \frac{G(x+h) - G(x)}{h} = \frac{1}{(m-1)!} \int_0^x (x - t)^{m-1} f(t)\, dt$$

and thus (11.8) is an antiderivative of (11.9). Since

$$\underbrace{\int_0^x \cdots \int_0^x}_{m + 1 \text{ times}} f$$

is also an antiderivative of (11.9) and both functions are 0 at $x = 0$, they are equal. This shows that (11.7) holds for $m + 1$. Consequently, in view of the induction principle, (11.7) holds for every $n \in \mathbb{N}$. $\qquad \square$

EXERCISES 11.3

(1) Evaluate $\int_0^x \int_0^x \int_0^x \int_0^x f$ if $f(x) = \cos x$.
(2) Justify the use of the Dominated Convergence Theorem in the proof of Theorem 11.3.1.

11.4 Taylor's formula

From Cauchy's formula we obtain *Taylor's formula*, which is one of the most important formulas in calculus.

Theorem 11.4.1 (Taylor's formula). *Let $a < 0 < b$. If f has n continuous derivatives in (a, b), then*

$$f(x) = f(0) + \frac{f'(0)}{1!}x + \frac{f''(0)}{2!}x^2 + \cdots + \frac{f^{(n-1)}(0)}{(n-1)!}x^{n-1} + \underbrace{\int_0^x \cdots \int_0^x}_{n \text{ times}} f^{(n)} \quad (11.10)$$

for $x \in (a, b)$.

Proof. By Theorem 11.1.5 we may write

$$f(x) = f(0) + \int_0^x f'(t)\, dt,$$

so (11.10) holds for $n = 1$. For induction, assume that (11.10) holds for some n and all functions satisfying the conditions of the theorem. If we assume that $f^{(n+1)}$ is continuous in (a, b), we may replace f in (11.10) by f' and get

$$f'(x) = f'(0) + \frac{f''(0)}{1!}x + \cdots + \frac{f^{(n)}(0)}{n!}x^{n-1} + \underbrace{\int_0^x \cdots \int_0^x}_{n \text{ times}} f^{(n+1)}.$$

Integrating this equality from 0 to x we obtain

$$f(x) - f(0) = \frac{f'(0)}{1!}x + \frac{f''(0)}{2!}x^2 + \cdots + \frac{f^{(n)}(0)}{n!}x^n + \underbrace{\int_0^x \cdots \int_0^x}_{n+1 \text{ times}} f^{(n+1)}.$$

This proves that the formula holds for all $n \in \mathbb{N}$. $\qquad\square$

The expression

$$R_n = \underbrace{\int_0^x \cdots \int_0^x}_{n \text{ times}} f^{(n)},$$

is called the *remainder*. The remainder can be transformed in several ways, for example,

Integral form: $R_n = \dfrac{1}{(n-1)!} \displaystyle\int_0^x (x-t)^{n-1} f^{(n)}(t)\, dt,$

Lagrange's form: $R_n = \dfrac{x^n}{n!} f^{(n)}(\xi)$, where ξ is a number between 0 and x.

The integral form follows immediately from Cauchy's formula (Theorem 11.3.1). To prove Lagrange's form we will use the following theorem.

Theorem 11.4.2. *If f is continuous in $[a, b]$ and g is nonnegative and integrable over (a, b), then there exists a $\xi \in (a, b)$ such that*

$$\int_a^b f(t)g(t)\, dt = f(\xi) \int_a^b g(t)\, dt. \qquad (11.11)$$

Proof. Denote by m and M the minimum and the maximum of f in $[a, b]$, respectively. Then

$$mg(t) \le f(t)g(t) \le Mg(t)$$

for all $t \in (a, b)$. Integrating these inequalities from a to b we get

$$m \int_a^b g(t)\, dt \le \int_a^b f(t)g(t)\, dt \le M \int_a^b g(t)\, dt.$$

On the other hand, we have

$$m \int_a^b g(t)\, dt \le f(t) \int_a^b g(t)\, dt \le M \int_a^b g(t)\, dt,$$

for all $t \in [a, b]$. Since the function $f(t) \int_a^b g(t)\, dt$ is continuous in $[a, b]$, there exists a $\xi \in (a, b)$ for which (11.11) holds, by the intermediate value theorem (Theorem 3.2.4). $\qquad \square$

Applying the above theorem to the integral form of the remainder R_n we get Lagrange's form:

$$\frac{1}{(n-1)!} \int_0^x (x-t)^{n-1} f^{(n)}(t)\, dt = \frac{1}{(n-1)!} f^{(n)}(\xi) \int_0^x (x-t)^{n-1}\, dt = \frac{x^n}{n!} f^{(n)}(\xi)$$

for some ξ between 0 and x. It is important to remember that ξ depends on x.

Examples 11.4.3.

(a) $e^x = 1 + \frac{x}{1!} + \frac{x^2}{2!} + \cdots + \frac{x^{n-1}}{(n-1)!} + \frac{x^n}{n!} e^\xi$, for some ξ between 0 and x.

(b) $\ln(x+1) = \frac{x}{1} - \frac{x^2}{2} + \frac{x^3}{3} - \cdots + \frac{(-1)^{n-2}}{n-1} x^{n-1} + \frac{(-1)^{n-1}}{n} \frac{x^n}{(1+\xi)^n}$, for some ξ between 0 and x.

(c) $\sin x = \frac{x}{1!} - \frac{x^3}{3!} + \frac{x^5}{5!} - \cdots + (-1)^{n-1} \frac{x^{2n-1}}{(2n-1)!} + \frac{x^{2n}}{(2n)!} \sin(\xi + n\pi)$, for some ξ between 0 and x, because $(\sin x)^{(m)} = \sin\left(x + \frac{m\pi}{2}\right)$.

EXERCISES 11.4

(1) Find Taylor's formula with Lagrange's form of the remainder for the function $f(x) = (1+x)^{-1/2}$.

(2) Find Taylor's formula with Lagrange's form of the remainder for the function $f(x) = e^{-x^2}$ in the special case when $n = 4$.

(3) Prove that $(\sin x)^{(m)} = \sin\left(x + \frac{m\pi}{2}\right)$ for all $m \in \mathbb{N}$.

(4) Show that Theorem 11.4.2 does not hold if the assumption that g is non-negative is dropped.

11.5 The integral test for convergence of infinite series

The integral gives us a very convenient test for convergence of certain infinite series.

Theorem 11.5.1 (Cauchy's integral test). *Let f be a positive decreasing function defined on $[1,\infty)$ such that $\lim_{x\to\infty} f(x) = 0$. Then the series $\sum_{n=1}^{\infty} f(n)$ converges if and only if the function f is integrable over $[1,\infty)$.*

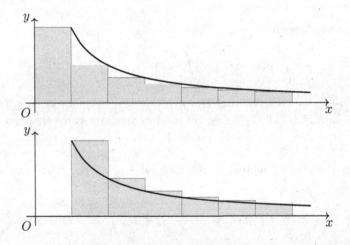

Fig. 11.5   Illustration for the proof of Cauchy's integral test.

Proof. The main idea of the proof is illustrated by Fig. 11.5. We have

$$\sum_{n=1}^{m} f(n+1) \leq \int_{1}^{m+1} f(x)\,dx \leq \sum_{n=1}^{m} f(n), \qquad (11.12)$$

which follows by induction from

$$f(n+1) \leq \int_{n}^{n+1} f(x)\,dx \leq f(n).$$

From (11.12), by letting $m \to \infty$, we obtain

$$\sum_{n=2}^{\infty} f(n) \leq \int_{1}^{\infty} f(x)\,dx \leq \sum_{n=1}^{\infty} f(n), \tag{11.13}$$

and the assertion follows easily. □

Corollary 11.5.2. *The series* $\displaystyle\sum_{n=1}^{\infty} \frac{1}{n^{\alpha}}$ *converges if and only if* $\alpha > 1$.

Proof. This is a direct consequence of Theorems 11.5.1 and 11.1.8. □

In particular, the series $\sum_{n=1}^{\infty} \frac{1}{n}$ is divergent while the series $\sum_{n=1}^{\infty} \frac{1}{n^2}$ is convergent. From (11.13) we obtain

$$1 + \frac{1}{2^2} + \frac{1}{3^2} + \cdots < \frac{4}{3},$$

because

$$\sum_{k=2}^{\infty} \frac{1}{k^2} \leq \int_{1}^{\infty} \frac{dx}{x^2} = \left[-\frac{1}{3}\frac{1}{x^3}\right]_{1}^{\infty} = \frac{1}{3}.$$

EXERCISES 11.5

(1) Test the series $\sum_{n=2}^{\infty} \frac{1}{n(\ln n)^2}$ for convergence.

(2) Test the series $\sum_{n=2}^{\infty} \frac{1}{n \ln n \ln(\ln n)}$ for convergence.

(3) Consider the series $\sum_{n=2}^{\infty} \frac{1}{n^p (\ln n)^q}$. Prove the following assertions.

    (a) If $p > 1$, then the series converges for arbitrary q.
    (b) If $p = 1$ and $q > 1$, then the series converges.
    (c) If $p < 1$, then the series diverges for arbitrary q.
    (d) If $p = 1$ and $q \leq 1$, then the series diverges.

(4) Complete the proof of Theorem 11.5.1.

11.6   Examples of trigonometric series

Theorems 8.6.1 and 11.5.1 can help us show that a series is convergent, but they will not help us find the sum of a convergent series. An approximate value of the sum can be found by adding a finite number of initial terms. This method can be used to approximate the sum of

$$1 + \frac{1}{2^2} + \frac{1}{3^2} + \cdots,$$

but by no means would it lead to the discovery of a magic connection between the series and the number π:

$$1 + \frac{1}{2^2} + \frac{1}{3^2} + \cdots = \frac{\pi^2}{6}.$$

The proof of this equality is much more difficult than the proof of convergence of the series. We will prove the equality using trigonometric series.

By a *trigonometric series* or *Fourier series* (Jean Baptiste Joseph Fourier 1768–1830) we mean any series of the form

$$\frac{1}{2}a_0 + \sum_{n=1}^{\infty}(a_n \cos nx + b_n \sin nx).$$

We will first consider some elementary cases. For example, we are going to show that

$$\sum_{n=1}^{\infty} \frac{\sin nx}{n} = \frac{\pi - x}{2} \quad \text{for } x \in (0, 2\pi), \tag{11.14}$$

and that the series is uniformly convergent on every closed interval of the form $[\varepsilon, 2\pi - \varepsilon]$ contained in $[0, 2\pi]$.

In view of the identity

$$2\cos nx \sin \frac{1}{2}x = -\sin\left(n - \frac{1}{2}\right)x + \sin\left(n + \frac{1}{2}\right)x,$$

the expression

$$2(\cos x + \cos 2x + \cdots + \cos nx)\sin \frac{1}{2}x$$

can be written as

$$-\sin \frac{1}{2}x + \sin \frac{3}{2}x - \sin \frac{3}{2}x + \sin \frac{5}{2}x + \cdots - \sin\left(n - \frac{1}{2}\right)x + \sin\left(n + \frac{1}{2}\right)x$$

which simplifies to

$$-\sin \frac{1}{2}x + \sin\left(n + \frac{1}{2}\right)x.$$

Hence

$$\sum_{k=1}^{n} \cos kx = -\frac{1}{2} + \frac{\sin\left(n + \frac{1}{2}\right)x}{2\sin \frac{1}{2}x}.$$

Integrating the above equality from π to x we get

$$\sum_{k=1}^{n} \frac{\sin kx}{k} = \frac{\pi - x}{2} + I_n(x)$$

where

$$I_n(x) = \int_{\pi}^{x} \frac{\sin\left(n + \frac{1}{2}\right)t}{2\sin \frac{1}{2}t}\, dt.$$

Integration by parts gives us

$$I_n(x) = -\frac{1}{2n+1} \left(\frac{\cos\left(n+\frac{1}{2}\right)x}{\sin\frac{1}{2}x} + \frac{1}{2}\int_\pi^x \frac{\cos\frac{1}{2}t\cos\left(n+\frac{1}{2}\right)t}{\left(\sin\frac{1}{2}t\right)^2}\,dt \right).$$

Hence

$$|I_n(x)| \le \frac{1}{2n+1} \left(\frac{1}{\sin\frac{1}{2}\varepsilon} + \frac{1}{2}\int_\varepsilon^\pi \frac{dt}{\left(\sin\frac{1}{2}t\right)^2} \right) \qquad \text{for } x \in [\varepsilon, 2\pi - \varepsilon]. \tag{11.15}$$

This inequality proves that the sequence of functions I_n converges to 0 uniformly on $[\varepsilon, 2\pi - \varepsilon]$. Since ε is arbitrary, (11.14) follows.

For $x = 0$ the series in (11.14) is not convergent to $\pi/2$, but to 0. Moreover, since all terms of the series are periodic, the series in (11.14) converges to a function whose graph is shown in Fig. 11.6.

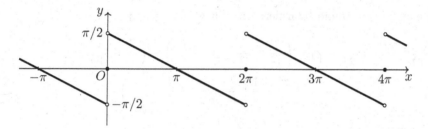

Fig. 11.6 Graph of the sum $\sum_{n=1}^\infty \frac{\sin(nx)}{n}$.

Now we consider the function

$$H(x) = -\sum_{n=1}^\infty \frac{\cos nx}{n^2}. \tag{11.16}$$

Note that this series converges uniformly on $[0, 2\pi]$. Since

$$H'(x) = \left(-\sum_{n=1}^\infty \frac{\cos nx}{n^2} \right)' = -\sum_{n=1}^\infty \left(\frac{\cos nx}{n^2} \right)' = \sum_{n=1}^\infty \frac{\sin nx}{n} = \frac{\pi - x}{2}, \tag{11.17}$$

we have

$$H(x) = -\frac{(\pi - x)^2}{4} + C \quad \text{for } x \in (0, 2\pi).$$

The constant C can be found by letting $x = \pi$, which gives us

$$C = \sum_{n=1}^\infty \frac{(-1)^{n-1}}{n^2}.$$

Hence

$$-\sum_{n=1}^\infty \frac{\cos nx}{n^2} = -\frac{(\pi - x)^2}{4} + \sum_{n=1}^\infty \frac{(-1)^{n-1}}{n^2} \qquad \text{for } x \in (0, 2\pi),$$

which, after changing signs, yields

$$\frac{\cos x}{1^2} + \frac{\cos 2x}{2^2} + \frac{\cos 3x}{3^2} + \cdots = \frac{(\pi - x)^2}{4} - \left(\frac{1}{1^2} - \frac{1}{2^2} + \frac{1}{3^2} - \cdots \right) \quad \text{for } x \in (0, 2\pi).$$

But the series on the left converges uniformly and its sum is a function continuous at every $x \in \mathbb{R}$. In particular, at $x = 0$ we have

$$\frac{1}{1^2} + \frac{1}{2^2} + \frac{1}{3^2} + \cdots = \frac{\pi^2}{4} - \left(\frac{1}{1^2} - \frac{1}{2^2} + \frac{1}{3^2} - \cdots \right). \tag{11.18}$$

Hence we easily find that

$$\frac{1}{1^2} + \frac{1}{3^2} + \frac{1}{5^2} + \cdots = \frac{\pi^2}{8}.$$

This equality was found by Euler. Using it we may write

$$\frac{1}{1^2} + \frac{1}{2^2} + \frac{1}{3^2} + \cdots = \left(\frac{1}{1^2} + \frac{1}{3^2} + \frac{1}{5^2} + \cdots \right) + \left(\frac{1}{2^2} + \frac{1}{4^2} + \frac{1}{6^2} + \cdots \right)$$
$$= \frac{\pi^2}{8} + \frac{1}{4} \left(\frac{1}{1^2} + \frac{1}{2^2} + \frac{1}{3^2} + \cdots \right)$$

and hence

$$\frac{1}{1^2} + \frac{1}{2^2} + \frac{1}{3^2} + \cdots = \frac{\pi^2}{6}. \tag{11.19}$$

This equality was also found by Euler. From (11.18) and (11.19) we easily obtain

$$\frac{1}{1^2} - \frac{1}{2^2} + \frac{1}{3^2} - \cdots = \frac{\pi^2}{12}.$$

Starting from the series $\sum\limits_{n=1}^{\infty} \frac{\cos nx}{n}$ and using a similar method we can obtain

$$\frac{1}{1^4} + \frac{1}{2^4} + \frac{1}{3^4} + \cdots = \frac{\pi^4}{90}. \tag{11.20}$$

Comparing (11.19) and (11.20) we are anxious to find for what number α will we have

$$\frac{1}{1^3} + \frac{1}{2^3} + \frac{1}{3^3} + \cdots = \frac{\pi^3}{\alpha}.$$

The answer is quite unexpected: α is probably an irrational number, but we do not know whether this has ever been proved. The approximate value of α is 25.79435017.

EXERCISES 11.6

(1) Prove that the series in (11.16) converges uniformly on the interval $[0, 2\pi]$.

(2) Provide a detailed proof for (11.15).

(3) Justify (11.17).

(4) Prove that $\frac{\sin x}{1} - \frac{\sin 2x}{2} + \frac{\sin 3x}{3} - \cdots = \frac{x}{2}$ for $x \in (-\pi, \pi)$.

(5) Prove that $\frac{\sin x}{1} + \frac{\sin 3x}{3} + \frac{\sin 5x}{5} + \cdots = \begin{cases} -\frac{\pi}{4}, & \text{for } x \in (-\pi, 0), \\ \frac{\pi}{4}, & \text{for } x \in (0, \pi). \end{cases}$

(6) Prove that

$$\frac{\sin x}{1^3} + \frac{\sin 2x}{2^3} + \frac{\sin 3x}{3^3} + \cdots = \frac{x(\pi - x)(2\pi - x)}{12}$$

for $x \in (0, 2\pi)$.

(7) Prove that $\frac{1}{1^3} - \frac{1}{3^3} + \frac{1}{5^3} - \cdots = \frac{\pi^3}{32}$.

(8) Prove that $\frac{1}{1^4} + \frac{1}{2^4} + \frac{1}{3^4} + \cdots = \frac{\pi^4}{90}$.

11.7 General Fourier series

The following theorem gives us the general form of coefficients of the Fourier series of a periodic function. Since all the functions involved are periodic with period 2π one can think of them as functions defined just on $[0, 2\pi]$.

Theorem 11.7.1. *If*

$$f(x) = \frac{1}{2}a_0 + \sum_{n=1}^{\infty} (a_n \cos nx + b_n \sin nx) \quad i.n. \quad on \ [0, 2\pi], \tag{11.21}$$

then

$$a_0 = \frac{1}{\pi} \int_0^{2\pi} f(x) \, dx, \tag{11.22}$$

and

$$a_n = \frac{1}{\pi} \int_0^{2\pi} f(x) \cos nx \, dx \quad and \quad b_n = \frac{1}{\pi} \int_0^{2\pi} f(x) \sin nx \, dx \tag{11.23}$$

for $n = 1, 2, \ldots$.

Proof. Note that (11.21) means that f is an integrable function on $[0, 2\pi]$ and

$$\lim_{n \to \infty} \int_0^{2\pi} \left| \frac{1}{2}a_0 + \sum_{k=1}^{n} (a_k \cos kx + b_k \sin kx) - f(x) \right| dx = 0.$$

Integrating (11.21) from 0 to 2π we get

$$\int_0^{2\pi} f(x) \, dx = \pi a_0,$$

proving (11.22). In order to find the remaining coefficients a_n we first multiply both sides of (11.21) by $\cos mx$. The series

$$\frac{1}{2}a_0 \cos mx + \sum_{n=1}^{\infty} (a_n \cos mx \cos nx + b_n \cos mx \sin nx) \tag{11.24}$$

converges in norm and thus we also have

$$\int_0^{2\pi} \cos mx\, f(x)\, dx = \frac{1}{2} \int_0^{2\pi} a_0 \cos mx\, dx$$

$$+ \sum_{n=1}^{\infty} \int_0^{2\pi} (a_n \cos mx \cos nx + b_n \cos mx \sin nx)\, dx. \tag{11.25}$$

Since

$$\int_0^{2\pi} (\cos mx)^2\, dx = \pi,$$

$$\int_0^{2\pi} \cos mx \cos nx\, dx = 0 \text{ for } m \neq n,$$

and

$$\int_0^{2\pi} \cos mx \sin nx = 0 \text{ for all } m, n \in \mathbb{N},$$

we get

$$\int_0^{2\pi} f(x) \cos mx\, dx = \pi a_m.$$

Similarly, multiplying (11.21) by $\sin mx$ and integrating from 0 to 2π we obtain

$$\int_0^{2\pi} f(x) \sin mx\, dx = \pi b_m.$$

$\square$

For every 2π-periodic and integrable function f it is possible to find its Fourier series, namely,

$$\frac{1}{2}a_0 + \sum_{n=1}^{\infty} (a_n \cos nx + b_n \sin nx), \tag{11.26}$$

where the a_n's and b_n's are defined by (11.22) and (11.23). But does the series (11.26) converge to f? The answer depends on properties of f and what is meant by the word "converge." For example, if the derivative of f is continuous, then the series (11.26) converges to f uniformly. Continuity of f is not sufficient for uniform convergence of its Fourier series. In fact it does not even imply pointwise convergence. On the other hand, if f is continuous, then the series (11.26) converges to f almost everywhere. This follows from the famous theorem proved in 1966 by the Swedish mathematician Lennart Carleson (born 1928). Carleson proved that, if f is a square integrable function on $[0, 2\pi]$, then its Fourier series converges to f almost everywhere. A function f on $[0, 2\pi]$ is called *square integrable* if it is integrable and the function f^2 is also integrable. Every continuous function on $[0, 2\pi]$ is square integrable. The theory of square integrable functions will not be presented in this book.

EXERCISES 11.7

(1) In the proof of Theorem 11.7.1 we say that the series (11.24) converges in norm. Justify that claim.

(2) Show that (11.24) implies (11.25).

(3) Define

$$f(x) = \begin{cases} 0 \text{ if } 0 \le x < \pi, \\ 1 \text{ if } \pi < x \le 2\pi. \end{cases}$$

Find the Fourier series of f and check its convergence.

(4) Define

$$f(x) = \begin{cases} x & \text{if } 0 \le x < \pi. \\ 2\pi - x & \text{if } \pi < x \le 2\pi. \end{cases}$$

Find the Fourier series of f and check its convergence.

11.8   Euler's constant

The series $\sum_{n=1}^{\infty} \frac{1}{n}$ is called the *harmonic series*. We have shown in Section 11.5 that it is divergent. This means that the sequence of partial sums $\frac{1}{1} + \cdots + \frac{1}{n}$ tends to infinity. It turns out that the sequence

$$c_n = \frac{1}{1} + \frac{1}{2} + \cdots + \frac{1}{n} - \ln n$$

is convergent. The limit $C = \lim_{n \to \infty} c_n$ is called *Euler's constant*. To show convergence of the sequence (c_n) we write

$$c_n = \alpha_1 + \cdots + \alpha_{n-1} + \frac{1}{n}$$

where

$$\alpha_n = \frac{1}{n} + \ln n - \ln(n+1).$$

The curve in Fig. 11.7 is the graph of the function $f(x) = \frac{1}{x}$. On this curve lie vertices of shaded rectangles $P_1, P_2, \ldots$. It is easily seen that the sum of the areas of all these rectangles is 1. In fact, when shifting horizontally all the rectangles to the square K, they would fill up the whole square. The part of the rectangle P_n which is above the curve has the area

$$\frac{1}{n} - \int_n^{n+1} \frac{dx}{x} = \alpha_n.$$

Consequently, for every $n \in \mathbb{N}$, $\alpha_n > 0$ and the sum $\alpha_1 + \cdots + \alpha_n$ is less than 1, because it is less than the sum of areas of rectangles P_n. This proves convergence of the sequence (c_n). Moreover, we know that $\lim_{n \to \infty} c_n = C < 1$. To estimate C from below note that the curve $y = \frac{1}{x}$ is concave up and, consequently, the parts of

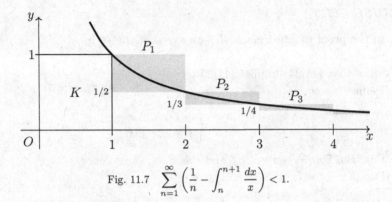

Fig. 11.7 $\displaystyle\sum_{n=1}^{\infty}\left(\frac{1}{n}-\int_n^{n+1}\frac{dx}{x}\right)<1.$

the rectangles above the curve, are greater than the area of half of the rectangle. Hence $C>1/2$ and thus

$$\frac{1}{2}<C<1.$$

In order to find a better estimate of C we consider the sequence

$$\gamma_n=c_n-\frac{1}{2n}+\frac{1}{12n^2}$$

that is also convergent to C. It is easy to check that

$$\gamma_n-\gamma_{n+1}=\frac{1}{6}\int_n^{\infty}\frac{dx}{x^3(x+1)^3}.$$

Since

$$\frac{1}{x^3(x+1)^3}=\frac{1}{5}\frac{5x^4+10x^3+5x^2}{x^5(x+1)^5}<\frac{1}{5}\frac{(x+1)^5-x^5}{x^5(x+1)^5}=\frac{1}{5}\left(\frac{1}{x^5}-\frac{1}{(x+1)^5}\right),$$

we have

$$\frac{1}{6}\int_n^{\infty}\frac{dx}{x^3(x+1)^3}<\frac{1}{30}\int_n^{\infty}\left(\frac{1}{x^5}-\frac{1}{(x+1)^5}\right)dx=\frac{1}{30}\int_n^{n+1}\frac{dx}{x^5}$$

and so

$$0<\gamma_n-\gamma_{n+1}<\frac{1}{30}\int_n^{n+1}\frac{dx}{x^5}.$$

Hence

$$0\le\gamma_n-C\le\frac{1}{30}\int_n^{\infty}\frac{dx}{x^5}=\frac{1}{120n^4}. \tag{11.27}$$

Conseqently, approximating C by

$$\gamma_n=\left(\frac{1}{1}+\frac{1}{2}+\cdots+\frac{1}{n}\right)-\frac{1}{2n}+\frac{1}{12n^2}-\ln n$$

we make an error which does not exceed $\frac{1}{120n^4}$. For example, when using γ_{10} we make an error that does not exceed $1/1,200,000$. In fact, from (11.27) it follows that

$$\gamma_{10} - \frac{1}{1,200,000} < C < \gamma_{10}.$$

Notice that

$$\gamma_{10} = \left(\frac{1}{1} + \frac{1}{2} + \frac{1}{3} + \frac{1}{4} + \frac{1}{5} + \frac{1}{6} + \frac{1}{8} + \frac{1}{20} \right) + \frac{1}{7} + \frac{1}{9} + \frac{1}{1200} - \ln 10$$

$$= \frac{21}{8} + \frac{1}{7} + \frac{1}{9} + \frac{1}{1200} - \ln 10 = \frac{72,571}{25,200} - \ln 10 \approx 0.577216494.$$

In view of the inequality $1/1,200,000 < 1/10^6$, we see that

$$0.577215 < \gamma_{10} - \frac{1}{10^6} < \gamma_{10} - \frac{1}{1,200,000} < C < 0.577217.$$

This shows that

$$C \approx 0.57721.$$

A more exact value is

$$C \approx 0.5772156649.$$

EXERCISES 11.8

(1) In this section we use the equality $\gamma_n - \gamma_{n+1} = \frac{1}{6} \int_n^\infty \frac{dx}{x^3(x+1)^3}$. Verify it.

(2) Explain in detail why $\gamma_n - \gamma_{n+1} < \frac{1}{30} \int_n^{n+1} \frac{dx}{x^5}$ implies $0 \leq \gamma_n - C \leq \frac{1}{30} \int_n^\infty \frac{dx}{x^5}$.

11.9 Iterated integrals

This book is, in principle, concerned with functions of a single real variable. However we occasionally used functions of two variables, for example, in Cauchy's formula Section 11.3. The function which to a number x assigns the value x^α is actually a function of two variables: x and a parameter α. From the mathematical point of view there is no reason to make a distinctions between a function of one variable with a parameter and a function of two variables. In each case we mean a function which assigns to a pair of numbers, say x and y, a real number $f(x,y)$. In this section we will restrict ourselves to continuous functions for which x ranges over an interval I and y over an interval J, so that all pairs (x,y) represent points in the rectangle $I \times J = \{(x,y) : x \in I \text{ and } y \in J\}$.

Definition 11.9.1. A function f defined on a rectangle $I \times J$ is said to be *continuous at a point* $(x_0, y_0) \in I \times J$, if for every $\varepsilon > 0$ there exists a $\delta > 0$ such that

$$|f(x, y) - f(x_0, y_0)| < \varepsilon$$

for all points $(x, y) \in I \times J$ such that $|x - x_0| < \delta$ and $|y - y_0| < \delta$. A function that is continuous at every point of a rectangle is said to be *continuous* on that rectangle.

In other words, a function f defined on a rectangle $I \times J$ is continuous at a point $(x_0, y_0) \in I \times J$ if and only if $|f(x_n, y_n) - f(x_0, y_0)| \to 0$ as $n \to \infty$ whenever $(x_n, y_n) \in I \times J$, $|x_n - x_0| \to 0$ and $|y_n - y_0| \to 0$ as $n \to \infty$.

Let f be a function defined on a rectangle $I \times J$ and let $y_0 \in J$. Then, by fixing $y = y_0$ in $f(x, y)$, we obtain a function of a single variable defined on I, namely $f_{y_0}(x) = f(x, y_0)$. Similarly, if $x_0 \in I$, then by fixing $x = x_0$ in $f(x, y)$, we obtain a function of a single variable defined on J, namely $f_{x_0}(y) = f(x_0, y)$. If f is continuous on $I \times J$, then both functions f_{y_0} and f_{x_0} are clearly continuous on I and J, respectively. It might seem also that, conversely, if f_{y_0} and f_{x_0} are continuous for every $y_0 \in J$ and $x_0 \in I$, then f must also be continuous. This however is not true as shown by the following example.

Example 11.9.2. The function

$$f(x, y) = \begin{cases} \frac{xy}{x^2 + y^2}, & \text{if } (x, y) \neq (0, 0), \\ 0, & \text{if } x = y = 0, \end{cases}$$

is continuous with respect to each variable separately, that is, the functions f_{x_0} and f_{y_0} are continuous for every $x_0, y_0 \in \mathbb{R}$. On the other hand, for every $\delta > 0$ we have $|f(\delta, \delta) - f(0, 0)| = \frac{1}{2}$ which implies that f is not continuous at $(0, 0)$.

Theorem 11.9.3. *A function f that is continuous on a bounded, closed rectangle $[a, b] \times [c, d]$ is bounded, that is, there is a constant M such that $|f(x, y)| \leq M$ for all $(x, y) \in [a, b] \times [c, d]$.*

Proof. Suppose f is a continuous function on $[a, b] \times [c, d]$ that is not bounded. This means that for every $n \in \mathbb{N}$ there is a point $(x_n, y_n) \in [a, b] \times [c, d]$ such that $|f(x_n, y_n)| > n$. Since $a \leq x_n \leq b$ for all $n \in \mathbb{N}$, there exists a subsequence (x_{p_n}) which converges to some $x_0 \in [a, b]$, by the Bolzano–Weierstrass theorem. Now we consider the sequence (y_{p_n}). Since $c \leq y_{p_n} \leq d$ for all $n \in \mathbb{N}$, there exists a subsequence (y_{q_n}) of (y_{p_n}) which converges to some $y_0 \in [c, d]$. On the other hand, for every $n \in \mathbb{N}$ we have

$$|f(x_{q_n}, y_{q_n}) - f(x_0, y_0)| \geq |f(x_{q_n}, y_{q_n})| - |f(x_0, y_0)| > q_n - |f(x_0, y_0)|.$$

But this is not possible, since $|f(x_{q_n}, y_{q_n}) - f(x_0, y_0)| \to 0$, by continuity of f, and $q_n - |f(x_0, y_0)| \to \infty$ as $n \to \infty$. $\qquad\square$

Theorem 11.9.4. *If f is a continuous function on a bounded closed rectangle* $[a, b] \times [c, d]$, *then for every $\varepsilon > 0$ there is a $\delta > 0$ such that*

$$|f(x_1, y_1) - f(x_2, y_2)| < \varepsilon$$

for every $(x_1, y_1), (x_2, y_2) \in [a, b] \times [c, d]$ *such that* $|x_1 - x_2| < \delta$ *and* $|y_1 - y_2| < \delta$.

Proof. Suppose, on the contrary, that there is an $\varepsilon > 0$ such that for every $n \in \mathbb{N}$ there are $(s_n, t_n), (u_n, v_n) \in [a, b] \times [c, d]$ such that

$$|s_n - u_n| < \frac{1}{n}, \ |t_n - v_n| < \frac{1}{n}, \ \text{and} \ |f(s_n, t_n) - f(u_n, v_n)| \geq \varepsilon$$

for all $n \in \mathbb{N}$. As in the proof of Theorem 11.9.3, by the Bolzano–Weierstrass theorem, there is an increasing sequence of indices (p_n) such that $s_{p_n} \to s_0$ and $t_{p_n} \to t_0$ for some $s_0 \in [a, b]$ and $t_0 \in [c, d]$. Since $|s_n - u_n| < \frac{1}{n}$ and $|t_n - v_n| < \frac{1}{n}$, we must also have $u_{p_n} \to s_0$ and $v_{p_n} \to t_0$. Now we have

$$\varepsilon \leq |f(s_{p_n}, t_{p_n}) - f(u_{p_n}, v_{p_n})| \leq |f(s_{p_n}, t_{p_n}) - f(s_0, t_0)| + |f(s_0, t_0) - f(u_{p_n}, v_{p_n})|,$$

which is impossible since, by continuity of f, $|f(s_{p_n}, t_{p_n}) - f(s_0, t_0)| \to 0$ and $|f(s_0, t_0) - f(u_{p_n}, v_{p_n})| \to 0$ as $n \to \infty$. $\square$

The condition in the above theorem is called *uniform continuity* of f. The lemma says that every continuous function on a bounded closed rectangle $[a, b] \times [c, d]$ is uniformly continuous on that interval. We will use that property in the proof of the following theorem.

Theorem 11.9.5. *If a function f is continuous on a bounded closed rectangle* $[a, b] \times [c, d]$, *then the function*

$$F(x, y) = \int_c^y f(x, u) \, du$$

is continuous on the same rectangle.

Proof. Let $(x_0, y_0) \in [a, b] \times [c, d]$. For every $(x, y) \in [a, b] \times [c, d]$ we have

$$|F(x, y) - F(x_0, y_0)| = \left| \int_c^y f(x, u) \, du - \int_c^{y_0} f(x_0, u) \, du \right|$$

$$= \left| \int_c^y f(x, u) \, du - \int_c^y f(x_0, u) \, du + \int_{y_0}^y f(x_0, u) \, du \right|$$

$$\leq \int_c^y |f(x, u) - f(x_0, u)| \, du + \left| \int_{y_0}^y f(x_0, u) \, du \right|.$$

Let $\varepsilon > 0$. Since f is continuous, by Theorem 11.9.3, it is bounded by some constant M. Then

$$\left| \int_{y_0}^y f(x_0, u) \, du \right| \leq \int_{y_0}^y M \, du = M|y - y_0|.$$

On the other hand, by Theorem 11.9.4, there is a $\delta_0 > 0$ such that such that

$$|f(x, u) - f(x_0, u)| < \frac{\varepsilon}{2(d - c)}$$

for every $u \in [c, d]$ and every $x \in [a, b]$ such that $|x - x_0| < \delta_0$. Hence

$$\int_c^y |f(x, u) - f(x_0, u)|\, du < \frac{\varepsilon}{2} \quad \text{for } |x - x_0| < \delta_0.$$

Now, if we define $\delta = \min\{\delta_0, \frac{\varepsilon}{2M}\}$, then $|F(x, y) - f(x_0, y_0)| < \varepsilon$ whenever $|x - x_0| < \delta$ and $|y - y_0| < \delta$. This proves continuity of F at (x_0, y_0). $\square$

Under the hypotheses of the above theorem the function whose value at x is given by

$$F(x, d) = \int_c^d f(x, y)\, dy,$$

is continuous on $[a, b]$. Therefore, it is integrable over $[a, b]$ and we have

$$\int_a^b F(x, d)\, dx = \int_a^b \left(\int_c^d f(x, y)\, dy \right) dx.$$

To avoid inconvenient parentheses, this iterated integral is usually written in the form

$$\int_a^b dx \int_c^d f(x, y)\, dy. \qquad (11.28)$$

Similarly, we can consider the integral

$$\int_c^d dy \int_a^b f(x, y)\, dx. \qquad (11.29)$$

Integrals (11.28) and (11.30) differ only by the order of integration. We are going to show that they are equal. To this end we first prove a theorem about differentiation under the integral sign.

When considering derivatives of functions of two variables it is always necessary to indicate the variable with respect to which the differentiation is performed. We will use the following notation:

$$D_x f(x, y) = \lim_{h \to 0} \frac{f(x + h, y) - f(x, y)}{h}$$

and

$$D_y f(x, y) = \lim_{k \to 0} \frac{f(x, y + k) - f(x, y)}{k}.$$

These derivatives are called *partial derivatives*.

Theorem 11.9.6. *If a function F is continuous on a rectangle $[a, b] \times [c, d]$ and $D_y F$ is continuous and bounded on $[a, b] \times (c, d)$, then the function*

$$I(y) = \int_a^b F(x, y) \, dx$$

is differentiable on (c, d) and we have

$$I'(y) = D_y \left(\int_a^b F(x, y) \, dx \right) = \int_a^b D_y F(x, y) \, dx$$

for all $y \in (c, d)$.

Proof. Let $y_0 \in (c, d)$. Then

$$\left| \frac{I(y_0 + k) - I(y_0)}{k} - \int_a^b D_y F(x, y_0) \, dx \right|$$

$$\leq \int_a^b \left| \frac{F(x, y_0 + k) - F(x, y_0)}{k} - D_y F(x, y_0) \right| dx$$

$$= \int_a^b |D_y F(x, y_0 + \xi) - D_y F(x, y_0)| \, dx,$$

where ξ is some number between 0 and k that depends on both x and k. If $k \to 0$, then also $\xi \to 0$ and $D_y F(x, y_0 + \xi) \to D_y F(x, y_0)$ for every $x \in (a, b)$, because of continuity of $D_y F$. Since $D_y F$ is bounded, the last integral tends to 0, by the Dominated Convergence Theorem. $\square$

Now we are ready to prove the first results on iterated integrals.

Theorem 11.9.7. *If a function f is continuous on a rectangle $[a, b] \times [c, d]$, then*

$$\int_a^b dx \int_c^d f(x, y) \, dy = \int_c^d dy \int_a^b f(x, y) \, dx. \tag{11.30}$$

Proof. Consider the functions

$$\Phi(y) = \int_a^b dx \int_c^y f(x, u) \, du \quad \text{and} \quad \Psi(y) = \int_c^y du \int_a^b f(x, u) \, dx$$

defined for $y \in (c, d)$. Since f is continuous and bounded on $[a, b] \times [c, d]$, so is the function

$$F(x, y) = \int_c^y f(x, u) \, du$$

by Theorem 11.9.5. We can apply Theorem 11.9.6 to Φ to get

$$\Phi'(u) = \int_a^b D_y F(x, u) \, dx = \int_a^b f(x, u) \, dx.$$

Since

$$\Psi'(u) = \int_a^b f(x, u) \, dx,$$

Φ and Ψ have the same derivative and thus $\Phi = \Psi + C$ for some constant C. From the definitions of Φ and Ψ we clearly have $\Phi(c) = \Psi(c)$. Therefore $\Phi(u) = \Psi(u)$ for every $u \in [c, d]$. In particular $\Phi(d) = \Psi(d)$, which is the desired equality (11.30). $\square$

Conditions in Theorem 11.9.7 can be relaxed.

Theorem 11.9.8. *If a function f is continuous on $(a, b) \times (c, d)$, bounded or unbounded, and one of the iterated integrals*

$$\int_a^b dx \int_c^d |f(x, y)| \, dy \quad or \quad \int_c^d dy \int_a^b |f(x, y)| \, dx$$

exists, then both integrals in the equality

$$\int_a^b dx \int_c^d f(x, y) \, dy = \int_c^d dy \int_a^b f(x, y) \, dx$$

exist and the equality holds.

Proof. Suppose f is continuous on the open rectangle $(a, b) \times (c, d)$ and, without loss of generality, that the one of the integrals that exists is

$$\int_a^b dx \int_c^d |f(x, y)| \, dy.$$

More precisely, we assume that the integral $F(x) = \int_c^d |f(x, y)| \, dy$ exists almost everywhere in (a, b) and that the function F is integrable over (a, b). First we prove that the integral $\int_c^d dy \int_a^b |f(x, y)| \, dx$ exists and that

$$\int_a^b dx \int_c^d |f(x, y)| \, dy = \int_c^d dy \int_a^b |f(x, y)| \, dx. \tag{11.31}$$

Let $[\alpha_n, \beta_n]$ be a sequence of subintervals of (a, b) such that

$$[\alpha_n, \beta_n] \subset [\alpha_{n+1}, \beta_{n+1}] \subset (a, b), \quad \alpha_n \to a, \quad and \quad \beta_n \to b.$$

Similarly, let $[\gamma_n, \delta_n]$ be a sequence of subintervals of (c, d) such that

$$[\gamma_n, \delta_n] \subset [\gamma_{n+1}, \delta_{n+1}] \subset (c, d), \quad \gamma_n \to c, \quad and \quad \delta_n \to d.$$

Then, for all $m, n \in \mathbb{N}$, we have

$$\int_{\alpha_m}^{\beta_m} dx \int_{\gamma_n}^{\delta_n} |f(x, y)| \, dy = \int_{\gamma_n}^{\delta_n} dy \int_{\alpha_m}^{\beta_m} |f(x, y)| \, dx \tag{11.32}$$

by Theorem 11.9.7. For $m, n \in \mathbb{N}$ we define functions

$$f_{m,n}(x, y) = \begin{cases} f(x, y), & \text{for } (x, y) \in [\alpha_m, \beta_m] \times [\gamma_n, \delta_n], \\ 0, & \text{elsewhere,} \end{cases}$$

and

$$f_n(x, y) = \begin{cases} f(x, y), & \text{for } (x, y) \in (a, b) \times [\gamma_n, \delta_n], \\ 0, & \text{elsewhere.} \end{cases}$$

Then, by multiple uses of the Dominated Convergence Theorem and the Monotone Convergence Theorem, we obtain

$$\int_a^b dx \int_c^d |f(x,y)|\, dy = \lim_{n\to\infty} \int_a^b dx \int_c^d |f_n(x,y)|\, dy \tag{11.33}$$

$$= \lim_{n\to\infty} \lim_{m\to\infty} \int_a^b dx \int_c^d |f_{m,n}(x,y)|\, dy \tag{11.34}$$

$$= \lim_{n\to\infty} \lim_{m\to\infty} \int_c^d dy \int_a^b |f_{m,n}(x,y)|\, dx \tag{11.35}$$

$$= \lim_{n\to\infty} \int_c^d dy \int_a^b |f_n(x,y)|\, dx \tag{11.36}$$

$$= \int_c^d dy \int_a^b |f(x,y)|\, dx. \tag{11.37}$$

Now, since

$$\left| \int_c^d f_{m,n}(x,y)\, dy \right| \le \int_c^d |f(x,y)|\, dy$$

and

$$\left| \int_a^b f_{m,n}(x,y)\, dx \right| \le \int_a^b |f(x,y)|\, dx$$

for all $m, n \in \mathbb{N}$, we obtain, after multiple uses of the Dominated Convergence Theorem,

$$\lim_{n\to\infty} \lim_{m\to\infty} \int_a^b dx \int_c^d f_{m,n}(x,y)\, dy = \int_a^b dx \int_c^d f(x,y)\, dy \tag{11.38}$$

and

$$\lim_{n\to\infty} \lim_{m\to\infty} \int_c^d dy \int_a^b f_{m,n}(x,y)\, dx = \int_c^d dy \int_a^b f(x,y)\, dx. \tag{11.39}$$

Moreover, since

$$\int_a^b dx \int_c^d f_{m,n}(x,y)\, dy = \int_c^d dy \int_a^b f_{m,n}(x,y)\, dx$$

for all $m, n \in \mathbb{N}$, the desired equality follows. □

Theorems 11.9.7 and 11.9.8 are similar. However in Theorem 11.9.8 the integrand may be unbounded which significantly increases the applicability of the theorem. The proof is much more complicated because existence of the iterated integrals has to be proven, while in the first theorem it follows immediately from the fact that the function is continuous and the domain of integration is closed and bounded. Theorem 11.9.8 is a weak version of the famous Fubini's theorem (Guido Fubini 1879–1943). In that theorem the condition of continuity of f is replaced by integrability over the rectangle $(a, b) \times (c, d)$, bounded or unbounded.

EXERCISES 11.9

(1) Show that a function f defined on a rectangle $I \times J$ is continuous at a point $(x_0, y_0) \in I \times J$ if and only if $|f(x_n, y_n) - f(x_0, y_0)| \to 0$ as $n \to \infty$ whenever $(x_n, y_n) \in I \times J$, $|x_n - x_0| \to 0$ and $|y_n - y_0| \to 0$ as $n \to \infty$.

(2) Let $f(x, y) = \frac{x^2 y}{x^2 + y^2}$ if $x \neq 0$ and $y \neq 0$, and $f(x, y) = 0$ if $x = y = 0$. Prove that f is continuous at $(0, 0)$.

(3) Let f and g be continuous functions at (x_0, y_0). Prove the following:

(a) $f + g$ is continuous at (x_0, y_0),

(b) λf is continuous at (x_0, y_0) for any $\lambda \in \mathbb{R}$,

(c) fg is continuous at (x_0, y_0),

(d) f/g is continuous at (x_0, y_0) provided $g(x_0, y_0) \neq 0$.

(4) Let f be a continuous function on $[a, b] \times [c, d]$ and let $x_0 \in (a, b)$. Prove that $f(x, y)$ converges to $f(x_0, y)$ as $x \to x_0$ uniformly on $[c, d]$, that is, for every $\varepsilon > 0$ there exists $\delta > 0$ such that for all $y \in [c, d]$ we have $|f(x, y) - f(x_0, y)| < \varepsilon$ whenever $|x - x_0| < \delta$.

(5) Supply the details to verify (11.33), (11.34), (11.35), (11.36), (11.37), (11.38), and (11.39).

11.10   Euler's gamma and beta functions

By Euler's *gamma function* we mean

$$\Gamma(p) = \int_0^\infty x^{p-1} e^{-x} \, dx, \quad p > 0. \tag{11.40}$$

In Section 11.2 we proved that the integral exists for any $p > 0$. We have

$$\Gamma(1) = \int_0^\infty e^{-x} \, dx = 1.$$

Integration by parts

$$\int_0^\infty x^{p-1} e^{-x} \, dx = \left[\frac{1}{p} x^p e^{-x} \right]_0^\infty + \frac{1}{p} \int_0^\infty x^p e^{-x} \, dx = \frac{1}{p} \int_0^\infty x^p e^{-x} \, dx$$

leads to the following important result

$$p\Gamma(p) = \Gamma(p+1),$$

and hence, by induction,

$$\Gamma(p+1) = p!$$

for all natural numbers p. We see that the gamma function gives a continuous extension to all positive real numbers of the factorial $p!$ defined only for natural numbers.

Euler's *beta function* is defined by

$$B(p, q) = \int_0^1 x^{p-1} (1 - x)^{q-1} \, dx, \quad p, q > 0. \tag{11.41}$$

Here the integrand is continuous for $x \in [\varepsilon, 1 - \varepsilon]$ for any $0 < \varepsilon < \frac{1}{2}$. It is bounded for $x \in (0, \frac{1}{2}]$ by $x^{p-1} k_q$ and for $x \in [\frac{1}{2}, 1)$ by $m_p (1 - x)^{q-1}$ where k_q and m_p are properly chosen numbers depending on q and p, respectively.

The beta function can also be defined by

$$B(p, q) = \int_0^\infty \frac{t^{p-1}}{(1+t)^{p+q}} \, dt \tag{11.42}$$

which is obtained from (11.41) by substitution $x = \frac{t}{1+t}$. Indeed, if

$$f(x) = x^{p-1}(1 - x)^{q-1} \quad \text{and} \quad g(t) = \frac{t}{1+t},$$

then we have

$$f(g(t))g'(t) = \frac{t^{p-1}}{(1+t)^{p-1}} \frac{1}{(1+t)^{q-1}} \frac{1}{(1+t)^2},$$

$g(0) = 0$ and $g(\infty) = 1$.

Theorem 11.10.1. *For all $p, q > 0$ we have*

$$B(p, q) = \frac{\Gamma(p)\Gamma(q)}{\Gamma(p + q)}. \tag{11.43}$$

Proof. Substituting $x = ty$ in (11.40) we obtain

$$\frac{\Gamma(p)}{t^p} = \int_0^\infty y^{p-1} e^{-ty} \, dy, \quad t > 0.$$

Now writing $p + q$ instead of p and $1 + t$ instead of t we get

$$\frac{\Gamma(p + q)}{(1+t)^{p+q}} = \int_0^\infty y^{p+q-1} e^{-(1+t)y} \, dy.$$

After multiplying both sides by t^{p-1} and integrating with respect to t from 0 to ∞ we obtain

$$\Gamma(p + q) \int_0^\infty \frac{t^{p-1}}{(1+t)^{p+q}} \, dt = \int_0^\infty \int_0^\infty y^{p+q-1} e^{-(1+t)y} \, dy \, t^{p-1} \, dt.$$

The integral on the left-hand side equals $B(p, q)$. On the right-hand side we interchange the order of integration (applying Theorem 11.9.8) and get

$$\Gamma(p + q) B(p, q) = \int_0^\infty \int_0^\infty t^{p-1} e^{-ty} \, dt \, y^{p+q-1} e^{-y} \, dy = \Gamma(p) \int_0^\infty y^{q-1} e^{-y} \, dy.$$

Hence (11.43) follows. $\square$

This beautiful proof is due to the German mathematician Dirichlet (Peter Gustav Lejeune Dirichlet 1805–1859).

From (11.43) we have $B\left(\frac{1}{2}, \frac{1}{2}\right) = \Gamma\left(\frac{1}{2}\right) \Gamma\left(\frac{1}{2}\right)$. But

$$B\left(\frac{1}{2}, \frac{1}{2}\right) = \int_0^1 \frac{dt}{\sqrt{t(1 - t)}} = \int_0^1 \frac{dt}{\sqrt{\frac{1}{4} - (t - \frac{1}{2})^2}}$$

$$= [\arcsin(2t - 1)]_0^1 = \frac{\pi}{2} + \frac{\pi}{2} = \pi.$$

Since $\Gamma(\frac{1}{2})\Gamma(\frac{1}{2}) = \pi$, we have $\Gamma(\frac{1}{2}) = \sqrt{\pi}$, that is,

$$\int_0^\infty \frac{e^{-t}}{\sqrt{t}}\, dt = \sqrt{\pi}.$$

Substituting $t = x^2$ we get $2\int_0^\infty e^{-x^2}\, dx = \sqrt{\pi}$ and hence

$$\int_{-\infty}^\infty e^{-x^2}\, dx = \sqrt{\pi}.$$

This integral is called the *Euler–Poisson integral* (Siméon Denis Poisson 1781–1840).

11.11   The Riemann integral

In this section we discuss briefly Riemann integrable functions and their relation to Lebesgue integrable functions. We define Riemann integrable functions in terms of step functions that were introduced in Section 10.4. This is not the standard definition of Riemann integrable functions, which uses the so-called Riemann sums. Our definition is equivalent to the definition based on Riemann sums, but it is simpler.

Definition 11.11.1. A function f is called *Riemann integrable* if for every $\varepsilon > 0$ there exist step functions g and h such that

$$g(x) \le f(x) \le h(x) \quad \text{for every } x \in \mathbb{R}$$

and

$$\int (h - g) < \varepsilon.$$

We are using the integral $\int(h - g)$ in the definition of Riemann integrable functions, but it is not necessary to use the Lebesgue integral here. We use the simple integral of step functions introduced in Section 10.4.

Note that a Riemann integrable function is bounded and it vanishes outside a bounded interval.

In the following lemma we use LUB and GLB to denote the least upper bound and the greatest lower bound, respectively.

Lemma 11.11.2. *Let f be a Riemann integrable function. Then*

$$\underline{\int} f = \mathrm{LUB}\left\{\int g : g \text{ is a step function and } g(x) \le f(x) \text{ for all } x \in \mathbb{R}\right\}$$

and

$$\overline{\int} f = \mathrm{GLB}\left\{\int h : h \text{ is a step function and } f(x) \le h(x) \text{ for all } x \in \mathbb{R}\right\}$$

exist (are finite) and they are equal.

Proof. First note that, since the set

$$\left\{ \int g : g \text{ is a step function and } g(x) \leq f(x) \text{ for all } x \in \mathbb{R} \right\}$$

is bounded from above by $\int h$ for any step function h such that $f(x) \leq h(x)$ for all $x \in \mathbb{R}$, the set has a finite least upper bound. A similar argument shows that $\overline{\int} f$ is a well-defined finite number. Clearly, $\overline{\int} f \geq \underline{\int} f$. Suppose that $\overline{\int} f - \underline{\int} f > \varepsilon$ for some $\varepsilon > 0$. Let f_l and f_u be step function such that $f_l(x) \leq f(x) \leq f_u(x)$ for every $x \in \mathbb{R}$ and $\int (f_u - f_l) < \varepsilon$. Then

$$\varepsilon < \overline{\int} f - \underline{\int} f \leq \int f_u - \int f_l < \varepsilon.$$

This contradiction shows that we must have $\overline{\int} f = \underline{\int} f$. □

The above lemma allows us to define the integral of a Riemann integrable function.

Definition 11.11.3. If f is a Riemann integrable function, then by the *Riemann integral* of f we mean the common value of $\overline{\int} f$ and $\underline{\int} f$, that is,

$$\mathcal{R}\!\!\int f = \overline{\int} f = \underline{\int} f.$$

From Lemma 11.11.2 we easily obtain the following useful property.

Corollary 11.11.4. *If f is a Riemann integrable function, then there exists a non-decreasing sequence of step functions g_n and a non-increasing sequence of step functions h_n such that such that*

$$g_n(x) \leq f(x) \leq h_n(x) \quad \text{for all } x \in \mathbb{R} \text{ and all } n \in \mathbb{N}$$

and

$$\mathcal{R}\!\!\int f = \lim_{n \to \infty} \int g_n = \lim_{n \to \infty} \int h_n.$$

Theorem 11.11.5. *If f is Riemann integrable, then it is Lebesgue integrable and the Riemann integral of f equals the Lebesgue integral of f.*

Proof. Let (g_n) be a non-decreasing sequence of step functions such that $\mathcal{R}\!\!\int f = \lim_{n \to \infty} \int g_n$. Then $g_n \to f$ a.e. and consequently, by the Monotone Convergence Theorem, f is Lebesgue integrable and we have

$$\mathcal{R}\!\!\int f = \lim_{n \to \infty} \int g_n = \int f.$$

□

In view of the above theorem, there is no need to use a special symbol to denote the value of the Riemann integral of a Riemann integrable function. The difference is not in the value of the integral, but in the condition of integrability. Not every Lebesgue inetgrable function is Riemann integrable. The function $f(x) = \frac{1}{1+x^2}$ is Lebesgue integrable, but it is not Riemann integrable, since there is no step function h such that $f(x) \leq h(x)$ for all $x \in \mathbb{R}$ (the support of f is not bounded). As the next example shows, even if we consider only functions with bounded support, there are Lebesgue integrable functions that are not Riemann integrable.

Example 11.11.6. Let f be the characteristic function of the set of all rational numbers in the interval $[0, 1]$, that is,

$$f(x) = \begin{cases} 1, & \text{for } x \in \mathbb{Q} \cap [0, 1], \\ 0, & \text{otherwise.} \end{cases}$$

Clearly, f is Lebesgue integrable. Since $\int g \leq 0$ for every simple function g such that $g \leq f$ and $\int h \geq 1$ for every simple function h such that $f \leq h$, f is not Reimann integrable.

Our definition of Riemann integrable functions implies that every Riemann integrable function is bounded and has bounded support. This is often inconvenient. If f is a function with bounded or unbounded support and the function $f\chi_{[a,b]}$ is Riemann integrable, then we say that f is Riemann integrable on the interval $[a, b]$ and we write

$$\int f\chi_{[a,b]} = \int_a^b f(x)\,dx.$$

It can be shown, by a simple modification of the proof of Theorem 10.13.6, that every continuous function on a bounded and closed interval is Riemann integrable over that interval. To deal with integrals of unbounded functions or functions with unbounded support in the context of the Riemann integrals we introduce the so called *improper integrals*.

Definition 11.11.7. Let f be a function (bounded or unbounded) defined on an interval (a, b) (bounded or unbounded) and let $c \in (a, b)$. If f is Riemann integrable on every bounded and closed interval $[\alpha, \beta] \subset (a, b)$ and both limits

$$\lim_{\alpha \to a+} \int_\alpha^c f(x)\,dx \quad \text{and} \quad \lim_{\beta \to b-} \int_c^\beta f(x)\,dx$$

exist and are finite, then we say that the improper integral $\int_a^b f(x)\,dx$ converges and we define

$$\int_a^b f(x)\,dx = \lim_{\alpha \to a+} \int_\alpha^c f(x)\,dx + \lim_{\beta \to b-} \int_c^\beta f(x)\,dx.$$

Note that there is no necessity to consider improper integrals in the case of the Lebesgue integral. On the other hand, not every function that has a convergent improper Riemann integral is Lebesgue integrable. For example, the improper Riemann integral

$$\int_{-\infty}^{\infty} \frac{\sin x}{x} dx$$

is convergent, but the function $f(x) = \frac{\sin x}{x}$ is not Lebesgue integrable.

EXERCISES 11.11

(1) Show that a Riemann integrable function is bounded and it vanishes outside a bounded interval.

(2) Show that the property in Lemma 11.11.2 characterizes Riemann integrability, that is, a function f is Riemann integrable if and only if $\overline{\int} f = \underline{\int} f$.

(3) Provide a formal proof for Corollary 11.11.4.

(4) In the proof of Theorem 11.11.5 we claim that $g_n \to f$ a.e. Justify that claim.

(5) Shown that every continuous on a bounded and closed interval is Riemann integrable over that interval.

(6) Show that every non-decreasing and bounded function on a bounded interval is Riemann integrable on that interval.

(7) Show that the improper Riemann integral $\int_{-\infty}^{\infty} \frac{\sin x}{x} dx$ is convergent, but the function $f(x) = \frac{\sin x}{x}$ is not Lebesgue integrable.

Chapter 12

CONSTRUCTION OF REAL NUMBERS FROM NATURAL NUMBERS

12.1 Introduction

At the beginning of this book we introduced real numbers as a set $\mathbb{R}$ with operations of addition and multiplication and an inequality relation. These were treated as *primitive notions*, that is, they were accepted without definitions. We assumed that those primitive notions satisfied certain *axioms*, that is, properties that are accepted without proofs:

1^+ $a + b = b + a$;

2^+ $(a + b) + c = a + (b + c)$;

3^+ The equation $a + x = b$ is solvable;

1^* $ab = ba$;

2^* $(ab)c = a(bc)$;

3^* The equation $ax = b$ is solvable whenever $a \neq 0$;

$+*$ $a(b + c) = ab + ac$;

$1^<$ For every number a, one and only one of the following relations holds:

$$a = 0 \quad \text{or} \quad 0 < a \quad \text{or} \quad 0 < -a;$$

$2^<$ If $0 < a$ and $0 < b$, then $0 < a + b$ and $0 < ab$;

D Every nonempty set bounded from below has a greatest lower bound.

Then we derived other properties of real numbers from the axioms, by providing proofs. We also introduced new definitions and proved properties of those new notions. Everything in the first eleven chapters of this book has been carefully derived from the above primitive notions and axioms.

We have not addressed the question whether a set $\mathbb{R}$ satisfying the above conditions exists. Our prior experience with numbers helps us accept the first nine axioms without much resistance, but the last axiom is far from being intuitively obvious.

In this chapter we introduce a set of axioms of natural numbers and then show that one can build a set of real numbers satisfying all axioms listed above starting from natural numbers. This construction allows us to base mathematical analysis on fewer, simpler, and more intuitive axioms. It also provides an additional insight into the structure of real numbers.

Ancient mathematicians knew integers and rational numbers well, but they had difficulties with irrational numbers. For instance, in the Pythagorean school they did not know of any number which could be correctly assigned as the length of the diagonal of a unit square. They were forbidden to speak to other people about it. Not until the nineteenth century was a model for real numbers constructed by Dedekind. Later on, other models were constructed. We use Cantor's method (Georg Cantor 1845–1918) of equivalence classes which is also needed when correctly introducing integers and rational numbers.

In the construction of $\mathbb{R}$ we do not use any results from the first eleven chapters of this book. From the logical point of view it would make sense to make this chapter the first chapter of the book, since the properties of real numbers introduced in Chapter 1 are derived in this chapter. However, such order introduces some rather difficult ideas and proofs very early in the course. For pedagogical reasons we decided to postpone this important construction and present it at the end of the book.

12.2   Axioms of natural numbers

We are already familiar with natural numbers (or positive integers) by which we mean $1, 2, 3, \ldots$, and with the properties $1 < 2$, $2 < 3$ and so on, as well as many others of their properties. What we may not realize is just how few of those properties one can use to derive all the other properties of natural numbers. In this chapter all the familiar properties of natural numbers are derived from just four of their properties.

To accomplish this, an abstract set $\mathfrak{N}$ and a sign $\prec$ are taken as primitive notions, that is, they are accepted without definitions. The sign $\prec$ is put between two elements of $\mathfrak{N}$ and is read "is less than." Although the set $\mathfrak{N}$ is, at this point in time, merely some abstract set, in reality it turns out to be the set of all natural numbers. Later on, the sign $\prec$ is observed to be $<$, the sign you have been familiar with for many years.

The set $\mathfrak{N}$ and the sign $\prec$ are not the only possible choices for primitive notions. Instead of the "less than" relation we could take, for instance, the "addition" operation as a primitive notion. However the "less than" relation seems to be simpler. A small child answers at once such questions as "Which is smaller?", even if she or he cannot count. Similarly, the concept of being the greatest or the biggest is almost innate in man. The smallest child, having a choice of sweets, usually takes the biggest one.

Now we formulate the axioms of natural numbers. We will refer to these properties as *principles*.

AXIOM I (ASYMMETRY PRINCIPLE) If $x, y \in \mathfrak{N}$ and $x \prec y$, then it is not true that $y \prec x$.

Before we can formulate the second axiom we need to define what it means for an element of a subset of $\mathfrak{N}$ to be a least element. While it is easy to guess, a formal definition is desirable.

Definition 12.2.1. An element $x \in X$ is said to be a *least element* of a subset $X \subset \mathfrak{N}$, if we have $x \prec y$ or $x = y$ for every $y \in X$.

Now we state the second axiom of our set $\mathfrak{N}$.

AXIOM II (MINIMUM PRINCIPLE) Each nonempty subset of $\mathfrak{N}$ has a least element.

The next axiom is about greatest elements of bounded subsets of $\mathfrak{N}$.

Definition 12.2.2. An element $x \in X$ is a *greatest element* of a subset $X \subset \mathfrak{N}$, if we have $y \prec x$ or $y = x$ for every $y \in X$.

The symbol $y \preceq x$ can be used to express "$y \prec x$ or $y = x$."

Definition 12.2.3. A subset $X \subset \mathfrak{N}$ is called *bounded* if there is an element $x \in \mathfrak{N}$ such that $y \preceq x$ for every $y \in X$. Such an x is called a *bound* for the set X.

Now we ready to state the third axiom of our set $\mathfrak{N}$.

AXIOM III (MAXIMUM PRINCIPLE) Each nonempty bounded subset of $\mathfrak{N}$ has a greatest element.

The relation $y \prec x$ is often called the strong inequality in contrast to the weak inequality $y \preceq x$. (For instance, back in the familiar realm, both inequalities $1 \leq 1$ and $1 \leq 2$ hold, because $1 = 1$ and $1 < 2$. On the other hand, $2 \leq 1$ does not hold, because neither $2 = 1$ nor $2 < 1$.) The inequality $y \preceq x$ can be read in four different ways: y is not greater than x, x is greater than or equal to y, x is not less than y, and y is less than or equal to x.

EXERCISES 12.2

(1) Are axioms I–III satisfied when $\mathfrak{N}$ and $\prec$ are replaced, respectively, by the familiar set $\mathbb{N}$ of all natural numbers and the inequality $\leq$, defined as usual?

(2) Are axioms I–III satisfied when $\mathfrak{N}$ and $\prec$ are replaced, respectively, by the familiar set $\mathbb{Z}$ of all integers and the inequality $<$, defined as usual?

(3) Are axioms I–III satisfied when $\mathfrak{N}$ and $\prec$ are replaced, respectively, by the familiar set $\mathbb{Q}$ of all rational numbers and the inequality $<$, defined as usual?

(4) Are axioms I–III satisfied when $\mathfrak{N}$ and $\prec$ are replaced, respectively, by the set $\mathbb{J}$ of all rational numbers x satisfying $0 \leq x \leq 1$ and the inequality $<$, defined as usual?

(5) Are axioms I–III satisfied when $\mathfrak{N}$ and $\prec$ are replaced, respectively, by the set $\{-1, 0, 1\}$ and the inequality $<$, defined as usual?

(6) Which of axioms I–III are satisfied when $\mathfrak{N}$ is replaced by the set $\mathbb{A} = \{1, 2, 3\}$ and $1 \prec 2$ is the only true instance of $x \prec y$? List all bounded subsets of $\mathfrak{N}$ in this example and give a bound for each. Do some bounded sets have more than one bound?

12.3 Some simple consequences of Axioms I–III

In this section we are going to derive some familiar properties of natural numbers from the three axioms stated in the first section.

12.3.1 (Trichotomy). *For each pair of elements x and y in $\mathfrak{N}$, one and only one of the following relations holds:*

$$x \prec y \quad or \quad x = y \quad or \quad y \prec x.$$

Proof. If $x \neq y$, then we have $x \prec y$ or $y \prec x$, by axiom II. These two inequalities exclude each other, by axiom I. If $x = y$, then neither of the inequalities can hold, again by axiom I. $\square$

We are so accustomed to thinking in terms of the above trichotomy that we hardly feel a need to prove it. However it is interesting to observe that this property can be rigorously deduced from our axioms.

12.3.2. *Each nonempty subset of $\mathfrak{N}$ has a unique least element.*

Proof. Existence of a least element is guaranteed by axiom II.

Suppose that there are two such least elements, x and y. Then $x \prec y$ cannot hold, because then y would not be a least element. Similarly, $y \prec x$ cannot hold. Therefore, by 12.3.1, $x = y$. $\square$

12.3.3. *Each nonempty bounded subset of $\mathfrak{N}$ has a unique greatest element.*

The proof is very similar to the last proof, as expected.

Proof. Existence of a greatest element is guaranteed by axiom III.

Suppose that there are two such least elements, x and y. Then $x \prec y$ cannot hold, because then x would not be a greatest element. Similarly, $y \prec x$ cannot hold. Therefore, by 12.3.1, $x = y$. $\square$

12.3.4 (Transitivity). *If $x \prec y$ and $y \prec z$, then $x \prec z$.*

Proof. The first two inequalities imply that neither y nor z is the least element of the set $\{x, y, z\}$. So x must be the least element, by axiom II, and consequently we have $x \preceq z$. If we had $x = z$, then we would have $y \prec x$ which contradicts $x \prec y$, by axiom I. Therefore $x \prec z$.   □

In axioms II and III we have the expression "nonempty subset of $\mathfrak{N}$," which means that there is at least one element in the set.

S: A set is a collection of objects. Therefore, if there are no objects, there is no collection and, consequently, there is no set. Thus the expression *empty set* does not make sense.

T: That is true. But traditionally one uses that expression for a set without elements.

S: What for?

T: To avoid exceptions. For instance, the intersection of two sets would lose its sense if the sets were disjoint. On the other hand, instead of using the jargon *the intersection of the sets X and Y is an empty set* one could simply say that X and Y have no elements in common or that they are disjoint.

Should we consider the empty set bounded? Intuitively it seems strange to say that it is unbounded, that is, not bounded. On the other hand, if we follow exactly the definition of a bounded set, we cannot claim that the empty set is bounded. Note that from axioms I–III it does not, in fact, follow that $\mathfrak{N}$ is not empty. If $\mathfrak{N}$ were empty, then any subset of $\mathfrak{N}$ would be unbounded. This sounds like nonsense, but it is a logical consequence of the definition. In order to avoid confusion we will use the word infinite for each subset of $\mathfrak{N}$ which is not empty and not bounded.

EXERCISES 12.3

(1) Does trichotomy property hold for the set $\mathbb{A} = \{1, 2, 3\}$ with $1 \prec 2$ being the only true instance of $x \prec y$?

(2) Is it possible to prove 12.3.4 using axiom III instead of axiom II?

(3) Let X be a nonempty bounded subset of $\mathfrak{N}$. Prove that, if the least element of X and the greatest element of X are equal, then X has exactly one element.

(4) Does the empty set satisfy axioms I–III?

(5) Explain why, if we follow exactly the definition of a bounded set, we cannot claim that the empty set is bounded.

12.4 The Infinity Principle

The properties of natural numbers described in 12.3.1-12.3.4 follow from axioms I, II, and III. Thus, every set satisfying I, II, and III has those properties. It is easy to give examples of such sets that are finite. For instance, the set $\{a, b, c\}$ consisting of elements a, b, and c (with $a \prec b$, $b \prec c$, and $a \prec c$) satisfies I, II, and III. On the other hand, the existence of infinite sets is not obvious at all. It even somewhat contradicts our intuition. However, in order to obtain all the interesting properties of natural numbers and real numbers, we now adopt the following axiom:

AXIOM IV (INFINITY PRINCIPLE) The set $\mathfrak{N}$ is infinite.

In other words, we assume that the set $\mathfrak{N}$ is not empty and is not bounded. Now, since we have added axiom IV to our properties of $\mathfrak{N}$, we see that the empty set is a bounded subset of $\mathfrak{N}$.

Property IV cannot be deduced from axioms I–III because there are sets satisfying I, II, and III, but not IV. We say that axiom IV is independent of I–III. Axiom IV plays the central role in the theory of natural numbers and even in the theory of real and complex analysis. It is surprising that axioms I–IV fully characterize natural numbers. One can also say that the set of natural numbers is defined by axioms I–IV. Giving substance to that claim is the next item on the agenda.

We denote the least element of $\mathfrak{N}$ by the symbol 1. If $x \in \mathfrak{N}$, then the least element of the set of all natural numbers greater than x is denoted by $x + 1$. The existence and uniqueness of such an element follow from II, IV, and 12.3.2.

Theorem 12.4.1 (Induction Principle). *Let X be a subset of $\mathfrak{N}$ such that*

(a) $1 \in X$,
(b) *If $x \in X$, then $x + 1 \in X$.*

Then $X = \mathfrak{N}$.

In other words, no proper subset of $\mathfrak{N}$ satisfies both (a) and (b).

Proof. Assume that X satisfies (a) and (b). Suppose that, on the contrary, there are elements in $\mathfrak{N}$ which do not belong to X and let y be the least of them. The existence of y follows from II. Let x be the greatest element of X such that $x \prec y$. There is such an element by (a) and III. In view of (b), we have $x + 1 \in X$. By 12.3.1, one of the relations

$$x + 1 \prec y \quad \text{or} \quad x + 1 = y \quad \text{or} \quad y \prec x + 1$$

holds. However the first relation is excluded by the assumption that x is the greatest element of X such that $x \prec y$. The second relation drops out because $x + 1 \in X$ and $y \notin X$. There remains the third relation which implies $x \prec y \prec x + 1$. But this contradicts the definition of $x + 1$. $\qquad\square$

The Induction Principle is not only a very important tool for proving theorems but, as we will see in the following sections, can be used to define addition and multiplication of natural numbers.

EXERCISES 12.4

(1) Find a set $\mathfrak{M}$ and define a relation $\lhd$ in $\mathfrak{M}$ such that, when $\mathfrak{N}$ and $\prec$ are replaced, respectively, by $\mathfrak{M}$ and $\lhd$, then axioms II, III, and IV are satisfied but I is not.

(2) Find a set $\mathfrak{M}$ and define a relation $\lhd$ in $\mathfrak{M}$ such that, when $\mathfrak{N}$ and $\prec$ are replaced, respectively, by $\mathfrak{M}$ and $\lhd$, then axioms I, III, and IV are satisfied but II is not.

(3) Find a set $\mathfrak{M}$ and define a relation $\lhd$ in $\mathfrak{M}$ such that, when $\mathfrak{N}$ and $\prec$ are replaced, respectively, by $\mathfrak{M}$ and $\lhd$, then axioms I, II, and IV are satisfied but III is not.

(4) Find a set $\mathfrak{M}$ and define a relation $\lhd$ in $\mathfrak{M}$ such that, when $\mathfrak{N}$ and $\prec$ are replaced, respectively, by $\mathfrak{M}$ and $\lhd$, then axioms I, II, and III are satisfied but IV is not.

(5) Does the set of all even numbers $\{2, 4, 6, \ldots\}$ with the usual definition of $<$ satisfy axioms I–IV? Does it contradict the claim that "axioms I–IV fully characterize the natural numbers"?

(6) Prove the following version of the Induction Principle: *Let X be a subset of $\mathfrak{N}$ such that*

  (a) $n \in X$;
  (b) If $x \in X$, then $x + 1 \in X$.

  Then every element of $\mathfrak{N}$ greater than n belongs to X.

(7) Show that the following version of the Induction Principle can be proved without the Infinity Principle: *Let X be a subset of $\mathfrak{N}$ such that*

  (a) $1 \in X$;
  (b) If $x \in X$ and x is not the greatest element of $\mathfrak{N}$, then $x + 1 \in X$.

  Then $X = \mathfrak{N}$.

12.5 Addition in $\mathfrak{N}$

Addition is defined by induction as follows:

(+) $x + 1 =$ the least element of $\mathfrak{N}$ greater than x;

(++) $x + (y + 1) = (x + y) + 1$.

The result of addition of two elements x and y is called the *sum* of x and y. Part (+) tells us how to add 1 to an element of $\mathfrak{N}$. Part (++) tells us how to add $y + 1$

if we already know how to add y. We need to prove that for any elements x and y in $\mathfrak{N}$ the sum $x + y$ is uniquely defined. We will use the Induction Principle.

Denote by X the set of all elements $z \in \mathfrak{N}$ for which the sum $x + z$ is uniquely defined for every element $x \in \mathfrak{N}$. By (+) and 12.3.2, we have $1 \in X$. (12.3.2 is used here to ensure the uniqueness of $x + 1$ for each fixed x). Now suppose $z \in X$. Then (++) tells us what $x + (z + 1)$ is and (+) ensures its uniqueness for every $x \in \mathfrak{N}$. Hence $z + 1 \in X$. From the Induction Principle we obtain $X = \mathfrak{N}$. But this means that the sum $x + y$ is uniquely defined for every $x, y \in \mathfrak{N}$.

12.5.1. *For all* $x, y \in \mathfrak{N}$ *we have* $x \prec x + y$.

Proof. We will use the Induction Principle. Denote by X the set of all elements $z \in \mathfrak{N}$ such that $x \prec x + z$ for all $x \in \mathfrak{N}$. By (+) , we have $1 \in X$. Now suppose $z \in X$. If $x \prec x + z$ for some $x \in \mathfrak{N}$, then

$$x \prec x + z \prec (x + z) + 1 = x + (z + 1),$$

by 12.3.4, (+), and (++). Hence, if $x \prec x + z$ for all $x \in \mathfrak{N}$, then $x \prec x + (z + 1)$ for all $x \in \mathfrak{N}$. In other words, if $z \in X$, then $z + 1 \in X$. By the Induction Principle, we conclude $X = \mathfrak{N}$, which means that $x \prec x + y$ for all $x, y \in \mathfrak{N}$. $\square$

It is also worthwhile to note that

12.5.2. *If* $x \prec y$, *then* $x + 1 \preceq y$.

which follows directly from (+). Moreover, for all $x, y, z \in \mathfrak{N}$ we have

12.5.3 (Associativity). $x + (y + z) = (x + y) + z$.

12.5.4 (Commutativity). $x + y = y + x$.

12.5.5 (Cancellation law). *If* $x + z = y + z$, *then* $x = y$.

Proof of 12.5.3. Denote by X the set of all elements $z \in \mathfrak{N}$ such that 12.5.3 holds for all $x, y \in \mathfrak{N}$. By (++), $1 \in X$. Moreover, if $z \in X$, then

$$x + (y + (z + 1)) = x + ((y + z) + 1) = (x + (y + z)) + 1$$
$$= ((x + y) + z) + 1 = (x + y) + (z + 1),$$

which means that $z + 1 \in X$. By the Induction Principle, we have $X = \mathfrak{N}$ and thus 12.5.3 holds for all elements $x, y, z \in \mathfrak{N}$. $\square$

Proof of 12.5.4. We are going to proceed in two steps. First we prove that

$$1 + y = y + 1 \tag{12.1}$$

holds for all $y \in \mathfrak{N}$. Denote by Y the set of all y for which (12.1) holds. Clearly $1 \in Y$. If $y \in Y$, then $y + 1 \in Y$, since

$$1 + (y + 1) = (1 + y) + 1 = (y + 1) + 1,$$

by (++). This proves, by the Induction Principle, that (12.1) holds for all $y \in \mathfrak{N}$.

In the second step we denote by X the set of all $x \in \mathfrak{N}$ such that 12.5.4 holds for each $y \in \mathfrak{N}$. By what we just proved, $1 \in X$. If $x \in X$, then $x + 1 \in X$, since

$$(x + 1) + y = x + (1 + y) = x + (y + 1) = (x + y) + 1 = (y + x) + 1 = y + (x + 1),$$

by 12.5.3 and (12.1). By the Induction Principle, we conclude that 12.5.4 holds for all $x, y \in \mathfrak{N}$. $\qquad\square$

Proof of 12.5.5. Denote by X the set of all z for which 12.5.5 holds. We first show that $1 \in X$. Assume $x + 1 = y + 1$. By 12.3.1 we have $x \prec y$ or $x = y$ or $y \prec x$. The first inequality implies $x \prec y \prec x + 1$, which is impossible. Similarly, the inequality $y \prec x$ is impossible. So we must have $x = y$.

Now assume that $z \in X$ and that $x + (z + 1) = y + (z + 1)$. Then, by 12.5.3 and 12.5.4, we have $(x + 1) + z = (y + 1) + z$. Since $z \in X$, we get $x + 1 = y + 1$ and then, by the first step of this proof, $x = y$. Thus $z + 1 \in X$.

By the Induction Principle, we conclude that 12.5.5 holds for all $x, y, z \in \mathfrak{N}$. $\quad\square$

Here is another important property of $\mathfrak{N}$.

12.5.6. *If $x, y \in \mathfrak{N}$ and $x \prec y$, then there exists a unique $z \in \mathfrak{N}$ such that $x + z = y$.*

Proof. Let $x, y \in \mathfrak{N}$ and assume that $x \prec y$. Denote by X the set of all $u \in \mathfrak{N}$ such that $x + u \preceq y$. This set is not empty, because it contains the element 1, by 12.5.2. Moreover, it is bounded by y, because $u \prec u + x = x + u$, which implies $u \prec y$. Let z be the greatest element in X. Then $x + z \prec y$ or $x + z = y$ or $y \prec x + z$. The first inequality is excluded by 12.5.2, because otherwise we would have $x + z + 1 \preceq y$, contrary to the definition of z. The last inequality is also excluded, because $z \in X$ and thus $x + z \preceq y$. Consequently $x + z = y$.

Now suppose $x + z_1 = y$ and $x + z_2 = y$. Then $x + z_1 = x + z_2$, and hence, by 12.5.5, $z_1 = z_2$. This proves the uniqueness. $\qquad\square$

The unique z such that $x + z = y$ is called the *difference* of y and x and is denoted by $y - x$. Property 12.5.6 guarantees that the difference $y - x$ exists in $\mathfrak{N}$ provided $x \prec y$.

The following corollary to 12.5.1 and 12.5.6 is often useful (for instance, see Exercise 1.5.4).

12.5.7. *$x \prec y$ if and only if there is some z such that $x + z = y$.*

S: This fact about natural numbers is known to every child in elementary school and is at least as evident as axioms I–IV. Why do we need to prove such an obvious fact?

T: We don't have to. We could accept it without proof.

S: Why then do we prove it?

T: For fun. Just like solving a cross-word puzzle or completing a jigsaw puzzle. Some people find fun and pleasure in deducing new facts from some given, quite different, facts.

S: I don't see any fun in it.

T: Besides fun there is another reason to play with axioms, theorems and proofs. It reveals the very method of mathematics, its core and spirit. Although we begin with childishly easy arguments, we gradually pass to more and more difficult ones. It is surprising that starting from the mere axioms I–IV one can build the noble edifice of calculus and real and complex analysis. This edifice, supported by four tiny columns, turns out to be immovably strong and absolutely precise at every stage.

EXERCISES 12.5

(1) Prove that $(a + b) + (c + d) = (a + c) + (b + d)$.

(2) Prove that $((a + b) + c) + d = a + (b + (c + d))$.

(3) Elements a, b, c, and d of $\mathfrak{N}$ must be added two at a time. Unless we prove it makes no difference how we meaningfully introduce parentheses into the expression $a + b + c + d$, then that expression without parentheses makes no sense. How many different ways can parentheses be meaningfully inserted into $a + b + c + d$ so that an expression that makes sense will result? Prove that all of those resulting expressions represent the same element of $\mathfrak{N}$.

(4) Prove that, if $a \prec b$ and $c \prec d$ then $a + c \prec b + d$.

12.6 Multiplication in $\mathfrak{N}$

The result of multiplication of two elements $x, y \in \mathfrak{N}$ is called the *product* of x and y and is denoted by xy (or sometimes by $x \cdot y$). The product of x and y is defined by induction as follows:

$(\cdot)$ $x \cdot 1 = x$;

$(\cdot\cdot)$ $x(y + 1) = (xy) + x$.

Let X be the set of all elements $y \in \mathfrak{N}$ such that the product xy is uniquely defined for every $x \in \mathfrak{N}$. By $(\cdot)$ we see that $1 \in X$. If we assume that $y \in X$, then xy is uniquely defined and so is $(xy) + x$ because of the uniqueness of addition. Thus, by $(\cdot\cdot)$, the product $x(y + 1)$ is uniquely defined, which proves that $y + 1 \in X$. By the Induction Principle, we conclude that $X = \mathfrak{N}$ so that the product xy is uniquely defined for all pairs of elements of $\mathfrak{N}$.

To simplify the notation we always assume that in an algebraic expression without parentheses with both addition and multiplication, the multiplication is performed first. For instance $(xy) + z$ can be written without parentheses: $xy + z$.

For all $x, y, z \in \mathfrak{N}$ we have

12.6.1 (Distributivity). $(x + y)z = xz + yz.$

12.6.2 (Commutativity). $xy = yx.$

12.6.3 (Associativity). $x(yz) = (xy)z.$

12.6.4 (Cancellation law). *If $xz = yz$, then $x = y$.*

Proof of 12.6.1. Denote by X the set of all $z \in \mathfrak{N}$ such that 12.6.1 holds for each pair $x, y \in \mathfrak{N}$. We have $1 \in X$, by $(\cdot)$. If $z \in X$, then

$$(x + y)(z + 1) = (x + y)z + (x + y)$$
$$= (xz + yz) + (x + y)$$
$$= ((xz + yz) + x) + y$$
$$= (xz + (yz + x)) + y$$
$$= (xz + (x + yz)) + y$$
$$= ((xz + x) + yz) + y$$
$$= (xz + x) + (yz + y)$$
$$= x(z + 1) + y(z + 1),$$

by $(\cdots)$, assumption $z \in X$, 12.5.3, 12.5.3, 12.5.4, 12.5.3, 12.5.3, and $(\cdots)$. Hence, by the Induction Principle, we conclude that $X = \mathfrak{N}$, which means that 12.6.1 holds for all $x, y, z \in \mathfrak{N}$. □

Proof of 12.6.2. We first prove that

$$1 \cdot x = x \cdot 1 \tag{12.2}$$

for all $x \in \mathfrak{N}$. Denote by X the set of all $x \in \mathfrak{N}$ for which (12.2) holds. Obviously, $1 \in X$. Moreover, if $x \in X$, then

$$1 \cdot (x + 1) = 1 \cdot x + 1 = x \cdot 1 + 1 = x + 1 = (x + 1) \cdot 1,$$

by $(\cdots)$, (12.2), $(\cdot)$, and again by $(\cdot)$. Thus $x + 1 \in X$. By the Induction Principle it follows that $X = \mathfrak{N}$, which means that (12.2) holds for all $x \in \mathfrak{N}$.

Assume now that 12.6.2 holds for some $y \in \mathfrak{N}$ and all $x \in \mathfrak{N}$. Then

$$x(y + 1) = xy + x = yx + x = yx + 1 \cdot x = (y + 1)x,$$

by $(\cdots)$, the assumption on y, $(\cdot)$, and 12.6.1. Therefore 12.6.2 continues to hold when y is replaced by $y + 1$. By the Induction Principle, we conclude that 12.6.2 holds for all $x, y \in \mathfrak{N}$. □

Proof of 12.6.3. The equality holds for $z = 1$ and all $x, y \in \mathfrak{N}$, by $(\cdot)$. If 12.6.3 holds for some $z \in \mathbb{N}$ and all $x, y \in \mathfrak{N}$, then

$$x(y(z + 1)) = x(yz + y) = (yz + y)x = (yz)x + yx$$
$$= x(yz) + xy = (xy)z + xy = (xy)(z + 1),$$

by $(\cdots)$, 12.6.2, 12.6.1, 12.6.2, the assumption on z, and $(\cdots)$. Thus 12.6.3 holds with $z + 1$ instead of z. This proves 12.6.3 for all $x, y, z \in \mathfrak{N}$, by the Induction Principle. □

Proof of 12.6.4. Suppose $xz = yz$ and $x \prec y$. By 12.5.6, there exists a $u \in \mathfrak{N}$ such that $x + u = y$. Since, by 12.6.1,

$$xz + uz = (x + u)z = yz = xz,$$

we obtain a contradiction with 12.5.1. The inequality $x \prec y$ is thus excluded. The inequality $y \prec x$ is excluded in a similar way. Hence, the only remaining possibility is $x = y$. $\qquad\qquad\qquad\qquad\qquad\qquad\qquad\qquad\qquad\qquad\qquad\qquad\qquad\square$

EXERCISES 12.6

(1) Prove that $(a + b)(c + d) = ac + ad + bc + bd$.

(2) Prove that $a \prec b$ implies $ac \prec bc$.

(3) Prove that $ac \prec bc$ implies $a \prec b$.

(4) Prove that $a \prec b$ and $c \prec d$ implies $ac \prec bd$.

(5) Prove that $a \prec ab$ whenever $1 \prec b$.

12.7   Addition table and multiplication table

In this section we intend to derive the addition and the multiplication tables.

We adopt the following notation

$$
\begin{aligned}
2 &= 1 + 1, \\
3 &= 2 + 1, \\
4 &= 3 + 1, \\
5 &= 4 + 1, \\
6 &= 5 + 1, \\
7 &= 6 + 1, \\
8 &= 7 + 1, \\
9 &= 8 + 1, \\
10 &= 9 + 1.
\end{aligned}
\tag{12.3}
$$

Each of the symbols 1, 2, 3, 4, 5, 6, 7, 8, 9 represents an element of $\mathfrak{N}$ and is called a *digit*. The symbol 0 (which does not represent an element of $\mathfrak{N}$) is also a digit. The symbol 10 also represents a member of $\mathfrak{N}$ and is called a two-digit number.

The multiplication of two elements of $\mathfrak{N}$, each represented by digits or numerals, will be always denoted by placing a point between them, for example $2 \cdot 3$ or $5 \cdot 10$. The symbol 75 will denote $7 \cdot 10 + 5$. Generally, if $\mathbb{A}$ and $\mathbb{B}$ denote any non-zero digits, we agree to understand that

$$\mathbb{A}\mathbb{B} = \mathbb{A} \cdot 10 + \mathbb{B} \tag{12.4}$$

and

$$\mathbb{A}0 = \mathbb{A} \cdot 10. \tag{12.5}$$

Elements of $\mathfrak{N}$ of the form (12.4) or (12.5) are called *two-digit numbers*.

Table 12.1 is organized so that each of its entries is the sum of the first number in the corresponding row and the top number in the corresponding column. It is the *addition table*.

Table 12.1   The addition table.

+	1	2	3	4	5	6	7	8	9
1	2	3	4	5	6	7	8	9	10
2	3	4	5	6	7	8	9	10	11
3	4	5	6	7	8	9	10	11	12
4	5	6	7	8	9	10	11	12	13
5	6	7	8	9	10	11	12	13	14
6	7	8	9	10	11	12	13	14	15
7	8	9	10	11	12	13	14	15	16
8	9	10	11	12	13	14	15	16	17
9	10	11	12	13	14	15	16	17	18

Almost everyone has known the whole addition table since childhood, but very few people think of deducing its entries rigorously from axioms I–IV. Here is how it can be done.

The entries of the second column are justified by (12.3). Next note that, if $\mathbb{A}$ is an arbitrary digit different from 0, then

$$10 + \mathbb{A} = 1\mathbb{A} \qquad\qquad (12.6)$$

according to (12.4). For example: $10 + 7 = 17$.

In addition to (12.6) we will still need the formula

$$(x + 1) + (y - 1) = x + y, \qquad\qquad (12.7)$$

for $x = 1, 2, 3, 4, 5, 6, 7, 8, 9$ and $y = 2, 3, 4, 5, 6, 7, 8, 9$. Observe that both sides of (12.7) are equal to $x + (1 + (y - 1))$, by associativity of addition and the definition of the difference.

Now we are ready to justify all entries of the addition table using (12.3), (12.4), and (12.7):

$$2 + 2 = 3 + 1 = 4,$$
$$3 + 2 = 4 + 1 = 5,$$
$$3 + 3 = 4 + 2 = 5 + 1 = 6,$$
$$4 + 3 = 5 + 2 = 6 + 1 = 7,$$
$$4 + 4 = 5 + 3 = 6 + 2 = 7 + 1 = 8,$$
$$5 + 4 = 6 + 3 = 7 + 2 = 8 + 1 = 9,$$
$$5 + 5 = 6 + 4 = 7 + 3 = 8 + 2 = 9 + 1 = 10,$$
$$6 + 5 = 7 + 4 = 8 + 3 = 9 + 2 = 10 + 1 = 11,$$
$$6 + 6 = 7 + 5 = 8 + 4 = 9 + 3 = 10 + 2 = 12,$$
$$7 + 6 = 8 + 5 = 9 + 4 = 10 + 3 = 13,$$
$$7 + 7 = 8 + 6 = 9 + 5 = 10 + 4 = 14,$$
$$8 + 7 = 9 + 6 = 10 + 5 = 15,$$
$$8 + 8 = 9 + 7 = 10 + 6 = 16,$$
$$9 + 8 = 10 + 7 = 17,$$
$$9 + 9 = 10 + 8 = 18.$$

Using the above equalities we can fill in all the entries of the addition table on its main diagonal and below. In view of commutativity of addition, the remaining entries can be filled in too. This completes the justification of the entire addition table.

Table 12.2 is organized similarly, but instead of sums there are products of numbers appearing at the beginning of the corresponding column and row. It is the *multiplication table*.

Table 12.2   The multiplication table.

·	1	2	3	4	5	6	7	8	9
1	1	2	3	4	5	6	7	8	9
2	2	4	6	8	10	12	14	16	18
3	3	6	9	12	15	18	21	24	27
4	4	8	12	16	20	24	28	32	36
5	5	10	15	20	25	30	35	40	45
6	6	12	18	24	30	36	42	48	54
7	7	14	21	28	35	42	49	56	63
8	8	16	24	32	40	48	56	64	72
9	9	18	27	36	45	54	63	72	81

The second column is justified by (·). Since

$$x \cdot 2 = x(1 + 1) = x + x,$$

the third column of the multiplication table is identical with the main diagonal in the addition table. In order to find the remaining entries we first prove two auxiliary properties.

12.7.1. *Each number in the multiplication table (except the first two columns) is the sum of the number next on the left and the number at the beginning of the same row.*

Proof. Assume that a number x stands in a row whose first element is $\mathbb{A}$ and in the column with the top element $\mathbb{B}$ ($\mathbb{B} \neq 1$). Then $x = \mathbb{A} \cdot \mathbb{B}$. The number on the left next to x is $\mathbb{A} \cdot (\mathbb{B} - 1)$, because it is in the row with the initial number $\mathbb{A}$ and in the column with the top number $\mathbb{B} - 1$. We thus have to prove that $\mathbb{A} \cdot \mathbb{B} = \mathbb{A} \cdot (\mathbb{B} - 1) + \mathbb{A}$. Indeed, we have

$$\mathbb{A} \cdot \mathbb{B} = \mathbb{A} \cdot ((\mathbb{B} - 1) + 1) = \mathbb{A} \cdot (\mathbb{B} - 1) + \mathbb{A} \cdot 1 = \mathbb{A} \cdot (\mathbb{B} - 1) + \mathbb{A}.$$

$\square$

12.7.2. *If $\mathbb{A}$, $\mathbb{B}$, $\mathbb{C}$, and $\mathbb{D}$ denote digits, $\mathbb{D}$ being different from 0 and 9, then*

$$\mathbb{D}\mathbb{A} + \mathbb{B} = \begin{cases} \mathbb{D}\mathbb{C} & \text{if } \mathbb{A} + \mathbb{B} = \mathbb{C}, \\ \mathbb{E}\mathbb{C} & \text{if } \mathbb{A} + \mathbb{B} = 1\mathbb{C}, \end{cases}$$

where $\mathbb{E} = \mathbb{D} + 1$.

For instance: $56 + 3 = 59$, since $6 + 3 = 9$, or $56 + 8 = 64$, since $6 + 8 = 14$.

Proof. If $\mathbb{A} + \mathbb{B} = \mathbb{C}$, then

$$\mathbb{D}\mathbb{A} + \mathbb{B} = (\mathbb{D} \cdot 10 + \mathbb{A}) + \mathbb{B} = \mathbb{D} \cdot 10 + (\mathbb{A} + \mathbb{B}) = \mathbb{D} \cdot 10 + \mathbb{C} = \mathbb{D}\mathbb{C}.$$

If $\mathbb{A} + \mathbb{B} = 1\mathbb{C}$ and $\mathbb{E} = \mathbb{D} + 1$, then

$$\begin{aligned} \mathbb{D}\mathbb{A} + \mathbb{B} &= (\mathbb{D} \cdot 10 + \mathbb{A}) + \mathbb{B} = \mathbb{D} \cdot 10 + (\mathbb{A} + \mathbb{B}) \\ &= \mathbb{D} \cdot 10 + 1\mathbb{C} = \mathbb{D} \cdot 10 + (1 \cdot 10 + \mathbb{C}) \\ &= (\mathbb{D} \cdot 10 + 1 \cdot 10) + \mathbb{C} = (\mathbb{D} + 1) \cdot 10 + \mathbb{C} \\ &= \mathbb{E} \cdot 10 + \mathbb{C} = \mathbb{E}\mathbb{C}. \end{aligned}$$

$\square$

Applying 12.7.1 and 12.7.2 and using the addition table and already established entries of the multiplication table we successively find:

$$3 \cdot 3 = 6 + 3 = 9, \qquad\qquad 7 \cdot 7 = 42 + 7 = 49,$$
$$4 \cdot 3 = 8 + 4 = 12, \qquad\qquad 8 \cdot 3 = 16 + 8 = 24,$$
$$4 \cdot 4 = 12 + 4 = 16, \qquad\qquad 8 \cdot 4 = 24 + 8 = 32,$$
$$5 \cdot 3 = 10 + 5 = 15, \qquad\qquad 8 \cdot 5 = 32 + 8 = 40,$$
$$5 \cdot 4 = 15 + 5 = 20, \qquad\qquad 8 \cdot 6 = 40 + 8 = 48,$$
$$5 \cdot 5 = 20 + 5 = 25, \qquad\qquad 8 \cdot 7 = 48 + 8 = 56,$$
$$6 \cdot 3 = 12 + 6 = 18, \qquad\qquad 8 \cdot 8 = 56 + 8 = 64,$$
$$6 \cdot 4 = 18 + 6 = 24, \qquad\qquad 9 \cdot 3 = 18 + 9 = 27,$$
$$6 \cdot 5 = 24 + 6 = 30, \qquad\qquad 9 \cdot 4 = 27 + 9 = 36,$$
$$6 \cdot 6 = 30 + 6 = 36, \qquad\qquad 9 \cdot 5 = 36 + 9 = 45,$$
$$7 \cdot 3 = 14 + 7 = 21, \qquad\qquad 9 \cdot 6 = 45 + 9 = 54,$$
$$7 \cdot 4 = 21 + 7 = 28, \qquad\qquad 9 \cdot 7 = 54 + 9 = 63,$$
$$7 \cdot 5 = 28 + 7 = 35, \qquad\qquad 9 \cdot 8 = 63 + 9 = 72,$$
$$7 \cdot 6 = 35 + 7 = 42, \qquad\qquad 9 \cdot 9 = 72 + 9 = 81.$$

In this way all the entries on and below the diagonal have been found. The remaining part of the multiplication table is to be filled up symmetrically with respect to the main diagonal, which is justified by the commutativity of multiplication.

It is now clear that all the properties of the familiar natural numbers can be deduced from axioms I–IV. For this reason, from this point on, we will refer to $\mathfrak{N}$ and $\prec$ as $\mathbb{N}$ and $<$, respectively. Also, as is customary, the elements of $\mathbb{N}$ will be called natural numbers and the symbols $+$, $-$, and $\cdot$ will, as usual, denote the operations addition, subtraction, and multiplication, respectively.

EXERCISES 12.7

The notation described in this section is based on the number ten. There is no mathematical reason for using that particular number. Other natural numbers can be used for the same purpose. In the following two problems you are asked to construct addition and multiplication tables in the notation based on two other numbers.

(1) Because we use only two digits, 0 and 1, the representations of natural numbers developed in this exercise are called binary representations. We have $1 + 1 = 10$, and, in general,

$$\mathbb{AB} = \mathbb{A} \cdot (1 + 1) + \mathbb{B}, \quad \mathbb{ABC} = \mathbb{A} \cdot (1 + 1) \cdot (1 + 1) + \mathbb{B} \cdot (1 + 1) + \mathbb{C},$$

and so on. Construct the addition table and the multiplication table including all one-, two-, and three-digit binary numbers.

(2) We use the digits $0, 1, 2, 3, 4, 5, 6, 7, 8, 9, \mathbb{T}, \mathbb{E}$ where $0, 1, 2, 3, 4, 5, 6, 7, 8, 9$ have the meaning given in (12.3) and

$$\mathbb{T} = 9 + 1, \mathbb{E} = \mathbb{T} + 1, 10 = \mathbb{E} + 1.$$

Construct the addition table and the multiplication table for all the one-digit numbers in this base twelve system of enumeration.

12.8 Equivalence classes

The method of equivalence classes enables us to construct new mathematical objects from given ones. It is a very important and powerful tool in mathematics so we discuss it in its full generality.

Definition 12.8.1. A relation $\sim$ defined on a set X is called a *weak equivalence* if it has the following properties:

Symmetry: $x \sim y$ implies $y \sim x$;

Transitivity: $x \sim y$ and $y \sim z$ implies $x \sim z$.

The above definition of "weak equivalence" is different from the standard definition of an equivalence relation. What is different is this: for a relation to be an *equivalence relation*, it is assumed to have a reflexivity property in addition to the properties we require. A relation is *reflexive* if $x \sim x$ for all $x \in X$. We believe the decision not to require this additional property makes our construction of the real numbers considerably easier.

If $x \sim y$, then we say x is *equivalent* to y. In view of (a), we may also say y is equivalent to x.

Definition 12.8.2. The set of all elements in X that are equivalent to an element $x \in X$ is called the *equivalence class* of x and denoted by $[x]$.

It may happen that an element x is not equivalent to any element in X, not even itself. Then $[x]$ is empty. On the other hand, if $[x]$ is not empty, then $x \sim x$ and therefore $x \in [x]$. Indeed, if $y \in [x]$, then $x \sim y$ and $y \sim x$, and hence $x \sim x$. An element of an equivalence class will be called a *representative* of that class.

Theorem 12.8.3. *Let $\sim$ be a weak equivalence in X.*

(a) *If $x \sim y$, then $[x] = [y]$, that is, the classes $[x]$ and $[y]$ have the same elements;*

(b) *If the relation $x \sim y$ does not hold, then the classes $[x]$ and $[y]$ are disjoint, that is, they have no common elements.*

Proof. To prove (a), suppose $x \sim y$. If $z \in [x]$, then $z \sim x$. Thus $z \sim y$, by transitivity, which implies $z \in [y]$. Therefore $[x] \subset [y]$. Similarly we prove that $[y] \subset [x]$.

To prove (b) suppose that the relation $x \sim y$ does not hold and that there exists an element z belonging to both $[x]$ and $[y]$. Then $z \sim x$ and $z \sim y$ and, by symmetry and transitivity, $x \sim y$ contrary to the hypothesis. □

Forming equivalence classes is like sticking some elements together to form bigger grains. Those elements which are not equivalent to any element drop out in the construction.

EXERCISES 12.8

(1) Give an example of a weak equivalence relation for which some equivalence classes are empty.
(2) Determine if the defined relation is an equivalence relation, a weak equivalence relation, or neither? If it is an equivalence relation or a weak equivalence relation, determine how many distinct equivalence classes this relation produces.

   (a) In $(\mathbb{Z}_4, \otimes)$ (see exercises in Section 1.1) define a relation Δ as follows: $a\Delta b$ if $2 \otimes a = 2 \otimes b$.

   (b) Let $\prec$ be a relation in $\mathbb{N}$ defined as follows: $x \prec y$ if there exists $z \in \mathbb{N}$ such that $xz = y$.

   (c) Let $\simeq$ be a relation defined in $\mathbb{N}$ as follows: $x \simeq y$ if there exist $m, n, k \in \mathbb{N}$, $k \neq 1$, such that $x = mk$ and $y = nk$.

   (d) Let n be a fixed natural number. We define a relation $=_n$ in $\mathbb{N}$ by $a =_n b$ if n divides $b - a$. In other words, $a =_n b$ if there exists an integer k such that $b - a = kn$.

   (e) Let $\mathcal{X}$ be the set of all ordered pairs of natural numbers. Define a relation $\square$ in $\mathcal{X}$ as follows: $(s, t) \square (u, v)$ if $s + t = u + v$.

(3) True or false: $y \sim z$ if and only if $[y] = [z]$?
(4) Let $\mathcal{A}$ be a collection of disjoint subsets of X. In X we define a relation $\sim$ as follows: $x \sim y$ if there is a set $S \in \mathcal{A}$ such that both x and y are in A. Show that $\sim$ is a weak equivalence in X.

12.9　Defining integers by natural numbers

If m and n are natural numbers and $n < m$, then $m-n$ is a natural number. However it is not so if $m \leq n$. Within the context of the natural numbers, when $m \leq n$, the indicated difference $m - n$ has no more meaning than an ordered pair (m, n) of natural numbers. If we try to extend the notion of numbers by merely considering different indicated differences as different numbers then, for instance, the indicated differences $1 - 2$, $5 - 6$, $8 - 9$, would represent different numbers in contrast to what we have learned at school. The identification of indicated differences can be correctly carried out by using the method of equivalence classes.

We say that two indicated differences $m_1 - n_1$ and $m_2 - n_2$ are equivalent and we write $m_1 - n_1 \sim m_2 - n_2$ if and only if $m_1 + n_2 = m_2 + n_1$. One can easily see that this relation is reflexive and symmetric. To check that it is also transitive, assume that

$$m_1 - n_1 \sim m_2 - n_2 \quad \text{and} \quad m_2 - n_2 \sim m_3 - n_3.$$

Then

$$m_1 + n_2 = m_2 + n_1 \quad \text{and} \quad m_2 + n_3 = m_3 + n_2.$$

Summing the corresponding parts of these two equalities we obtain

$$m_1 + n_3 + (m_2 + n_2) = m_3 + n_1 + (m_2 + n_2)$$

and in view of the cancellation law,

$$m_1 + n_3 = m_3 + n_1.$$

This proves $m_1 - n_1 \sim m_3 - n_3$.

The equivalence classes of indicated differences of natural numbers are a model of the set of integers. This model contains all natural numbers if we agree to make the following identification:

$$n = [(n + 1) - 1].$$

Now we introduce addition, multiplication and the property of being positive for integers:

$$[m_1 - n_1] + [m_2 - n_2] = [(m_1 + m_2) - (n_1 + n_2)],$$
$$[m_1 - n_1][m_2 - n_2] = [(m_1 m_2 + n_1 n_2) - (m_1 n_2 + m_2 n_1)],$$
$$[m - n] > 0 \text{ if and only if } m > n.$$

In the inequality $[m - n] > 0$, the number 0 is the equivalence class $[1 - 1]$.

Of course, we need proofs showing that these operations are well-defined and that the usual properties of addition, multiplication and inequalities are satisfied. This matter will be discussed together with rational numbers.

EXERCISES 12.9

In the following exercises, be careful to use only the properties of natural numbers.

(1) Let m and n be natural numbers. Prove that $[m - n]$ is not empty.
(2) To show that addition of integers is well-defined, that is, suppose $(m_1 - n_1) \sim (k_1 - l_1)$ and $(m_2 - n_2) \sim (k_2 - l_2)$, and then prove that $(m_1 + m_2) - (n_1 + n_2) \sim (k_1 + k_2) - (l_1 + l_2)$.
(3) To show that multiplication of integers is well-defined, that is, suppose $(m_1 - n_1) \sim (k_1 - l_1)$ and $(m_2 - n_2) \sim (k_2 - l_2)$, and then prove that $(m_1 m_2 + n_1 n_2) - (m_1 n_2 + m_2 n_1) \sim (k_1 k_2 + l_1 l_2) - (k_1 l_2 + k_2 l_1)$.

(4) To show that the property of being positive is well-defined for integers, suppose $(m-n) \sim (k-l)$, and then prove that $m > n$ if and only if $k > l$.

(5) Prove that for any pair of integers a and b the equation $a + x = b$ has a solution.

(6) Prove that for integers $a = [3-1]$ and $b = [2-1]$ the equation $ax = b$ has no integer $[m-n]$ as a solution.

(7) Prove that for any integers a, b, and c we have $a(b+c) = ab + ac$.

(8) Prove that, for any pair of integers a and b, if $a > 0$ and $b > 0$, then $a + b > 0$ and $ab > 0$.

(9) Suppose $m, n \in \mathbb{N}$. Then we can ask if $m < n$ as natural numbers or if $m < n$ as integers, that is, $[(m+1)-1] < [(n+1)-1]$. Prove that the two conditions are equivalent.

(10) Suppose $m, n \in \mathbb{N}$. Does $m + n$ mean the same thing when m and n are viewed as natural numbers as $m + n$ means when m and n are considered as integers. Justify your response.

(11) Suppose $m, n \in \mathbb{N}$. Does $m \cdot n$ mean the same thing when m and n are viewed as natural numbers as $m \cdot n$ means when m and n are considered as integers. Justify your response.

12.10 Defining rational numbers by integers

The construction of rational numbers is similar to the construction of integers, but the relation of equivalence is different. We consider the set of all pairs (p, q), where p and q are integers and $q \neq 0$. Since those pairs correspond to fractions we write $\frac{p}{q}$ instead of (p, q).

Recall that $\frac{6}{15}$ and $\frac{8}{20}$ are equivalent fractions because $6 \cdot 20 = 8 \cdot 15$. In general, two pairs $\frac{p}{q}$ and $\frac{r}{s}$ are called equivalent if $ps = rq$. We then write $\frac{p}{q} \sim \frac{r}{s}$. It is easy to check that this relation is reflexive, symmetric, and transitive, so it is an equivalence. The set of equivalence classes of this relation is a model of the set of rational numbers. The equivalence class containing a pair $\frac{p}{q}$ is denoted by $\left[\frac{p}{q}\right]$.

We introduce the following definitions of addition, multiplication and the property of being positive:

$$\left[\frac{p}{q}\right] + \left[\frac{r}{s}\right] = \left[\frac{ps + rq}{qs}\right];$$

$$\left[\frac{p}{q}\right]\left[\frac{r}{s}\right] = \left[\frac{pr}{qs}\right];$$

$$\left[\frac{p}{q}\right] > 0 \text{ if } pq > 0.$$

The set of all rational numbers contains all the integers if we agree to make the identification $p = \left[\frac{p}{1}\right]$.

Now we have a model of rational numbers. This model satisfies all properties $\mathfrak{R}$

except for **D**. Consider, for example, property 2^+. We first prove this property for integers. We have

$$([p-q] + [r-s]) + [t-u] = [(p+r) - (q+s)] + [t-u]$$
$$= [((p+r)+t) - ((q+s)+u))]$$

and

$$[p-q] + ([r-s] + [t-u]) = [p-q] + [(r+t) - (s+u)]$$
$$= [(p+(r+t)) - (q+(s+u))].$$

It is now apparent, from the associative property for addition of natural numbers, that 2^+ is satisfied for integers.

The brackets have been used here for equivalence classes in the sense of the preceding section. Using our result for integers, we will now prove 2^+ for arbitrary rational numbers. In the following argument, brackets will be used in the new sense as equivalence classes of fractions.

We have

$$\left(\left[\frac{p}{q}\right] + \left[\frac{r}{s}\right]\right) + \left[\frac{t}{u}\right] = \left[\frac{ps+rq}{qs}\right] + \left[\frac{t}{u}\right] = \left[\frac{(ps+rq)u+tqs}{qsu}\right]$$

and

$$\left[\frac{p}{q}\right] + \left(\left[\frac{r}{s}\right] + \left[\frac{t}{u}\right]\right) = \left[\frac{p}{q}\right] + \left[\frac{ru+ts}{su}\right] = \left[\frac{psu+(ru+ts)q}{qsu}\right].$$

In both cases the result can be written as

$$\left[\frac{psu+rqu+tqs}{qsu}\right],$$

so 2^+ holds for rational numbers.

S: This proof is very long and not interesting, actually boring.

T: Even so it is not complete.

S: Why not?

T: Because in the calculation with rational numbers we have used more properties of integers. In fact, first of all we should prove that the definitions of addition and multiplication of integers are correct, that is, if $[m_1 - n_1] = [p_1 - q_1]$ and $[m_2 - n_2] = [p_2 - q_2]$, then $[(m_1 + m_2) - (n_1 + n_2)] = [(p_1 + p_2) - (q_1 + q_2)]$ and $[(m_1 m_2 + n_1 n_2) - (m_1 n_2 + m_2 n_1)] = [(p_1 p_2 + q_1 q_2) - (p_1 q_2 + p_2 q_1)]$. Next, we have to prove properties 2^+, 1^*, 2^*, and $+*$ for integers, because they have also been used in the proof presented above. Finally we need to check that the definition of addition of rational numbers is correct. Only then would the proof be complete.

S: So we should first prove all these properties for integers! There should not be any serious problem with that. However the proofs would be awfully boring and the complete proof of 2^+ for rational numbers would be at least twice as long.

T: This is right. The proofs of the first nine properties in $\mathfrak{R}$ are like a big puzzle for children. One needs nothing but patience.

S: Moreover, as those properties have been familiar since the elementary school, we strongly believe that they must be true.

T: In our case we do not need belief, because we have shown a way which enables us to logically check the entire theory.

S: So, should every genuine mathematician check all the items in $\Re$?

T: No, but every genuine mathematician should feel able to provide a proof for any property in $\Re$. I think that no mathematician in the world has checked all the proofs, except perhaps those who were writing a book on arithmetic.

EXERCISES 12.10

In the following exercises, be careful to use only the properties of integers.

(1) Let p and q be integers with $q \neq 0$. Prove that $\left[\frac{p}{q}\right]$ is not empty.

(2) Let $\frac{p_1}{q_1} \sim \frac{r_1}{s_1}$ and $\frac{p_2}{q_2} \sim \frac{r_2}{s_2}$. Prove that $\frac{p_1 q_2 + p_2 q_1}{q_1 q_2} \sim \frac{r_1 s_2 + r_2 s_1}{s_1 s_2}$.

(3) Let $\frac{p_1}{q_1} \sim \frac{r_1}{s_1}$ and $\frac{p_2}{q_2} \sim \frac{r_2}{s_2}$. Prove that $\frac{p_1 p_2}{q_1 q_2} \sim \frac{r_1 r_2}{s_1 s_2}$.

(4) Let $\frac{p}{q} \sim \frac{r}{s}$. Prove that $pq > 0$ if and only if $rs > 0$.

(5) Let $b > 0$ be a rational number. Prove that there is some natural number n such that $n > b$. This property of rational numbers is called the *Archimedean property*.

(6) Let p and q be integers. Then p and q can also be considered as rational numbers $\left[\frac{p}{1}\right]$ and $\left[\frac{q}{1}\right]$. Is $p + q$ the same whether you consider p and q as integers or as rational numbers? Justify your response.

(7) Let p and q be integers. Then p and q can also be considered as rational numbers, $\left[\frac{p}{1}\right]$ and $\left[\frac{q}{1}\right]$. Is $p \cdot q$ the same whether you consider p and q as integers or as rational numbers? Justify your response.

12.11   Defining real numbers by rational numbers

Real numbers, like integers and rational numbers, will be introduced by the method of equivalence classes. However it will not be sufficient to consider pairs. Instead we will consider infinite sequences of rational numbers.

Before we define what it means to say that two sequences of rational numbers are equivalent, lets remind ourselves of some experience we already have with real numbers and also what we are trying to accomplish.

Every real number can be approximated arbitrarily closely by rational numbers, that is, every real number is the limit of an appropriately chosen sequence of rational numbers. For example, the number π is the limit of a sequence that might start like this: 3, 3.1, 3.14, 3.141, 3.1415, 3.14159, 3.141592, ... Of course there are other sequences of rational numbers that will converge to π. Another might start like this: 4, 3.2, 3.15, 3.142, 3.1416, 3.1416, 3.141593, ... Our goal is to lump together all the sequences of rational numbers that converge to π into a single equivalence class and, in the model we construct, that equivalence class will be our model of π. Now, of course, if all we have so far are rational numbers, then we cannot refer to π in the construction of our model. We have to be much more clever than that. Loosely speaking, our approach will be to declare that two sequences are to be put into the same equivalence class if their terms get and stay arbitrarily close when the terms are sufficiently far out in their respective lists. The following definition makes this idea precise.

Definition 12.11.1. Two sequences of rational numbers (a_n) and (b_n) are *equivalent*, denoted by $(a_n) \sim (b_n)$, if for any rational number $\varepsilon > 0$ there exists a natural number K such that $-\varepsilon < a_m - b_n < \varepsilon$ for all $m, n > K$.

The defined relation is evidently symmetric. To see that it is also transitive, assume that $(a_n) \sim (b_n)$ and $(b_n) \sim (c_n)$. Then, for any $\varepsilon > 0$, there exist natural numbers K_1 and K_2 such that

$$-\frac{\varepsilon}{2} < a_m - b_n < \frac{\varepsilon}{2} \quad \text{for all } m, n > K_1,$$

and

$$-\frac{\varepsilon}{2} < b_n - c_k < \frac{\varepsilon}{2} \quad \text{for all } k, n > K_2.$$

Adding corresponding terms of these inequalities we obtain

$$-\varepsilon < a_m - c_k < \varepsilon \quad \text{for all } m, k > K,$$

where K is the greater of K_1 and K_2. This proves that $\sim$ is a weak equivalence.

For instance, the sequences $(1/n)$ and (0) are equivalent, where (0) denotes the sequence (a_n) for which $a_n = 0$ for all $n \in \mathbb{N}$. The sequence (n) is not equivalent to any sequence, not even to itself.

The equivalence class containing a sequence (a_n) will be denoted by $[(a_n)]$. According to this definition the symbol $[(n)]$ represents the empty set, because there is no sequence of rational numbers equivalent to the sequence (n). We will follow a convention that if the symbol $[(a_n)]$ is used then it represents a nonempty class. The set of all those nonempty equivalence classes is a model of real numbers. Therefore, in this model, a real number is a nonempty equivalence class of sequences of rational numbers.

Note that if $[(a_n)]$ represents a real number, then the sequence (a_n) is equivalent to itself; (see the paragraph immediately following Definition 12.8.2). Also conversely, if $(a_n) \sim (a_n)$, then $[(a_n)]$ represents a real number, because (a_n) belongs to $[(a_n)]$.

Evidently, for any rational number a the constant sequence (a) belongs to an equivalence class, since it is equivalent to itself or, for example, to the sequence $(a + \frac{1}{n})$. This equivalence class is denoted by $[(a)]$. This suggests the identification $a = [(a)]$, so that each rational number is a real number.

Having so introduced real numbers, we have to define operations on them, and then prove that all the properties in $\mathfrak{R}$ hold. We adopt the following definitions of addition and multiplication:

$$[(a_n)] + [(b_n)] = [(a_n + b_n)] \quad \text{and} \quad [(a_n)][(b_n)] = [(a_n b_n)].$$

Note that there is reason to question whether the definitions for sum and product are really well-defined. We leave verification of that as exercises.

To see that the real numbers satisfy condition $\mathbf{2}^+$, notice that

$$([(a_n)] + [(b_n)]) + [(c_n)] = [(a_n + b_n)] + [(c_n)] = [((a_n + b_n) + c_n)],$$

and

$$[(a_n)] + ([(b_n)] + [(c_n)]) = [(a_n)] + [(b_n + c_n)] = [(a_n + (b_n + c_n))].$$

The fact that addition of rational numbers is associative may now be employed to observe that the real numbers also possess property $\mathbf{2}^+$. We see that the proof is a little simpler than the one for rational numbers (of course we use here former results for rational numbers). The detailed verification of all the axioms in $\mathfrak{R}$ would be an idle task for the same reason as for rational numbers. The only interesting question is about axiom $\mathbf{D}$.

To address axiom $\mathbf{D}$, one first needs to define inequalities for real numbers. That definition is somewhat more sophisticated than the preceding definitions. Before we give the rigorous definition, lets talk about it first. Consider some real number $[(a_n)]$. It might be tempting to think that in order for the real number $[(a_n)]$ to be positive, each term of the sequence (a_n) should be positive. However, the limit of a sequence bears little resemblance to the first few terms so the first few terms might be negative even though the limit of the sequence is positive. This might suggest that we modify our thinking to say that the terms of (a_n) should eventually be positive. That seems like an improvement, but there is another hitch. The sequence $\left(\frac{1}{n}\right)$ converges to 0 so $\left[\left(\frac{1}{n}\right)\right]$ is, in fact, the real number 0, even though every single one of its terms is positive. So a further modification of our thinking is required. In order for the real number $[(a_n)]$ to be positive, there should be a positive rational number ε such that the terms of (a_n) are eventually bigger than ε. The next definition makes this precise.

Definition 12.11.2. We write $[(a_n)] > 0$, and say that $[(a_n)]$ is *positive*, if there exist a positive rational number ε and a natural number K such that $a_n > \varepsilon$ for all $n > K$. Similarly, we write $[(a_n)] < 0$, and say that $[(a_n)]$ is *negative*, if there exists a positive rational number ε and a natural number K such that $a_n < -\varepsilon$ for all $n > K$.

Again, it is reasonable to ask whether these notions of "greater than" and "less than" are well-defined. We leave verification of that as an exercise.

Clearly, the inequality $[(a_n)] < 0$ is equivalent to $[(-a_n)] > 0$.

Now we will prove property $\mathbf{1}^<$. First note that the relations $[(a_n)] = 0$, $0 < [(a_n)]$, and $0 < -[(a_n)]$ exclude one another, that is, it cannot happen that two of them are simultaneously satisfied. It remains to be shown that at least one of them must hold. To this end, suppose $[(a_n)]$ is a real number that is neither positive nor negative. Since $(a_n) \sim (a_n)$, for every $\varepsilon > 0$ there exists a number K such that

$$-\frac{\varepsilon}{2} < a_m - a_n < \frac{\varepsilon}{2} \quad \text{for all } m, n > K.$$

Moreover, since $[(a_n)]$ is not positive, then there exists an index $n > K$ such that $a_n < \frac{\varepsilon}{2}$. Hence, $a_m < \varepsilon$ for every $m > K$. Similarly, since $[(a_n)]$ is not negative we get $a_m > -\varepsilon$ for every $m > K$. Thus, whenever $[(a_n)]$ is neither positive nor negative, we have

$$-\varepsilon < a_n < \varepsilon \quad \text{for all } n > K.$$

Since ε can be chosen arbitrarily small, this proves that $(a_n) \sim (0)$, and consequently, $[(a_n)] = [(0)] = 0$. Property $1^<$ is proved.

To prove $2^<$ assume that $[(a_n)] > 0$ and $[(b_n)] > 0$. There exist rational numbers $\varepsilon_1 > 0$ and $\varepsilon_2 > 0$ and numbers K_1 and K_2 such that $a_n > \varepsilon_1$ for $n > K_1$ and $b_n > \varepsilon_2$ form $n > K_2$. Then

$$a_n + b_n > \varepsilon_1 + \varepsilon_2 \quad \text{and} \quad a_n b_n > \varepsilon_1 \varepsilon_2 \quad \text{for } n > K,$$

where K is the greater of the numbers K_1 and K_2. Since $\varepsilon_1 + \varepsilon_2 > 0$ and $\varepsilon_1 \varepsilon_2 > 0$, property $2^<$ is proved.

Definition 12.11.3. By $\alpha < \beta$ or $\alpha \le \beta$, where α and β are real numbers, we mean that $\beta - \alpha > 0$ and $\beta - \alpha \ge 0$, respectively.

Theorem 12.11.4. *For every real number $\alpha > 0$ there is a rational number ε such that $0 < \varepsilon < \alpha$.*

Proof. If $\alpha = [(a_n)] > 0$, them there exists a rational number $\varepsilon > 0$ and a natural number K such that $a_n > 2\varepsilon$ for $n > K$. Hence $a_n - \varepsilon > \varepsilon$ for $n > K$, and thus $[(a_n - \varepsilon)] > 0$. But since ε is a rational number, we have $[(a_n)] - \varepsilon = [(a_n - \varepsilon)] > 0$, that is, $\alpha = [(a_n)] > \varepsilon$. $\square$

Theorem 12.11.5. *If $a_1, a_2, \ldots$ and b are rational numbers such that*

$$a_n \le a_{n+1} < b \quad \text{for all } n \in \mathbb{N}, \tag{12.8}$$

then the sequence (a_n) represents a real number and we have $[(a_m, a_m, a_m, \ldots)] \le [(a_n)]$ for each $m \in \mathbb{N}$. In other words, every non-decreasing sequence of rational numbers bounded from above represents a real number which is greater than or equal to every term of the sequence.

Proof. To prove that the sequence (a_n) represents a real number it suffices to show that $(a_n) \sim (a_n)$, that is, for every rational number $\varepsilon > 0$ there exists a number K such that

$$-\varepsilon < a_m - a_n < \varepsilon \quad \text{for } m, n > K.$$

Suppose this is not true. Then there exists a rational number $\varepsilon > 0$ and two increasing sequences (p_n) and (q_n) of natural numbers (indices) such that

$$a_{p_n} - a_{q_n} \geq \varepsilon$$

for all $n \in \mathbb{N}$. We may assume, without loss of generality, that $q_n < p_n < q_{n+1}$ for all $n \in \mathbb{N}$. Then

$$
\begin{aligned}
a_{p_n} - a_{q_1} &= (a_{p_n} - a_{q_n}) + (a_{q_n} - a_{p_{n-1}}) + (a_{p_{n-1}} - a_{q_{n-1}}) + (a_{q_{n-1}} - a_{p_{n-2}}) \cdots \\
&\quad + (a_{q_2} - a_{p_1}) + (a_{p_1} - a_{q_1}) \\
&\geq (a_{p_n} - a_{q_n}) + (a_{p_{n-1}} - a_{q_{n-1}}) + \cdots + (a_{p_2} - a_{q_2}) + (a_{p_1} - a_{q_1}) \\
&\geq n\varepsilon,
\end{aligned}
$$

because in the sum each of the $2n - 1$ differences is nonnegative (the sequence (a_n) is non-decreasing) and n of them are greater than ε. Thus $a_{p_n} \geq a_{q_1} + n\varepsilon$ for all $n \in \mathbb{N}$. By the Archimidean property of the rational numbers, the preceding statement contradicts the assumption that the sequence is bounded above. Thus, $(a_n) \sim (a_n)$ as claimed.

Now we prove that $[(a_m, a_m, a_m, \dots)] \leq [(a_n)]$. Suppose, conversely, that there is some $m_0 \in \mathbb{N}$ for which $[(a_{m_0}, a_{m_0}, a_{m_0}, \dots)] > [(a_n)]$. Then, for that m_0, we have $[(a_{m_0} - a_n)] > 0$, that is, $a_{m_0} > a_n$ for every n greater than some n_0, which contradicts the assumption that the sequence is increasing. $\qquad\square$

Now we are ready to prove that axiom **D** holds in our model of the real numbers.

Theorem 12.11.6. *Every nonempty set of real numbers bounded from below has a greatest lower bound.*

Proof. Let P be a nonempty set of real numbers, bounded from below. To prove that P has a greatest lower bound we plan to actually construct the greatest lower bound a for P by producing a non-decreasing sequence of rational numbers whose equivalence class is precisely a. To this end, let p_1 be the greatest integer such that $\frac{p_1}{2} \leq x$ for all $x \in P$. Next, we denote by p_2 the greatest integer such that $\frac{p_2}{4} \leq x$ for all $x \in P$. Clearly $\frac{p_2}{4} \geq \frac{p_1}{2}$. In general, for all $n \in \mathbb{N}$, we define p_n to be the greatest integer such that $\frac{p_n}{2^n} \leq x$ for all $x \in P$.

Clearly, the sequence $\left(\frac{p_n}{2^n}\right)$ is bounded and non-decreasing. By Theorem 12.11.5, the sequence represents a real number $a = \left[\left(\frac{p_n}{2^n}\right)\right]$. It is easy to see that a is a lower bound for P. To prove that a is the greatest lower bound of P suppose that b is a real number such that $a < b \leq x$ for all $x \in P$. By Theorem 12.11.4, there is a rational number $\varepsilon > 0$ such that $\varepsilon < b - a$. Hence $a + \varepsilon < b$. Let n_0 be the least natural number such that $\frac{1}{\varepsilon} < 2^{n_0}$. Then, by Theorem 12.11.5,

$$\frac{p_{n_0} + 1}{2^{n_0}} = \frac{p_{n_0}}{2^{n_0}} + \frac{1}{2^{n_0}} \leq \frac{p_{n_0}}{2^{n_0}} + \varepsilon \leq a + \varepsilon < b.$$

Therefore $\frac{p_{n_0} + 1}{2^{n_0}} \leq x$ for all $x \in P$, contrary to the definition of p_{n_0}. This proves that a is the greatest number satisfying $a \leq x$ for all $x \in P$. $\qquad\square$

This concludes our construction of a set equipped with addition, multiplication, and inequality that satisfy the ten axioms of real numbers. The construction started with the four axioms of natural numbers. Then integers and rational numbers where constructed. Finally, we introduced a set of equivalence classes of sequences of rational numbers as our model of real numbers. This does not mean that from now on we should think of real numbers as equivalence classes of sequences of rational numbers. In fact, one could argue that the question "What is a real number?" is not that important. What we need is the properties of real numbers. As we can see in Chapters 1–10, in the development of mathematical analysis we only use the ten properties of real numbers and never refer to the construction presented in this chapter.

S: I noticed that Cauchy sequences are introduced in other books where the method of equivalence classes is used in the construction of real numbers. We have not even mentioned them here.

T: We don't need Cauchy sequences.

S: Why?

T: Because we use the weak equivalence instead of the standard equivalence relation.

S: Why does it help?

T: The traditional method consists of two steps. First one introduces Cauchy sequences. Then the equivalence relation is defined only for Cauchy sequences. The reason is that the relation defined for pairs of sequences of rational numbers is not an equivalence relation if all sequences are considered. In our approach, the first step becomes superfluous. All the "undesirable" sequences like $1, 2, 3, \ldots$ just drop out in the construction.

EXERCISES 12.11

(1) Prove that $\left(\frac{1}{n}\right) \sim \left(\frac{1}{n}\right)$.

(2) Use the Archimedean property of the rational numbers to prove that $\left(\frac{1}{n}\right) \sim (0)$.

(3) Prove that $\left(\frac{1}{n}\right) \sim \left(\frac{n}{n^2+1}\right)$.

(4) Prove that $\left(\frac{2n+1}{n}\right) \sim \left(\frac{4n-3}{2n+1}\right)$.

(5) Prove that $[(n)]$ is empty.

(6) Prove that $[((-1)^n)]$ is empty.

(7) Prove that addition of real numbers is well-defined, that is, $(a_n) \sim (b_n)$ and $(c_n) \sim (d_n)$ implies $(a_n + c_n) \sim (b_n + d_n)$.

(8) Prove that multiplication of real numbers is well-defined, that is, if $(a_n) \sim$

(b_n) and $(c_n) \sim (d_n)$, then $(a_n c_n) \sim (b_n d_n)$.

(9) Prove that the inequality for real numbers is well-defined, that is, if $[(a_n)] \sim [(b_n)]$ and there exist a positive rational number ε_a and a natural number K_a such that $a_n > \varepsilon_a$ for all $n > K_a$, then there exist a positive rational number ε_b and a natural number K_b such that $b_n > \varepsilon_b$ for all $n > K_b$.

(10) Let a and b be two rational numbers. Let $\alpha = [(a, a, a, \dots)]$, $\beta = [(b, b, b, \dots)]$, and $\gamma = [(a + b, a + b, a + b, \dots)]$. Prove that $\alpha + \beta = \gamma$.

(11) Let a and b be two rational numbers. Let $\alpha = [(a, a, a, \dots)]$, $\beta = [(b, b, b, \dots)]$, and $\gamma = [(ab, ab, ab, \dots)]$. Prove that $\alpha\beta = \gamma$.

(12) Let a be a rational number and let $\alpha = [(a, a, a, \dots)]$. Prove that $a > 0$ if and only if $\alpha > 0$.

(13) Explain the importance of the properties described in exercises 10, 11, and 12.

(14) Prove: If $a_1, a_2, \dots$ and b are rational numbers such that $a_n \geq a_{n+1} > b$ for all $n \in \mathbb{N}$, then the sequence (a_n) represents a real number and we have $a_m \geq [(a_n)]$ for each $m \in \mathbb{N}$.

(15) Prove that $(a_n) \sim ((-1)^n a_n)$ if and only if $[(a_n)] = 0$.

(16) True or false: If $a_n > 0$ for all $n \in \mathbb{N}$, then $[(a_n)] > 0$?

(17) Define

$$a_1 = 1 \quad \text{and} \quad a_{n+1} = \frac{3a_n + 4}{2a_n + 3} \quad \text{for } n = 1, 2, 3 \dots.$$

Prove that (a_n) represents a real number.

(18) Prove: $[(a_n)] > 0$ if and only if there exist a sequence (b_n) and a positive rational number ε such that $(a_n) \sim (b_n)$ and $b_n > \varepsilon$ for all $n \in \mathbb{N}$.

(19) A sequence (a_n) of rational numbers is called a *Cauchy sequence* if for any positive ε there exists a number n_0 such that

$$-\varepsilon < a_m - a_n < \varepsilon \quad \text{for all } m, n \geq n_0.$$

(a) Prove that $\left(\frac{n}{n+1}\right)$ is a Cauchy sequence.

(b) Show that $\left(\frac{n^2+1}{n+1}\right)$ is not a Cauchy sequence.

(c) Prove that every Cauchy sequence is bounded.

(d) Let (a_n) and (b_n) be Cauchy sequences. Show that $(a_n + b_n)$ and $(a_n b_n)$ are Cauchy sequences.

(e) Let (a_n) and (b_n) be sequences of rational numbers. Prove that $(a_n) \sim (b_n)$ if and only if $(a_1, b_1, a_2, b_2, a_3, b_3, \dots)$ is a Cauchy sequence.

(f) Prove that a subsequence of a Cauchy sequence is a Cauchy sequence.

Index

Printed by Book-logo-Chosen
by Bookmasters

Printed in the United States
By Bookmasters